Ten Laws for Security

Eric Diehl

Ten Laws for Security

Eric Diehl
Sony Pictures Entertainment
Culver City, CA
USA

ISBN 978-3-319-82625-7 ISBN 978-3-319-42641-9 (eBook)
DOI 10.1007/978-3-319-42641-9

Softcover reprint of the hardcover 1st edition 2016

Printed on acid-free paper

This Springer imprint is published by Springer Nature
The registered company is Springer International Publishing AG
The registered company address is: Gewerbestrasse 11, 6330 Cham, Switzerland

Foreword

Twenty-six years ago, after a series of exhausting interviews, Eric Diehl agreed to hire me as a member of his team in Thomson Consumer Electronics. At the time, I could not imagine the huge impact that the encounter with Eric would have on my future career. Under his mantle, I learned what information security is about. Eric taught me the fundamentals of Pay TV scrambling, hacking, smart card protocols, and hardware security. He encouraged out-of-the-box thinking and constantly strove to perfection, clarity, and precision.

My work with Eric taught me to think about security from a variety of angles. Security can be approached according to the attack's timeline: Predictive security detects a coming attack (future), defensive security measures attempt to stop an ongoing attack, while reactive security comes after the attack and attempts to restore security. Security can also be seen from a threat-source perspective: accounting for the different motivations and means of hackers, agencies, academics, and criminals. The traditional way to approach security consists in addressing security by function: confidentiality, availability, integrity, etc. Eric's holistic approach consists in comprehensively approaching security by answering systemic clarifying questions such as: Where do we compute (device security)? With whom do we compute (network security)? What computes (system security)? How do we compute (program security)? and What does computation mean (information and knowledge security)?

As the years passed, Eric felt that recording his experience and passing it to the coming generation of security engineers became a necessity. His first book became a reference textbook in engineering laboratories and academia around the world. This second book is a practically oriented continuation allowing readers to understand both the commonsense foundations and the general principles underlying security.

Such a comprehensive approach to security has become a necessity: A brief look at any "good" cryptographic paper reveals that cryptographers rarely consider the *meaning or even the structure of protected data*. When a message is signed, hashed, or encrypted, data is considered as raw bits fed into functions. Interestingly, cryptographers consider this low-level treatment as a *virtue rather than a limitation* because cryptographic algorithms *do not assume anything* about the structure of the data that they process. Information security specialists work at a higher abstraction level and devise methods to protect *structured information*. For instance, SQL injections target *database entries*, Java bytecode verifiers check *type* semantics, and antiviruses analyze *executable programs*. As we write these lines, humanity is moving into an era of ontology and *knowledge* where protecting data and information starts to become insufficient. Ontologies already allow autonomous cars to make driving decisions and entrust computers with the authority to make important financial decisions. Hence, it appears necessary to start formalizing the *foundations of ontological security* and approach security in a holistic way.

This precious book sheds light on the fundamental underlying principles of information security. I hope that you will enjoy reading it as much as I did.

David Naccache
Professor at Ecole normale supérieure
Paris, France

Preface

First of all, I would like to thank my wife Laila for her constant support for this work and her invaluable patience. I would also like to thank my son Chadi for his illustrations.

I would like to thank many colleagues and friends who carefully reviewed portions of the manuscript. Their comments made this book more readable and attractive. First, I thank my colleagues at Sony Pictures Entertainment, including Bryan Blank, Mike Melo, and Tim Wright. Second, my gratitude goes to friends, including Patrice Auffret (La Poste), Mahesh Balasubramanian (Disney), Martin Bergenwall (Inside Secure), Olivier Brique (Kudelski Security), Eric Filiol (ESIEA), Eric Freyssinet (French Interior Ministry), Julien Iguchi (Laboratoire Cristal, Université de Lille), Helena Handschuh (Cryptography Research), Olivier Heen (Technicolor), Jean-Louis Lanet (INRIA), Jeff Lotspiech, Andrew McLennan (Inside Secure), Jean-Jacques Quisquater, and Rei Savafi-Naini (University of Calgary).

I would like to give a special thanks to my friend David Naccache (Ecole Normale Supérieure) who wrote the Foreword for this book.

Finally, I am grateful to my editor Ronan Nugent and Springer, who showed a keen interest in this second book.

Culver City, USA | Eric Diehl

Contents

Abbreviations and Acronyms

AIK	Attestation Identity Key
API	Application programming interface
ATM	Automated teller machine
BIOS	Basic Input/Output System
BYOC	Bring Your Own Cloud
BYOD	Bring Your Own Device
C&C	Command and control
CA	Certification authority
CAS	Conditional access system
CCC	Chaos Computer Club
CEO	Chief executing officer
CERT	Computer Emergency Response Team
CIA	Central Intelligence Agency
CMOS	Complementary metal–oxide semiconductor
CobiT	Control Objectives for Information and Related Technology
CRC	Cyclic redundancy code
CSIRT	Computer Security Incident Response Team
CSS	Content scramble system
CVV	Card verification value
DARPA	Defense Advanced Research Projects Agency
DDoS	Distributed denial of services
DES	Data Encryption Standard
DLP	Data loss prevention
DMCA	Digital Millennium Copyright Act
DMZ	Demilitarized zone
DNS	Domain name service
DoS	Denial of services
DRAM	Dynamic random-access memory
DRM	Digital Rights Management
DSA	Digital signature algorithm

DVB	Digital Video Broadcasting
EAL	Evaluation Assurance Level
EC	European Commission
ECC	Elliptic curve cryptography
ECM	Entitlement control message
EEPROM	Electrically erasable programmable read-only memory
EFI	Extensible Firmware Interface
EK	Endorsement key
EMM	Entitlement management message
FBI	Federal Bureau of Investigation
FTC	Federal Trade Commission
FTP	File Transfer Protocol
GPS	Global Positioning System
HIPAA	Health Insurance Portability and Accountability Act
HRoT	Hardware Root of Trust
HSM	Hardware Secure Module
HTTP	Hypertext Transfer Protocol
HTTPS	Hypertext Transfer Protocol Secure
IaaS	Infrastructure as a Service
IDS	Intrusion detection systems
IoT	Internet of Things
IP	Internet Protocol
IT	Information technology
MaaS	Malware as a Service
MAC	Medium access control (or message authentication code)
NBS	National Bureau of Standards
NDA	Non-disclosure Agreement
NIST	National Institute of Standards and Technology
NMOS	Negative metal–oxide semiconductor
NSA	National Security Agency
OWASP	Open Web Application Security Project
PaaS	Platform as a Service
PCB	Printed circuit board
PCI-DSS	Payment Card Industry Data Security Standard
PII	Personally identifiable information
PKES	Passive Keyless Entry and Start
PLL	Phase-locked loop
PMOS	Positive metal–oxide semiconductor
POS	Point of sale
PP	Protection Profile
PRNG	Pseudorandom number generator
QKD	Quantum key distribution
RAM	Random-access memory
RAT	Remote Administration Tool
RF	Radio frequency

RFID	Radio frequency identification
RNG	Random number generator
ROI	Return on investment
ROM	Read-only memory
RSA	Rivest–Shamir–Adleman
SaaS	Software as a Service
S-box	Substitution box
SCADA	Supervisory Control and Data Acquisition
SIM	Security information management
SMTP	Small Mail Transfer Protocol
SOX	Sarbanes–Oxley Act
SQL	Structured Query Language
SRK	Storage root key
SSID	Service Set Identifier
SSL	Secure Socket Layer
SSN	Social Security Number
STB	Set-top box
TCG	Trusted Computing Group
TCP	Transport Control Protocol
TLS	Transport Layer Security
TOE	Target Of Evaluation
UDP	User Datagram Protocol
UK	United Kingdom
URL	Uniform resource locator
US	United States
VERIS	Vocabulary for Event Recording and Incident Sharing
VHDL	VHSIC Hardware Description Language
VPN	Virtual private network
WAN	Wide area network
WPS	Wi-fi Protected Setup
WWI	World War I
WWII	World War II
XSS	Cross-site scripting

List of Figures

Introduction

Security is an extensive, complex discipline that encompasses many domains, many technologies, and multiple methodologies and requires a broad range of skills. The variety of topics covered is astonishing. For instance, the standard ISO 27002:2005 (previously known as ISO 17999) provides a checklist of 11 different items that should be audited when evaluating the security of an enterprise [1]. The checklist is an excellent overview of various aspects of security. These topics include security policy; organization of information security; asset management; human resources security; physical and environmental security; communications and operations management; access control; information systems acquisition; development and maintenance; information security incident management; business continuity management; and compliance. All these features are mandatory to achieve a satisfactory result, i.e., a reasonably secure organization.

Is there any commonality between all these different, complementary aspects of security? Are there some general rules that widely apply to all the various facets of security? We believe so.

For 15 years, together with my previous security team, I defined and refined a set of ten laws for security [2]. These laws are simple but powerful. Over the years, when meeting other security experts, solution providers, potential customers, and students, I discovered that these laws were an excellent communication tool. These rules allowed us to benchmark quickly whether both parties shared the same vision for security. Many meetings successfully started by me introducing these laws, which helped build reciprocal respect and trust between teams. Over time, I found that these laws were also an excellent educational tool. Each law can introduce different technologies and principles of security. They constitute an entertaining way to present security to new students or to introduce security to non-experts [3]. Furthermore, these laws are mandatory heuristics that should drive any design of secure systems. There is no valid, rational reason for a system to violate one of these rules. The laws can be used as a checklist for a first-level sanity check. As such, this set of laws is also a reasonable first sieve for identifying snake-oil vendors.

Each chapter of this book addresses one law. The first part of the chapter always starts with examples. These anecdotes either illustrate an advantageous application of the law or outline the consequences of not complying with it. The second part of the chapter explores different security principles addressed by the law. Each chapter introduces, at least, one security technology or methodology that illustrates the law, or that is paramount. From each law, the last section deduces some associated rules that are useful when designing or assessing a security system.

The Devil Is in the Detail Security is complex. A minor implementation mistake may ruin a theoretically secure solution. As in my previous book [4], inserts, entitled "The Devil Is in the Detail," illustrate the gap between theory and real-world security. The book depicts some real hacks to highlight what went wrong. Each insert concludes with a lesson that the security practitioners have learned (sometimes the hard way). It should give a glimpse of the work of designers of security solutions.

Before continuing, I would like to introduce a few famous characters who will follow us in this book. They are often encountered in security-related publications.

- Alice and Bob are extremely talkative persons. They always exchange messages. They try to have secure communication. In other words, anybody other than themselves should not understand their messages and no one else should be able to modify the meaning of these messages.
- Eve wishes to learn what Alice and Bob exchange. So, she is always eavesdropping on them. She is a passive attacker, i.e., she only listens.
- Mallory also wants to learn what Alice and Bob exchange. Unlike Eve, Mallory is an active attacker. When spying on Alice and Bob, she may interfere with the exchanged communications. For instance, she may delete messages, inject forged messages, or modify exchanged messages.

This book uses another similar convention. When the book does not use the above characters, it assumes that the attacker is a female character, whereas the attacked person is a male character.[1] This convention allows easy identification in the same sentence of who is who and helps to distinguish clearly the attacker from her victim. Feminine pronouns refer to the attacker, whereas masculine pronouns refer to the target.

Now, it is time to present these ten laws.

- **Law 1: Attackers Will Always Find Their Way**
- **Law 2: Know the Assets to Protect**
- **Law 3: No Security Through Obscurity**

[1]The reader should not look for any stealthy meaning or sexist interpretation in this distribution of roles.

- **Law 4: Trust No One**
- **Law 5: Si Vis Pacem, Para Bellum**[2]
- **Law 6: Security Is No Stronger Than Its Weakest Link**
- **Law 7: You Are the Weakest Link**
- **Law 8: If You Watch the Internet, the Internet Watches You**
- **Law 9: Quis Custodiet Ipsos Custodes?**[3]
- **Law 10: Security Is Not a Product, Security Is a Process**

Although the book presents the ten laws in a given order, this order has no semantic significance. No law, except maybe "Law 1—Attackers Will Always Find Their Way," is more influential than the other ones. If one system does not comply with one law, then the entire system becomes more vulnerable. Nevertheless, the book puts an emphasis on the first law. This law should impart a sense of humility in security practitioners: Attackers will defeat them. It is not a message of desperation. Rather, this law provides a glimpse of the impossible, but thrilling challenge to all security defenders. Security is an extremely addictive cat-and-mouse game. Thus, the first law is my favorite law. The chapters are independent, and the reader may read them in his or her preferred order.

A cautionary note is necessary at this stage of the book. These laws do not have to be interpreted literally. Like everything, security is not black or white, and it is about shades of gray. Security is an exercise in finding a subtle equilibrium that does not accommodate a Manichean vision. Nevertheless, using strong punch lines is an efficient approach for communication, and they are shorter and easier to remember. The chapters nuance these punch lines, so we avoid oversimplification.

[2]"If you want peace, prepare for war."

[3]"Who guards the guardians?"

Chapter 1
Law 1: Attackers Will Always Find Their Way

> *The major difference between a thing that might go wrong and a thing that cannot possibly go wrong is that when a thing that cannot possibly go wrong goes wrong it usually turns out to be impossible to get at or repair.*
>
> ADAMS D., Mostly Harmless [5]

1.1 Examples

Attackers will always find their way. This statement is cruel and pessimistic! Unfortunately, it reflects the brutal bare reality of the world of security. No secure system is invulnerable. This is common knowledge and has been used well in the literature and the movie industry. One well-known Scottish saying is "Three can keep a secret if twa be awa."[1] This saying highlights that secrets never remain hidden. They will leak out eventually 1 day.

The belief that invulnerability is a myth was already known in antiquity. In his *Iliad*, the Greek author Homer provided an illustration of this law with the Greek mythological hero Achilles [6]. At the birth of Achilles, his mother, the nymph Thetis, bathed him in the Styx, one of Hell's rivers. Styx's water had the magical power to make someone invulnerable. No weapon could hurt Achilles. Unfortunately, Thetis had to hold her son by his heel during the bath to immerse him in the Styx. Therefore, a tiny part of Achilles's body, one heel, remained vulnerable. During the Trojan War, Achilles was the terror of the Trojans as he slaughtered many of them. His most famous victory was against Hector, son of the Trojan King Priam. As vengeance, Hector's brother, Paris, killed Achilles with an arrow guided by the God Apollo to the unique, vulnerable spot of Achilles's heel. Interestingly, the coward Trojan Paris defeated the mighty Greek warrior with the help of an insider: Apollo. Indeed, only a God could know the existence of this vulnerability. This classical scheme implies an insider is commonly encountered in the security arena. An insider, here the God

[1]Three can keep a secret if two are away (sometimes also found as "Three can keep a secret if two are dead").

E. Diehl, *Ten Laws for Security*, DOI 10.1007/978-3-319-42641-9_1

Apollo, will reveal the lethal secret or will partake in the plot of the attacker. Section 4.3.5 will explore this important scheme.

This myth of one single point of vulnerability appears in many other mythologies, such as with Sigurd in Norse mythology.[2]

Before the era of information technology (IT), many historical events demonstrated the truth of this law. The sinking of RMS Titanic is an iconic example. In 1912, the British company White Star Line built the largest cruiser of its time. Not only was RMS Titanic the largest ship, but she was also claimed to be the safest in naval history. White Star Line declared that Titanic was designed to be unsinkable with her sixteen watertight compartments. Her maiden journey from Southampton to New York began on April 10, 1912. During the night of April 14, 1912, the Titanic struck a colossal iceberg. The supposedly unsinkable ship sank in less than 3 h. Of the 2224 passengers and crewmembers, only 722 survived. The ship did indeed have sixteen watertight compartments. The purpose of watertight compartments is to isolate a section of the boat's hull from the rest of the vessel. If this part of the hull were to be ripped open, then water would fill only this compartment. The flooding would not reduce the boat's buoyancy significantly. Thus, at least in theory, sixteen watertight compartments should have prevented the ship from sinking. Unfortunately, the iceberg ripped off more compartments than foreseen by the naval designers. Therefore, the incoming quantity of water exceeded the buoyancy threshold, and the Titanic sank, like an ordinary boat.

Rule 1.1: Always Expect the Attacker to Push the Limits

The attacker never obeys the rules; nor does she comply with the behavior expected by the designer of the system. A common method of attack is to take the system out of its nominal cases with the hope either of provoking a failure or of gaining some advantages. A buffer overflow attack is such an example, where the attacker inputs information longer than expected. Fault injection attacks are another example, where the attacker changes the operating environment (Sect. 1.2.2) so that the system cannot operate properly anymore.

In 2014, Joseph Hemanth designed a simple denial of service (DoS) attack on the Pebble, a smartwatch [7]. Each time the Pebble receives a notification from the email service, Twitter, or any social network, it vibrates and displays the corresponding notification message on its small screen. Unfortunately, the Pebble does not check the length of the message that it displays. If an attacker sends a few hundred emails or Facebook notifications in five seconds to the targeted user, then the display buffer overflows. The display of the targeted user's Pebble becomes dull with many white lines. After a while, the overflown Pebble resets to factory defaults, deleting all configuration settings and stored messages. The attack does

[2]Sigurd killed the dragon Fafnir. Following the advice of the God Odin, he bathed in its blood to become invulnerable. Unfortunately, a leaf sticking on his shoulder created a weak point in this otherwise armored skin. Of course, his opponent, Gottrum, would defeat him through this unique vulnerable point.

not only DoS the device, but also wipes out the targeted device. Hemanth pushed the limits of the Pebble.

The number of examples that could illustrate this first law is vast. Every day, new exploits occur. The book will present many other illustrations.

1.2 Analysis

There is a strong asymmetry in this war between security defenders and attackers. The attacker has many advantages over the defender. First, the attacker needs to succeed only once, whereas the defender has to succeed every time. The attacker can afford many defeats. The defender cannot afford one. Furthermore, the attacker may reuse the winning strategy or technique against many different defenders. Second, the attacker benefits from all the security technologies and tools that the defender may use. She can twist these technologies to become weapons rather than armors. Furthermore, the attacker does not share the same constraints as the defender. She does not have to obey a bureaucracy, follow processes, and generally obey rules. She has the freedom to act outside of the established context and rules. Furthermore, nature favors the attacker. The second law of thermodynamics states that entropy tends not to decrease. It highlights that it is easier to break a system than to build it. Building a system increases the order and decreases chaos. Therefore, it reduces entropy. Breaking a system increases the chaos and thus increases entropy. The remainder of this chapter illustrates this sad reality.

1.2.1 Should Vulnerabilities Be Published?

These anecdotes may seem far from the IT world. Unfortunately, the current information era has a plethora of failure stories. The first example is from World War II (WWII): Enigma M4. For many centuries, nations have used some form of cryptography to protect their vital communication. Even Julius Caesar encrypted his messages.[3] During WWII, every country used its own proprietary encryption

[3]Julius Caesar's substitution algorithm was extremely rudimentary. The encrypted character is the original character shifted by a fixed value within the alphabet. For instance, if the shift value is 3, A becomes D, B becomes E, …, and X becomes A. With this key, "`WHQ ODZV IRU VHFXULWB`" is the cipher text of "`TEN LAWS FOR SECURITY`." Obviously, this type of encryption can be easily broken. The easiest method is to make a statistical analysis of the frequency of occurrence of the encrypted characters in the cipher text and then try to match their distribution with the Gaussian distribution of the supposed language. For instance, in English, the three most frequent characters are E, T, and A, whereas in French they are E, S, and A. The analysis is even more efficient when using pairs of characters or groups of three characters. If the encrypted message is long enough, then the identification of these most frequent characters is easy. The correspondence reveals the "key" and thus the original message. Al Kindi introduced this statistical method for cryptanalysis in the ninth century [8].

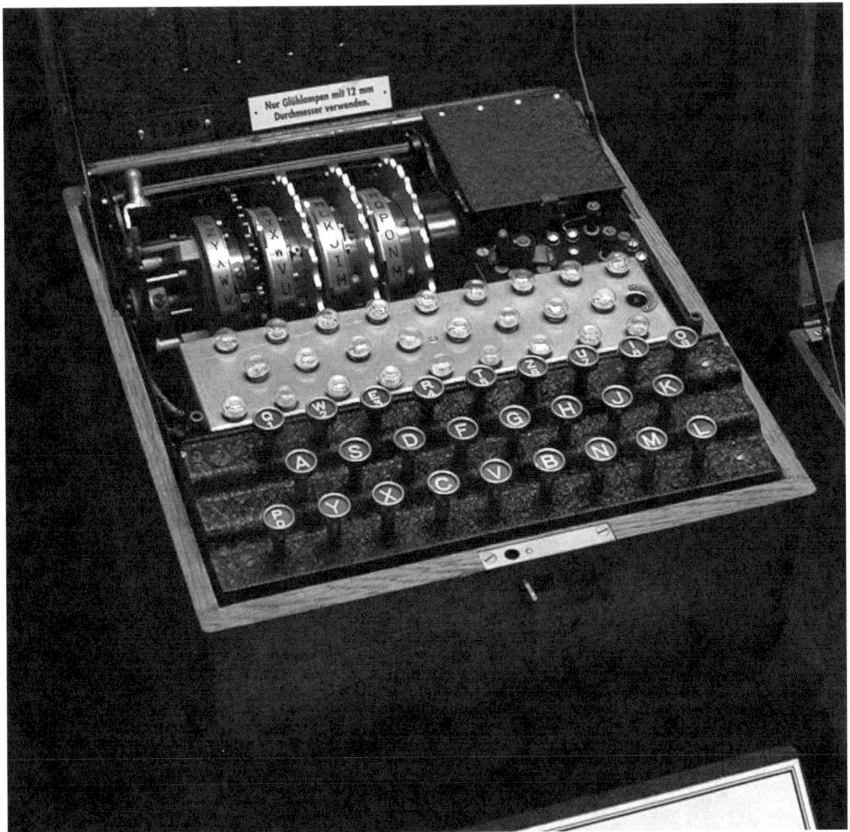

Fig. 1.1 Four-rotor Enigma machine

algorithm. Of course, each camp attempted, often successfully, to break the enemy's cipher [9]. They used cipher machines to encrypt and decrypt their confidential messages. Usually, these cipher machines were rotor-based. The Swiss Hagelin was the most renowned manufacturer of these cipher machines [10]. The best-known cipher machine is Enigma. Germans were inspired by the Hagelin machines to design their military, version Enigma G, by adding their own proprietary enhancements. Thus, they produced the Enigma M3 and later the Enigma M4 (with a fourth supplemental rotor), as illustrated in Fig. 1.1.

An Enigma machine looks like a typewriter with an extra set of rotors and switchboards. The rotors and switchboards serve for setting the secret key. The keyboard is used to enter the clear text to encrypt, or the ciphertext to decrypt. The entered text has no punctuation or spacing. Letters replace numbers. The nine letters

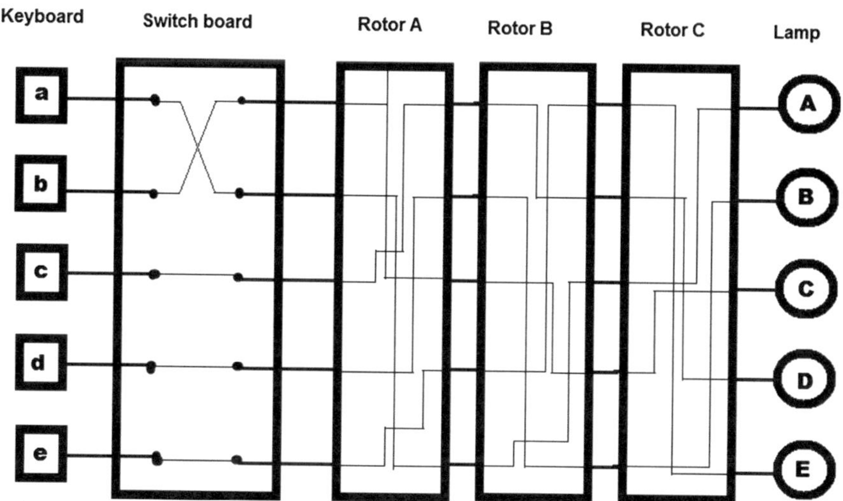

Fig. 1.2 Simplified view of Enigma M3

at the top (Q, W, E, R, T, Z, U, I, and O) also represent nine digits. 0 is associated with letter P. Therefore, Q may mean the letter Q or the digit 1. Lightbulbs, one per letter, display the decrypted or encrypted result. Figure 1.2 depicts a simplified Enigma machine using only five letters and three rotors. The real devices use the full 26-character Latin alphabet, no punctuation, no digits, and no space. By convention, the input has uppercase characters, whereas the output has lowercase characters. Each rotor has 26 entry points physically wired to 26 exit points (one point per character). Rotors A, B, and C have different wiring. Of course, the rotors are identical in Enigma machines of the same model. The physical connection lights the corresponding letter's bulb. Rotor A moves one position each time the operator dials a new character. After a full rotation of rotor A, i.e., after 26 characters, rotor B moves one position. After a full rotation of rotor B, rotor C moves one position. The switchboard allows a fixed reshuffling of the keyboard. To decrypt a message correctly, the operator has to reproduce the initial conditions of the encrypting machine, i.e., the same configuration of the switchboard and the same initial position of the rotors. The secret key was the combination of the switchboard's configuration and the rotors' position.

During the war, Germans believed that no enemy could break their cipher. Most countries accepted this statement, and not many nations tried to challenge this assumption. However, in 1932, Polish cryptanalysts had the conviction that mathematics could be used to solve the problem. Through reverse engineering, they had already understood the particular wiring of the four rotors. They developed an electromechanical device, called the Bomba, which allowed accelerating the code breaking. With the help of these tools, they found two crucial design flaws in Enigma. Nevertheless, continuous improvements of Enigma kept it ahead in the ensuing race. In 1939, when Germans invaded Poland, the Polish cryptanalysts sent all their work, with a few replicas of Enigma machine, to their British and French counterparts. This contribution would prove to be invaluable.

In 1940, the team established at Bletchley Park succeeded in breaking the cipher. Under the lead of Alan Turing and Gordon Welchman, the team partly automated the cracking process using their own electromechanical device: Colossus. The German forces continuously improved their Enigma machines, making each version harder to crack than the previous one. For instance, in 1942, the German naval forces deployed the four-rotor Enigma M4 on their submarines (called U-boats) in replacement of the three-rotor version. Nevertheless, during most of the WWII, Bletchley Park was able to decrypt German ciphertexts. This success seriously influenced the outcome of WWII. Some historians estimate that it reduced the duration of the war by 3 years and saved thousands of lives. For instance, the German Air Force lost the Battle of England in 1940 partly because it entrusted bomber-target information to the insecure Air Force Enigma [11]. Thanks to Bletchley Park, the British Royal Air Force knew in advance the targets that German bombardiers would bomb during the coming night raid. The German Navy lost a staggering number of submarines for the same reason because Bletchley Park could break German Navy Enigma-encrypted messages.

The way the Allied forces treated the decrypted messages is interesting. The decrypted messages were called Ultras. The handling of Ultras was restricted to a limited set of dedicated, trained, trusted officers. The German could never become aware that the Allies had cracked their Enigma codes. Thus, the exploitation of information collected from Ultras was extremely sensitive and tightly controlled. The Allied high command decided that military operations should never exploit Ultras unless there was another potential source of disclosure of the corresponding information. Because of this clever strategy, the German forces never discovered that the Allied forces could read their secret encrypted communications. Despite its losing many U-boats, the German naval high command always thought that its losses were due to the Allies' better radar and direction capabilities rather than a weak cipher. The Germans systematically incriminated the second source of information rather than questioning the secrecy of their communication.

The design of a robust cryptographic system requires, at least, two different skill sets: cryptography and cryptanalysis. Cryptography, practiced by cryptologists, is the art of designing the actual algorithms, whereas cryptanalysis, practiced by cryptanalysts, is the art of breaking the designed algorithms. Cryptology encompasses both cryptography and cryptanalysis. The real robustness of the encryption depends on the quality of both teams. A skilled cryptographer may develop a suitable algorithm. Nevertheless, only skilled cryptanalysts can give reasonable assurance that this algorithm is robust. According to American military experts, the German cryptographic systems were brilliantly conceived [11]. German cryptologists were brilliant. Unfortunately, the Allied cryptanalysts were better than the German cryptanalysts. At the end of the war, the German forces had new encryption devices ready for rollout. The Allied cryptanalysts would not have been able to crack them. Fortunately, until the end of the war, the German forces were always convinced that Enigma was robust, and they did not see the need to deploy the most robust version. Hence, the special treatment of Ultras was a winning strategy.

Unfortunately, this strategy is not a good one anymore. The designer of a secure system must request other skilled people to attack his system. Furthermore, it is essential to look continuously for evidence that the system can be eventually broken. This search of proof requires monitoring the Internet, dedicated forums such as Doom9 [12], security contests such as Pwn2Own, and hacking conferences such as Defcon or Black Hat. This continuous monitoring may reveal attacks applied to another system but also applicable to yours. Having researchers and ethical hackers publish exploits is essential to the process of designing more secure systems. This practice is called full disclosure policy. Some people claim that publishing vulnerabilities benefits attackers and is dangerous for consumers, and so disclosing vulnerabilities should not be allowed. According to them, software vulnerabilities should be kept secret. Indeed, once a vulnerability is publicly disclosed, the number of attacks based on this vulnerability may multiply by up to five orders of magnitude [13]. Jason Lanier, one of the fathers of virtual reality, called the full disclosure policy the ideology of violation [14]. According to him, the only purpose of the researchers was to seek glory. I disagree. I believe that refusing to disclose vulnerabilities is a poor strategy. Indeed, publishing exploits has three positive consequences.

- *Disclosure of the Vulnerability*: The designers of the vulnerable system can now implement a countermeasure that will close the hole. That a given vulnerability has not been publicly revealed does not mean that an attacker is not already exploiting it. Maybe an attacker already knows and exploits this vulnerability but keeps this knowledge to herself. In that case, this stealthy attacker may exploit the vulnerability without her victims being aware and with no chance of seeing this hole fixed.
  An unknown vulnerability is called a 0-day vulnerability. 0-day vulnerabilities are the quintessence of vulnerability for attackers, as there is yet no available defense. Thus, the attack will be taking place off the radar. Therefore, there is a black market for trading 0-day vulnerabilities. Their price is ranging from $5000 to $250,000 depending on the criticality of the exploit [15, 16]. Furthermore, it is claimed that some private companies, such as ReVuln [17], Vupen Security, FinFisher, and the Hacking Team, purchase such 0-day vulnerabilities from hackers. Later, these enterprises sell these vulnerabilities to interested customers [18]. The sale of these vulnerabilities is secret to give a competitive edge to the customer. The potential clients of such 0-day vulnerabilities are private corporate teams looking for new attack vectors or attempting to set up an advanced persistent threat (APT), as well as government agencies. For instance, in 2012, the US National Security Agency (NSA) paid Vupen Security for a 1-year supply of 0-day vulnerabilities. Vupen Security provided both the vulnerabilities and the pieces of software needed to exploit those security flaws to attack electronic systems [19].

Interestingly, many companies may reward the private disclosure of vulnerabilities in their products or services [20]. Since November 2010, Google fostered responsible disclosure through its vulnerability reward program [21]. The reward ranges from $100 for a new cross-site scripting (XSS) to $20,000 for unknown remote code execution. As an extremely smart incentive, it also publishes the name of the contributors in its security hall of fame. In September 2012, Google had rewarded about 250 persons. On June 26, 2013, Microsoft launched a similar program, called "mitigation bypass and blue hat defense," with bounties of up to $100,000 [22]. In January 2014, Facebook paid a $33,500 bounty to a Brazilian security researcher who found one remote code vulnerability [23]. These bounty programs have started to become popular, well attended, and well funded. In 2013, Facebook received 14,763 submissions. This number was an increase of about 250 % compared to the previous year [24]. Among them, only 687 submissions were valid vulnerabilities. About 40 of them were critical vulnerabilities. Google went even one step further. In 2015, Google launched the Vulnerability Research Grant program [25]. Researchers may apply for this grant program to look for potential vulnerabilities within sensitive Google applications. The grant ranges from $500 to $3133.70. The application clearly states that some researchers might not find any vulnerability. In other words, the grant is not contingent on the discovery of any actual vulnerability. Many forms of reward are in use. For example, United Airlines rewards the vulnerability discoverers with air miles [26].

Nevertheless, the bounty is not necessarily the primary driver of the disclosure, as many ethical hackers are motivated by fame rather than by monetary reward. Reputation is a powerful motivation. The Web site HackerOne.com is a hub for about 100 of such vulnerability discovering programs. The programs range from large companies or projects (Adobe, OpenSSL, Python, Twitter, or Yahoo) to small sites or projects. A little less than half of these projects offer pecuniary rewards. Furthermore, the site provides a federated hall of fame. Some security-oriented conferences also organize contests. For instance, the annual Pwn2Own contest is an organized challenge to crack the security of the four top browsers.[4]

Of course, responsible disclosure requires careful scheduling. Known, published, responsible disclosure policies exist, such as the Coordinated Vulnerability Disclosure [27] or the Zero Day Initiative [28]. Without such policies, the disclosure of vulnerabilities would subscribe to the ideology of violation. The ethical hacker must first disclose the vulnerability to the product owner/developer. The product owner should negotiate a reasonable delay with the ethical hacker or security researcher before the public disclosure. During this grace period, the product owner will fix the vulnerability and provide a security patch to its customers. Then, customers must apply the security patch, which

[4] In 2015, Jung Hoon Lee was rewarded $225,000 for three successful exploits on Chrome, Safari, and Internet Explorer 11.

may take many months. A recent study evaluated the average duration of one 0-day vulnerability in the field to 312 days. In extreme cases, it lasted up to 30 months [13]. Later, the ethical hacker publishes the exploit. Unfortunately, as exploits are often described immediately after the delivery of the corresponding patch, and as people do not immediately update their systems, a flurry of exploits may be launched during this transitional period [29]. Late in 2014, with its Project Zero, Google unilaterally decided that its researchers should disclose within 7 days the critical vulnerabilities they discovered [30] and that less critical vulnerabilities should be disclosed within 90 days. The claimed objective was to accelerate the availability of a patch or, at least, the publication of a security advisory bulletin. After some hiccups, Google added a potential grace period of 14 days if the software editor needed it to complete the security patch [31].

- *Availability of the Patch for Vulnerability*: Public disclosure gives a strong incentive to the software editor to fix the vulnerability. Avoiding bad mouthing that may ruin the reputation of a firm can be a powerful inducement for developing a new security patch. The more secure the deployed pieces of software or the Web services will be, the more secure the digital world will become [32]. Having all systems patched correctly is of general public interest. See Rule 6.2: Patch, Patch, Patch.
- *An Excellent Educational Tool for Students and Especially for Practitioners*: A new vulnerability may use a new type of attack and technique. It is hard to defend against an attack that you are not aware of. Knowledge is paramount to security. The disclosed attack may apply to targets other than the original one. Designers of such potential targets can preventively address these vulnerabilities. They do not need to wait until the exploits are deployed on these targets. Furthermore, designers can avoid the same mistakes in new designs.

Rule 1.2: Responsibly Publish Vulnerabilities

In the role of dissemination and education, the Computer Emergency Response Teams (CERTs) are key players. CERTs are primarily dedicated to helping enterprises and administrations rather than end users. Nevertheless, most of the time, the CERT information is available to everyone. Their primary missions are to analyze all reported threats, to maintain a database of all disclosed vulnerabilities, and to provide support and advice for implementing mitigation plans for these vulnerabilities. The first CERT was created in 1988 in Pittsburgh at Carnegie Mellon University under the lead of the US Defense Advanced Research Projects Agency (DARPA) following the appearance of the Morris worm. As CERT is a trademark of Carnegie Mellon University, the community should instead use the generic term Computer Security Incident Response Team (CSIRT). There are several hundreds of CSIRTs in the world. The CERT from Carnegie Mellon University remains the most famous CSIRT.

1.2.2 Jailbreaking and Secure Bootloaders

During World War II, the cryptanalysts at Bletchley Park succeeded in breaking the algorithms that the Enigma machines implemented. The next example describes another hack applied to a different type of device. This time, it is a consumer device rather than a military apparatus. Many modern devices execute only pieces of software approved by the device's manufacturer. This is the case, for instance, for game consoles, set-top boxes (STBs), and some smartphones. There are two main rationales for this design decision.

- Security: Controlling what software executes on a platform is a way to ensure that no malware contaminates the platform. This method is especially efficient for closed garden environments such as STBs and even smart cards. It ensures that the device behaves as expected or that no forged piece of software can extract secrets or misbehave. In the case of game consoles, it is a theoretical, copy protection method, as only signed, controlled pieces of software can execute on the platform. If usual ripping tools cannot make a pristine, complete copy of the genuine game, then it is not possible to produce illegal copies of genuine game. Usually, the corresponding physical media uses a logical format that is different from the format used by the media industry for movies, in order to thwart ripping tools. For instance, the logical format of a game DVD is different from the format of DVD-Video. Furthermore, it uses a proprietary format, whereas DVD-Video's one is public.
- Business: Enforcing what software executes on a platform is a way to control how the market distributes the corresponding software. The manufacturer is a gatekeeper who may charge a fee to the software editors for authorizing the execution of their applications on its platform. This is the case, for instance, with Apple with its Apple Store, or the manufacturers of game consoles. They create controlled vertical markets.

Whatever the reason to implement such restriction methods, these systems use the same enforcement mechanism: a secure bootloader. The secure bootloader is quasi the first piece of software executed by the device. It has two functions.

- The secure bootloader checks the integrity of the system. It typically verifies whether the operating system and the drivers have been tampered with. For that purpose, it compares the hash value[5] of the binary code of these elements with the corresponding stored reference values. Obviously, the stored reference hash values should not be rewritable by an attacker. Why use a hash function rather than a lightweight checksum, given the fact that the hash function requires much more calculation? Under the assumption that the stored reference hash values cannot be modified, if the attacker changes even one bit of the software's binary code, the corresponding hash value changes drastically. The attacker will not be

[5]Section 15.3 provides an explanation of the hash function.

able to forge a proper piece of code whose corresponding hash value will match the reference value. This feature is intrinsic to the hash function. With a checksum, it is easy to forge a piece of binary code that will present a corresponding checksum by appending a few dumb bytes. If the attacker modifies one bit of the software, she has just to adjust one of the dumb bytes to obtain an invariant checksum.

- This solution is applicable only to applications and drivers that the bootloader knows before runtime. For dynamically loaded applications, thus unknown to the bootloader beforehand, integrity checking needs another mechanism. Section 4.2.5 describes a potential solution to this problem.
- Once the integrity of the system is established, the bootloader checks the integrity and the origin of the piece of software to be executed. Usually, it verifies that the originator cryptographically signed this piece of software (Sect. 4.3.6). Of course, the bootloader has to store the public, signing, root key of the expected originator securely. This public root key has not to be secret. Nevertheless, it must not be rewritable by the attacker. If an attacker could change this public root key, then she could replace it with her own public root key and so execute arbitrary code signed by her corresponding private root key. It is beneficial to note that the signature empowers two verifications. The first verification is that nobody tampered with the software package. The second check is whether the authority who knows the private, signing, root key issued this software package. In other words, checking the signature verifies both the integrity and the origin of the piece of software before executing it.

Sometimes, the bootloader is more complex and has several stages. The two initial stages are the ones described above. The following stage loads a signed middleware. This middleware will load the dynamic application once it successfully checks that the signature of the application is valid. In this configuration, it is good practice to sign the dynamic applications with a different key hierarchy than the one used to sign the middleware. This differentiation allows distributing application-signing keys to external, trusted, software developers while keeping control and limiting the risk of forgery of its proprietary middleware. With such a key distribution mechanism, only in-house developers have access to the process that signs the middleware.

The secure bootloader should be a tamper-resistant, carefully crafted, piece of software. Ideally, to prevent any modification, the secure bootloader should be stored in a non-writable memory. With such a secure bootloader, it seems impossible for an attacker to execute an arbitrary piece of software. Unfortunately, Law 1: Attackers Will Always Find Their Way is always true. Attackers sometimes find a way to unlock such devices. This kind of attack is often called jailbreaking. The objective is to find a vulnerability that grants root access to the device. With these elevated privileges, the attacker can modify the secure loader so that it accepts any binary code, even if not signed by the manufacturer [33]. The next example examines such a jailbreaking exploit.

The Devil Is in the Detail A very sophisticated secure bootloader protects Microsoft's game console: Xbox 360 [34]. It uses a three-stage system with the initial bootloader being signed by RSA and being encrypted with RC4. The bootloader securely launches a hypervisor. The hypervisor starts the signed games once their signature is verified. Some early versions of this hypervisor had flaws that allowed executing unsigned code. Microsoft systematically issued a newly revised version of the hypervisor that was immune to the weakness. Microsoft forced the update of Xbox 360 to enforce the systematic use of the most current hypervisor that is not crippled with vulnerability. Every known hack of Xbox 360 uses one of the flawed hypervisors. The enforcement of the use of immune hypervisors should prevent the existing attacks. Thus, the goal of the attacker is to launch a defective hypervisor instead of a flawless new one. The bootloader checks whether the signature of the hypervisor is legitimate. The flawed hypervisors are genuine Microsoft pieces of software. Therefore, they are duly signed. If the bootloader would only verify the signature, it would not ban the defective hypervisors. Therefore, Microsoft modified the bootloader to check whether it attempts to execute one of these flawed hypervisors. In addition to validating the signature, it calculates the hash of the hypervisor and checks whether the calculated hash belongs to a blacklist of forbidden hashes. Hackers must bypass this additional control to install a flawed hypervisor.

Gligli and four hacking colleagues implemented a hardware attack based on hardware fault injection. Fault injection is a sophisticated but devastating class of attack [35]. The objective of fault injection is to make a piece of software act in a different way than expected by its designers. The source of the fault is external stimuli in the environment of the hardware component. Typical stimuli can be glitches in the external clock signal, variations in the voltage supply, tiny pulses in the reset signal, low temperature, and even X-rays and ion beams. These attacks are not intrusive, as they do not require modifying the piece of software or using a debugger or injecting signals inside the chip.[6] The stimulus should generate a fault in the processor, resulting in the program execution's derailing. It does not anymore follow the expected flow of execution.

The hackers found that a tiny pulse in the reset signal of the main processor while the C function `memcmp`[7] executes resulted in it always returning a true Boolean value, regardless of the compared values. Unfortunately, developers often use the standard C function `memcmp` to compare two hash values to check whether their signatures match. To trick the bootloader, the

[6]Some fault injection attacks are more intrusive as they require the depackaging of the component. Depackaging is the operation that removes the silicon die package and sometimes removes some physical layers. This is the case with white light and laser attacks.

[7]`memcmp` is a standard function of the `libc` library that compares two blocks of consecutive bytes. If the blocks are the same, the returned value is true; else the returned value is false.

attackers have to generate a glitch in the reset signal at the proper time. Even if the signature of the forged software is to differ from the expected one, the comparison will indicate that it matches the expected result. Consequently, the bootloader may authorize the execution of the banned, flawed hypervisor. The bootloader believes that the presented hash does not belong to the blacklist and thus permits execution of the defective hypervisor. The difficult trick is to find the exact timing of when to apply the glitch.

The Xbox 360 comes in two versions: flat and slim. On the flat version of Xbox 360, the attackers discovered that pulling up a pin of the processor reduces its speed by about 120 times. Thanks to this trick, they profiled the bootloader to find the precise moment of the occurrence of the `memcmp` function that compares the hash of the executing hypervisor with the reference values. After many trials, they designed a successful hack.

1. Wait for the beginning of the execution of the `memcmp` function that compares the two hash values.
2. Slow down the processor when it starts the hash comparison by pulling up the corresponding pin.
3. Wait for a precise duration and then apply a 100-nS pulse to the reset pin of the processor. The slowing down of the processor increases the reliability of the hack as it requires less accurate timings.
4. Let the processor return to nominal speed by releasing the pin's voltage.

For the slim version of the Xbox 360, the hackers did not find a similar pin that would slow down the clock of the processor. Nevertheless, they found that a programmable phase-locked loop (PLL) chip drove the clock using an I2C bus. The I2C bus is a cheap, two-line, standardized, serial bus widely used in hardware design for many decades. The I2C bus is not protected and is well documented. The previous hack evolved into:

1. Wait for the beginning of the execution of the `memcmp` function that compares the two hash values.
2. Through the I2C bus, command the PLL to slow down the clock of the processor.
3. Measure exact duration and then apply a 20-nS pulse to the reset pin of the processor.
4. Through the I2C bus, command the PLL to return the processor to its nominal speed.

Fault injection attacks are not easily reproducible. They often require many successive failed attempts until one is successful. Nevertheless, these attacks are devastating. Moreover, they are difficult to prevent. In the case of the described hack, the attackers claimed that their average success rate was about 25 %. For this reason, successfully booting an unsigned code may require a few minutes of automatic trials before the Xbox is successfully fooled. Furthermore, fast and accurate hardware is required to generate pulses as small as 20-nS. In other words, this attack is not easy and not practicable by newbies. Nevertheless, the outcome is that the attackers succeed in running an arbitrary piece of software on Xbox 360. Once the flawed hypervisor is running, it is possible to trick it to accept any arbitrary piece of code. This exploit is a physical attack as it requires access to the device and some physical equipment. At the time of writing this book, there was no known purely software-based attack.

Lesson: Attackers are resourceful. Even garage hackers may use sophisticated attacks. All sophisticated attacks do not require expensive, sophisticated tools. All attacks are not easy to reproduce repeatedly. Thus, some attacks may not have an enormous business impact and affect reputation more than finances.

Fault injections are extremely powerful attacks. They are often wrongly categorized as a type of side-channel attack. Side-channel attacks collect information leaking from side channels, and through careful analysis, they retrieve secret information. For instance, the side channel can be the duration needed to perform encryption, measurement of power consumption which gives an idea of how many transistors commute, or measurement of local electromagnetic radiations. Since the publication in 1996 of the famous, first timing attack by Paul Kocher [36], side-channel attacks became more sophisticated than that seminal one. They have extensively proven to be devastating and have forced manufacturers to design and implement dedicated countermeasures (Sect. 6.2.2).

The previous insert describes a hack dedicated to the Microsoft Xbox 360. Microsoft is not the only hacked game console manufacturer. At the time of this writing, every game console has been broken. The Nintendo Wii [37] and Nintendo DS have been broken for each new generation of product.

There was no protection in the initial versions of the Sony Play Station (PS). Their design allowed running any piece of software. Soon the hobbyist community became enthusiastic and extremely active. With its PS3, Sony decided to restrict the execution to approved pieces of software. The latest versions of Sony PS3 have also been successfully jailbroken. At the December 2010 Chaos Computer Club (CCC)[8] conference, George Hotz, by the nickname of GeoHot, disclosed the digital signature algorithm (DSA) private key used to sign the firmware of all PS3 devices [39]. Normally, it is impossible to guess a private key while knowing only its public key. The usual security assumption is that this private key never leaks out. Usually, private root keys are hosted in Hardware Secure Modules (HSMs) and stored in a safe. Strictly enforced, security policies govern access to the HSM. The absolute secrecy of the private key is the cornerstone assumption of most trust models.

[8]For over 25 years, the CCC has been the largest European hacker's group. The activities of the club extend from technical research and dissemination to political engagement [38]. Each year, its December conference gathers many of the most influential hackers worldwide. Many new exploits are disclosed during this event.

GeoHot could not get access to the HSM. Then, how did he find this private key? The mistake was a poor implementation of the signing software. In a proper implementation of the DSA signature, the signing process uses a different random value for each new signature. Unfortunately, Sony used the same fixed value within its DSA implementation. Every signature employed the same "random" value. This mistake, well known by the cryptographic community, allows guessing the private key [40] with a few valid signatures. Indeed, GeoHot explored this option and discovered that he could retrieve the signing key. Nevertheless, this was not the ultimate key as it protected "only" the firmware. In October 2012, hackers by the nickname of "The Three Musketeers" went a step further [41]. They guessed and released the private key used for the bootloader LV0. With this key, it was possible to sign and install any arbitrary software on the console. Unfortunately, the Musketeers did not disclose how they discovered this private key.

The game industry is under continuous attack from the hacking community. Sadly, the game industry made many mistakes with its security systems. Copy protection systems regularly applied to games anger consumers. Nevertheless, in Sect. 2.3.1, we will come back to the world of game consoles, but this time with a positive example.

Jailbreaking applies to devices other than game consoles. Indeed, the phone industry initially coined the word jailbreaking. Historically, jailbreaking a mobile phone has meant circumventing the system that enforces that the mobile phone uses only a given operator's network and subscription. Many operators subsidize the cost of the mobile phone. Therefore, they want to guarantee their return on investment by restricting the use of the subsidized phone to its own carrier network (and associated carrier networks when roaming). Jailbreaking removes this restriction. At the end of July 2010, the US Copyright Office and the Librarian of Congress announced six new exemptions to the Digital Millennium Copyright Act (DMCA)[9] [43, 44]. Exemptions authorize circumventing protection measures as defined by the DMCA under very specific conditions. One of these exemptions states that jailbreaking mobile phones is legal if the purpose is to enable the use of the phone on other carrier networks.

With the advent of smartphones, and especially the Apple iPhone, jailbreaking took on a second meaning: allowing the execution of arbitrarily downloaded applications on the smartphone. This operation is sometimes called "rooting." For instance, Apple iPhone, Apple iTouch, and Apple iPad can exclusively execute applications downloaded from the Apple App Store. Apple applies extremely tight control on these applications. As usual when restricting usage, this was a challenge for hackers. An arms race started between Apple and the hacking community. Until iOS 9.0.2,[10] hackers offered a solution to run arbitrary applications on these

[9]Since October 28, 1998, the DMCA [42] defines the US copyright laws. Normally, under the DMCA, it is illegal to circumvent any security measure. Nevertheless, there are some exemptions to this rule. Since its inception, five such amendments were issued in 2000, 2003, 2006, 2010, and 2014, defining new exemptions to the DMCA rules.

[10]iOS 9.0.2 was the latest version of iOS at the time of editing this chapter.

devices. They used pieces of software such as `redsn0w`, `Absinthe`, `JailBreakMe` [45], and `Pangu` [46]. Rooting enabled using an alternate application store: `Cydia` [47].

Other manufacturers also attempt to control their execution environment. Nevertheless, the advent of Android has tended to reduce this practice. Having access to a vast library of applications is an excellent selling argument.[11] Unfortunately, rooting introduces other security issues, such as opening the door to malware.

1.2.3 Flawed Designs

In the case of side-channel attacks and Xbox attacks, the success came from a new attack that was unknown to the designers. Unfortunately, often the attack succeeds because the design has serious security flaws [49]. Sometimes, poor design choices of the security designer are the cause of such defects. The following is a grotesque example. Wi-Fi routers currently use a secure protocol, WPA2-PSK, to protect the wireless communications against eavesdropping. This protocol uses a secret symmetric key that should be unique for each router. Usually, the router manufacturer stores on the device a random key generated by the factory. This key (or the passphrase that will generate this key) is often printed together with the network name (SSID[12]) on a label affixed to the router. The key is a long set of hexadecimal digits that the user might have to enter when joining a network for the first time.[13] Best security practice assumes that this WPA2-PSK key is random so that an eavesdropper cannot guess the key needed to join a network stealthily. On some of its models, the manufacturer Belkin did not respect this rule [51]. Rather than being truly random, the key was derived from the wide area network (WAN) medium access control (MAC) address of the router. By construction, the WAN MAC address is public information. The router broadcasts it. Thus, this address is available to any attacker. Unfortunately, the algorithm that derived the key from the WAN MAC address was far too simplistic. Each digit of the key was the result of a static substitution of one digit of the WAN MAC address. Once the substitution table was guessed, it was easy to calculate this pseudorandom key. Furthermore, Belkin's passphrase used only eight hexadecimal digits, i.e., 32 bits of entropy.

---

[11] Nevertheless, there are also "rooting" exploits available for Android. One example is `Towelroot`, designed by GeoHot [48].

[12] The Secure Set Identifier (SSID) is the alphanumeric string that is part of the header of the packets over wireless local area networks.

[13] A new protocol called Wi-fi Protected Setup (WPS) allows users to bypass this clumsy, complex phase of dialing long passwords. Unfortunately, once more, some weak implementations of this protocol undermined the security of some devices [50].

Exploring by brute force[14] 32 bits of entropy is currently just a matter of a few hours. For a symmetric key to be safe, this key should be at least 90 bits long [52]. This too short length was another significant design error.

It is interesting to consider why a manufacturer uses a deterministic algorithm rather than a true random number generator for creating the secret key. The reason is the ability to restore default values in the case of a problem. If the manufacturer were to use a random number generator, then it would have no way to provide the passphrase to Alice if ever the sticker of her router was destroyed or erased. The only solution would be to store all the generated passphrases in a database securely. The manufacturer or the operator has to offer through its hotline a sophisticated retrieval service. First, the operator must assess whether that the calling customer owns the device. Once this ownership is verified, the operator can deliver the value of the forgotten passphrase. This manual operation is expensive. Using a pseudo-random number generator (PRNG) with a known seed simplifies the logistics and reduces the cost. Nevertheless, this money-driven design choice requires a strong PRNG, which was not the case for Belkin (Chap. 3).

In the previous example, the motivation driving the error was the manufacturer's convenience and cost constraints. One of the leading causes of this flaw comes from a bad balance between user-friendly, marketable features and security constraints. For instance, for increased user-friendliness, car manufacturers introduced a new breed of car keys: the Passive Keyless Entry and Start (PKES) system. These car keys use radio frequency identification (RFID) wireless communication with the car over a secure protocol. The idea is that every automobile and its car key are paired. The vehicle automatically detects the presence of the right paired car key and acts correspondingly. For instance, when the car key is in the range of two meters of the car, the car will authorize the driver to open the doors with the handle (without his having to use the physical key). If the driver is inside the car, it authorizes his starting of the engine. As the car key remains in the driver's pocket or bag, the interaction with the car's access control is invisible to the driver. Obviously, this feature is extremely user-friendly.

Unfortunately, in 2011, three researchers from ETH Zurich, Aurélien Francillon, Boris Danev, and Srdjan Capkun, demonstrated a simple attack: a classical relay attack [53]. In a relay attack, the messages are retransmitted to make the communicating entities believe that they are closer than in reality. Thus, the weakness was not in the protocol itself, but in the initial concept. In PKES, the car initiates the challenge. The researchers exploit this specificity. They place a first antenna near the car to capture the emitted message of the car (as the antenna of the car key would do). Then, these messages are relayed to a second antenna close to the car key (8–10 m). Thus, the second antenna mimics the car emitter's behavior. This message relaying is independent of any logical protocol. A simple cable or RF

[14]Brute force attacks explore systematically every possible value of the key until one succeeds. Thus, in the case of 32 bits, it means at maximum 4,294,967,296 trials. Unfortunately, with current computers, exploring a 32-bit space is extremely fast. A brute force attack is the simplest attack. The defense is to increase the length of the key.

transmission links the two antennas. In other words, the system transmits messages between the car and the physical key a longer distance than intended. Consequently, if the attacker knows the location of the car, and if the attacker comes reasonably close to the car's owner, she may steal the signal of the car key. Near this targeted car, her accomplice can unlock the car and steal it. The researchers demonstrated this attack on ten car models from eight manufacturers. A few years later, high-tech thieves used this attack in the street [54].

The design is flawed because it assumes that the presence of the signal means that the emitting device is nearby. The researchers suggested one first simple countermeasure: Deactivate the car key with a physical switch. Unfortunately, this countermeasure has two issues. First, it does not take into account human factors. Without doubt, some people will forget to deactivate the key when leaving their car. Other users may not remember that they had disabled the car key and thus will struggle when trying to open the car, expecting it to automatically open despite their having deactivated the feature previously. The second issue is that adding a switch button would annul the perceived benefit of this system: being buttonless. This is most probably the heart of the problem. Unlocking the car is done without conscious action by the owner. Is it wise from a security point of view? The security designer should never neglect human factors (Chap. 7).

The researchers proposed a second countermeasure more complicated than the previous one. This solution requires accurately measuring the trip time, i.e., the duration needed for a message to reach the target point. The relay attack increases the trip time. Therefore, the car could detect its interception of the communication channel. Unfortunately, the trip time of wireless communication is not stable and precise. The dispersion of values is high. A class of secure protocols called distance bounding protocols attempts to prevent relay attacks [55, 56].

Sometimes, a feature of a system is an intrinsic vulnerability by design, as users may employ it in a way that its designers did not foresee. For example, Microsoft designed a proprietary file format, the Advanced Systems Format (ASF), for storing and playing digital media content. It is a container format used for Windows Media Audio (`.wma`) and Windows Media Video (`.wmv`). In addition to the audio and video streams, an ASF file may also contain text streams, Web pages, and script commands. One script command is extremely attractive to attackers: URLANDEXIT (code value 81). This command requires the player to open the Web page whose uniform resource locator (URL) is following the command code. The Web page is accessed automatically without any user interaction. The initial purpose was to be able to download transparently a new breed of codec that was not yet supported by the host. Unfortunately, in 2008, attackers used this documented feature to access sites that would distribute malwares [57]. For instance, the Trojan `Win32.ASF-Hijacker.A` searched the hard drive of the infected computer for ASF files [58]. Once the malware found them, it injected into these files the URLANDEXIT command pointing to a remote Web site controlled by its designers. The infected media file was ready to spread the infection. Many variants and malwares used this URLANDEXIT command.

Unfortunately, this dangerous feature is enabled by default. Microsoft released an update patch that allowed disabling this feature by manually updating some registry keys. Such operation requires some skill and is not user-friendly. Interestingly, the Trojan `Win32.ASF-Hijacker.A` disabled this feature for its infection to remain unnoticed by the computer that created the forged content. Not all players are vulnerable. For instance, the widely deployed, open source multimedia player VLC does not support the script commands of ASF.

As specified by Microsoft, this issue was not a security vulnerability but rather was a feature by design [59].

> When a content owner creates an audio or a video stream, that content owner can add script commands (such as URL script commands and custom script commands) that are embedded in the stream. When the stream is played back, the script commands can trigger events in an embedded player program, or they can start your Web browser and then connect to a particular Web page. This behavior is by design.

Sometimes, an excellent feature may turn into an exploitable hole. The Thunderbolt port is an Apple proprietary port that combines both the DisplayPort for displays and PCIe for connecting devices using one unique physical connector. At the December CCC 2014 conference, Trammel Hudson disclosed the first known proof of concept of a bootkit for Mac OS X [60] using the Thunderbolt port. Bootkits are a particular category of rootkits that stealthily infect the master boot record or the volume boot record. The master boot record is the piece of software that the Basic Input/Output System (e 1st occ. expansion given) loads to initiate the boot. The volume boot record holds in the first partition the piece of software that loads the OS. In other words, a bootkit is a rootkit that installs itself in the boot system of the machine before the installation of the OS. Thus, the loaded OS is unaware of the presence of the bootkit.

Hudson's exploit uses several weaknesses in the boot system of Mac OS X.

- A CRC32, rather than a cryptographic signature, protects the integrity of the boot Read-Only Memory (ROM).[15] Unfortunately, the purpose of a cyclic redundancy code (CRC) is to check that the data is not corrupted (i.e., there is no mistake due to transmission). The goal of CRC is not to verify whether a principal modified a piece of data. Forging a CRC that matches an altered piece of software is extremely straightforward. The attacker can modify the boot process software and bypass the control of integrity by just calculating the new CRC and replacing the previous value with the calculated one.
- The firmware that is to be upgraded with the Extensible Firmware Interface (EFI) is signed with RSA 2048. However, the verification of the signature is done by the boot software. The previous vulnerability may alter this bootloader. The attacker may load her forged firmware at boot time using EFI if she can

[15] In the case of Mac OS, the ROM is indeed an electrically erasable programmable Read-Only Memory (EEPROM). This allows a potential upgrade of after its deployment.

deliver it to the machine to infect. Nevertheless, this attack requires physical access to the machine.

- Hudson used an attack that was disclosed in 2012 [61]. At boot time, EFI queries external devices that are connected through PCIe about whether they have any Option ROMs to execute. An Option ROM or expansion ROM is a piece of firmware that the BIOS launches at boot time. The Apple Thunderbolt port also supports the PCIe interface. The Thunderbolt port allows this function to load an arbitrary firmware from a connected device that would announce it has an Option ROM.
- Hudson fooled the boot firmware by replacing Apple's public key with his public key. Thus, the Apple software checks the signature of his malware and decides that it is legitimate, as the replaced key pair, used to verify the signature, and signed the malware. The legitimate Apple software then executes the malicious Option ROM. Later, the attacker's public key is written down in the ROM, preventing any Apple-authorized firmware upgrade from occurring. The compromised device will only accept firmware updates signed with the attacker's private key.

The potential attack is to design a forged Thunderbolt device with the expected malware as an Option ROM. The attacker needs physical access to the target computer. Then, she boots the device with the connected forged Thunderbolt device. In a few minutes, the attacker owns the machine. The entire attack is fast. Apple prepared fixes that prevent Option ROM execution during a firmware upgrade.

In 2014, Luyi Xing and his colleagues disclosed a new class of attack, which they coined pileup [62]. The attack uses an update of the OS to create new vulnerabilities. The researchers used Android to demonstrate the concept. Nevertheless, the attack may apply to other OSs. Google issues a new version of Android about every year. Each update usually features new system applications, new permissions, and new attributes. An increment of the application programming interface (API) level index signals this evolution. An attacker knows the history of this development. She designs a malicious application that uses selected permissions and attributes that are only available in the version of Android with the highest API level. The versions of the OS with a lower API level do not know these permissions. Thus, at installation time, they do not ask the user to grant this unknown permission. The forged application is installed on a device running a version of Android with a lower API level. The malicious application cannot use these newer unsupported features. Later, the device may be updated to the latest version of Android. When upgrading, Android automatically grants the previously claimed permissions, privileges, or attributes without asking for confirmation from the user again. It would not be user-friendly for the OS to request permission from the end user for each application he previously installed. This systematic request for each application would make the upgrade process tedious. The researchers gave some demonstrations of pileup attacks. For instance, Android 2.3 (so-called Gingerbread with API level 9) does not have permission to receive Google

Voice SMS.[16] This permission has been only present since Android 4.0 (so-called Ice Cream Sandwich with API level 14). Using such a pileup attack, the malicious application could read SMS messages of Google Voice without the user being aware once he upgraded to a version of Android greater than 4.0.

Another method is to claim some package name that the system will use in a future upgraded version. Once the OS is upgraded, the previous data will be kept. If malicious, this data may contaminate the system application. Pileup attacks may be dangerous in an extremely fragmented environment such as Android's.

Modern designs are extremely complex. Exploring all the use cases is difficult, if not impossible. It is even harder to foresee scenarios that will benefit from the evolution of the product. Flaws are ineluctable in elaborate designs. The attackers will use such flawed designs to create exploits.

1.2.4 Advanced Persistent Threats

There are two categories of attacks: opportunistic attacks and targeted attacks. In the case of opportunistic attacks, the attacker randomly selects her targets. For each random target, she checks whether she can successfully apply her exploit. If it is the case, then she perpetrates the attack, intrudes on the system, and attempts to dive deeper into it. If she fails, she looks for another victim. Usually, this type of attacker does not invest much effort into breaking into targets. If a target does not easily surrender, then the attacker moves to another random target. For example, this routinely occurs on Internet routers. Attackers scan IP addresses to find open ports, usually using automated tools, such as `nmap` or the `Shodan` search engine. Then, the reported open ports are analyzed to check whether their corresponding protocols are vulnerable to widely known exploits. Perimetric defense tools prevent attackers from breaching the perimeter defined by a company. They include firewalls, intrusion detection systems, and antiviruses. As long as they are up to date, typical perimetric defense tools are usually sufficient to deter opportunistic attacks. Opportunistic attackers always look for the weakest preys.

In the case of targeted attacks, the attacker focuses on a precise target or at least on a category of targets. The attacker's goal is to fulfill a particular task. For that purpose, the attacker uses a broad range of techniques. The attack may be sophisticated and encompasses many steps. The first step collects maximum amount of information about the target. This step uses all available resources, such as published information, stolen or leaked proprietary information, and information gathered through social engineering [63]. Business intelligence tools such as Maltego may help the attacker to cross-check the collected information. During the second step, the attacker attempts to breach the target using the collected information. Depending on the actual security of the target, this second step can be

[16]The permission is `com.google.googlevoice.RECEIVE_SMS`.

Fig. 1.3 RSA SecurID

extremely complex and may involve multiple attacks spread over different systems. The new term advanced persistent threats (APTs) describes the most sophisticated attacks of this category. The most interesting recent ATP was the attack against RSA Ltd.

SecurID is a widely used two-factor authentication token. The system is usually used to authenticate a user, either to grant him access to a remote service or system or to establish a virtual private network (VPN) with his company's network. The token displays a six-digit number that changes every few seconds following a secret proprietary algorithm (Fig. 1.3). Each token is associated with one person. When receiving her token from her IT department, Alice is also attributed a four-digit personal identification number (PIN). To authenticate with SecurID, Alice provides her login identity (for instance, her email address), her four-digit PIN, and the six digits currently displayed by the SecurID token. The authenticating server checks whether these three parameters fit together. If they do, then the server authenticates Alice. The SecurID system is a two-factor authentication mechanism because it checks:

1. Something Alice knows, e.g., her PIN; this renders the theft of the SecurID token useless. The thief would not know her four-digit PIN.
2. Something that Alice owns, e.g., the token itself; the continuously changing value of the six digits proves that she currently holds it. This solution renders shoulder surfing useless because the validity of the code has an extremely short life (a few seconds).

On March 18, 2011, Art Coviello, the executive chair of RSA Ltd., announced to his customers that his company had been attacked, and some information leaked out [64]. The stolen information may have reduced the security of the SecurID tokens. RSA Ltd. did not disclose what information the intruders took. Nevertheless [65], we may infer that the intruders have now a method to impersonate Alice's SecurID token or a way to fool the authentication server into believing that the login user is Alice.

RSA Ltd. properly managed this crisis. The company had a positive, responsible reaction regarding the hack. It provided some details and timely visibility on the issue. Therefore, this APT teaches us many interesting lessons.

The Devil Is in the Detail The attackers penetrated the company RSA Ltd. through a targeted phishing email using a malicious Excel file [66]. Over 2 days, two small groups of EMC[17] employees received two spear phishing emails [67]. Phishing emails are emails that attempt to impersonate a known user or entity. Attackers specifically craft spear phishing emails to target an organization or specific individuals within the targeted organization. On March 19, 2011, one employee of EMC opened one of these emails, which were copied to three other employees. This email was sent supposedly by a recruiting Web site: beyond.com. This site is a certified partner of RSA Ltd., making the phishing email less suspicious and even plausible. The subject of the email, the 2011 recruiting plan, had an attached file, `2011 recruiting plan.xls`. The attack used a typical social engineering ploy: exploiting human curiosity and greed. It always works. The inquisitive employee opened the attached object.

The malicious, poisonous Excel file embeds an Adobe Flash object that Excel executes automatically when the file is opened. Using the vulnerability CVE-2011-0609 [68], the Flash object installs a malware: PoisonIvy Backdoor. This vulnerability allows remote attackers to execute arbitrary code using a crafted Flash object. The vulnerability was first disclosed on March 15, 2011, i.e., 4 days before the attack. The malevolent Flash object installed the backdoor. PoisonIvy practically gives the attacker full control of the infected computer [69]. For example, PoisonIvy can rename, delete, execute, upload, or download files, read and edit the Windows registry, manage services, and access remote drives. In the RSA attack, the instance of PoisonIvy connected back to the domain `good.mincesur.com`. For many years, the domain `mincesur.com` was known to be associated with illegal activities [70]. Ideally, a firewall or proxy server should have blocked this domain. The compromised account was not one of a critical IT staff member. Nevertheless, the attacker succeeded in escalating her rights by looking for other more privileged accounts. Then, the attacker was able to gain access to critical information concerning SecurID tokens. She compressed the collected data into several password-protected rar files. Using encrypted files may allow passing under the radar of eventual data loss prevention (DLP) tools that analyze the semantics and structure of exported files. Finally, the attacker

[17]EMC is the company that owns RSA Ltd.

exfiltrated the stolen data using one `File Transfer Protocol` (FTP) server.

Was RSA Ltd. guilty of not being up to date? When the attack occurred, RSA had no means to patch this vulnerability. Adobe announced that the corresponding patch would be available only on March 21, 2011, i.e., 2 days after the infection [71]. The problem was that the targeted employee opened the infected file. We would expect that employees of a security-dedicated company would be extremely careful about spam and phishing emails.

Lesson: Increase the security awareness of all employees. As there will always be 0-day vulnerabilities in the field, the defense of antiviruses is not sufficient. Perimetric defenses are necessary but not anymore sufficient. Only vigilance may thwart 0-day email attacks.

This is not the end of the story. At the end of May 2011, Lockheed Martin, one of the largest US defense contractors, was under a "significant and tenacious" online attack. Some attackers attempted to penetrate Lockheed Martin's network [72]. It seems they tried to enter through Lockheed Martin's VPN which was protected by RSA SecurID. Apparently, the attackers possessed the seeds, serial numbers of valid tokens, and the underlying algorithm used to secure SecurID. Lockheed Martin immediately detected the attack and took drastic, preventive measures. The company disconnected all remote accesses to its corporate network. All employees were requested to work only from the office for 1 week. Telecommuting was banned. The VPN was disabled. All SecurID tokens were replaced. Every Lockheed Martin employee had to change his network password. The proximity in time to the RSA Ltd. hack, the intrusion vector (RSA SecurID), and the target itself (a defense contractor) tend to indicate a high correlation between the two events. The first stage of the attack would have been the theft of the secret information of the SecurID. The second stage used this information to try to penetrate Lockheed Martin's systems stealthily. This attack is perhaps the first publicly widely known instance of an APT.

The term APT comes from the military defense domain. Initially, APT was defined as a cyberattack launched by nation-state-funded organizations to steal, over a prolonged period, critical nation information using sophisticated, stealthy techniques. Currently, APT transcends the specialized realm of warfare to extend into business intelligence [73]. APT is an advanced attack because it employs an extensive set of tools and techniques to perform attacks. The attackers are well informed, well equipped, highly skilled, and well funded. They may design custom tools to succeed in their attacks. APT is persistent because the threat is not opportunistic, but it is intended to fulfill a specifically defined objective. The attackers will use all the time necessary to succeed. The cost of the operation is not an issue. APT is a dangerous threat because human beings fully manage the attack rather than robots or botnets. Human attackers are more opportunistic and reactive than robots. Usually, the life cycle of an APT has eight steps.

1. Collecting the target company: Information concerns the employees as well as the IT infrastructure. The objective of this step is twofold.

   - A better understanding of the network topology, the materials used, and the different pieces of software employed by the targeted company. This cartography can be done using exploration tools such as SinFP [74] and Metasploit [75] and by gathering public information available on the Internet. Data stolen via social engineering may be another worthwhile source of information.
   - A better understanding of the organization of the company; it is valuable to identify people who may have the "nearest" possible access to the targeted data. Obviously, IT administrators are ideal targets. The social engineer uses all available sources. There are many potential sources of names. Social networks are excellent hunting places. Furthermore, professional social networks such as LinkedIn are even better locations to identify and collect information about potential, initial entry points [76]. It is easy to find the name of the chief information officer of an organization and the key members of the IT staff of a company via professional social networks.

2. The analysis of data collected during the previous step provides a list of potentially interesting targets. Through social engineering, the attacker targets the identified users. This step uses the full panoply of methods, such as spamming, spear phishing, and impersonating calls. See Chap. 7.
3. The attacker attempts to intrude the target. Once a gullible user found, the attacker uses vulnerabilities (often 0-day vulnerabilities) to install a Trojan horse on the victim's computer. This backdoor will be the entry point of the hack. Of course, the installation and the operation of the infecting malware have to be stealthy. This is true for modern Trojans.
   Spear phishing remains the favorite method for initial infection. New file-sharing sites, such as DropBox, have become fashionable in the hacking world [77]. The phishing email points to a document shared through such a service. Such cloud-based services can offer efficient anonymity to the attacker. Tracing back the owner of the account is hard for forensic investigators.
   A waterhole attack is another example of a method employed to inject a backdoor [78]. Once the attacker has profiled the victim in step 2, she can identify a list of Web sites the victim will most likely visit. Then, the attacker analyzes these Web sites to find vulnerable ones. In the vulnerable ones, she attempts to implement some malware that will try to infect the victim's computer. The attacker's malware waits until the victim visits the Web site, like a lion waiting for its prey to come to the waterhole. During the victim's visit to a compromised site, the trapped malware attempts to infect the visiting computer. Infected files are often powerful vectors of infection. Some sophisticated malwares identify the victim and affect only the targeted victim and not the visitors that the APT did not target. This trick reduces the risk of being detected as only

a limited number of computers will be infected. Thus, they reduce their own attack surface as malware detection tools do not see them.

4. Once inside the system, if the victim has no access to the servers that the attacker was targeting, she expands laterally to roam around the network. In other words, the attacker looks for other vulnerable computers on the internal network that would allow her to come nearer to her final target.
5. Escalate privileges (social engineering, access to administrators' encrypted passwords, etc.): As the targeted data is sensitive, we may expect them to be protected strongly. Only higher privileged principals should be granted access to them.
6. Compromise the system holding the targeted data.
7. Exfiltrate the data stealthily: Usually, the attacker encrypts the exfiltrated data to avoid detection by DLP tools, deep packet inspection, or any filtering proxies. Of course, the data transmission uses covert channels, i.e., hiding among legitimate communications. Sometimes, these covert channels are throttled down to avoid triggering detection of anomalies in the communication. The overall transfer takes longer, but it is less visible to monitoring than the open channel. APTs are persistent, and time is usually not an issue. The exfiltration occurs either by using the available communication channels or via the backdoor installed during step 3. Often, communication and control (C&C) servers are used in a similar way for botnet management.
8. Clean up to avoid being traced back: An APT should not be detectable. Stolen data has even more value when the victim is not aware of the leak. Therefore, the attacker clears the logfiles but also may remove the backdoors that she installed. Sometimes, she may keep under control the compromised computer that she used to infiltrate the system. This control allows her getting back to collect information later. For instance, the alleged Chinese APT group called APT1 or comment crew keeps control of the victims' system for an average of 200 days [79].

In summary, the attacker studies the system in steps 1 and 2. She gains access to it in step 3. She enhances this access in steps 4 and 5. She exploits this access in steps 6 and 7. Eventually, she closes this access in step 8. In the literature, these eight steps are sometimes condensed into five steps: scouting (steps 1–2), intrusion (step 3), discovery (steps 4–5), capture (step 6), and exfiltration (steps 7–8).

An APT may be more complex than described in the previous section. For instance, an APT may first need an APT against another target to acquire some critical information or find external access to the final target. This was probably the case with the Lockheed Martin attack, which required first stealing information from RSA Ltd. to penetrate Lockheed Martin's system. In 2012, Adobe announced that it discovered two pieces of malware that presented a valid Adobe signature [80]. Attackers compromised a build server of Adobe and through it requested Adobe to sign the pieces of malware. As one of the pieces of malware was not "publicly" available, the likelihood that this was part of a larger APT is extremely high.

Defending against APT is a complex and costly endeavor. The traditional perimetric defense tools are insufficient. More sophisticated tools are mandatory, such as [81]:

- Automated sandboxing that analyzes every received file in a safe, isolated environment,
- Whitelisting applications that exclusively authorize execution of known trusted applications on a given device,
- Deep network traffic inspection that checks whether the communication is compliant with established protocols and acceptable patterns.

The ultimate defense is the human analysis by experts who carefully monitor activities and logfiles looking for anomalies and swiftly react. See Chap. 9.

APT should not be a concern for consumers. APT usually affects companies that handle strategic or critical infrastructures or organizations that have high-value intellectual properties. Third parties that work for such targets may become ancillary victims.

1.3 Takeaway

The previous sections highlight that attackers will always find a way to break the system. Once this fact is accepted, the security designer has to take this reality into account in all his designs. In a white paper, Kaspersky Lab identify that "assuming everything is OK" is one of the ten enablers of the IT department facilitating cybercrime [82]. Overconfidence is the worst weakness of the security designer. The security designer must never forget that the attackers will defeat him. The next sections present how the design must reflect this persistent threat.

1.3.1 Design Your System for Renewability

"*Even if the Allies lost this battle, we should not have lost the war.*" The British Minister of Information, Duff Cooper, noted this sentence on May 28, 1940, after the defeat of the Battle of France. Every security designer should endorse this sentence. The system must be able to recover after any successful attack. In other words, the system must be able to renew itself. This section starts by examining a class of systems that were designed with renewability as a core feature.

In 1984, the French broadcaster Canal+ launched its first subscription-based Pay TV channel. A few months later, the first schematics of pirate decoders were published. In 1985, the US operator HBO used Videocipher II to protect its satellite programs. Once more, pirate decoders were available soon on the market. The piracy market was organized as quasi-industrial organizations [83]. These first commercial deployments of Pay TV highlighted the need for longer keys and renewability. Two European Conditional Access Systems (CASs), Eurocrypt and Videocrypt, built the foundations of modern content protection. The Union Européenne de Radiodiffusion (UER) established Eurocrypt as an international standard. The Eurocrypt specifications created the vocabulary that is still in use today in Pay TV systems [84]. News DataCom and Thomson codesigned Videocrypt [85]. These systems created a new generic scheme for content protection [86] that is still in use today.

The main significant improvement over previous Pay TV systems came from the use of smart cards. Smart cards are removable secure processors. Figure 1.4 provides an overview of the utilization of the smart card within a Pay TV system. The set-top box receives a scrambled video stream multiplexed with encrypted information: entitlement control messages (ECMs) and entitlement management messages (EMMs). An ECM contains the decryption key, the so-called control word (CW), used to scramble the video, whereas an EMM carries the credentials that manage user rights. The STB extracts the ECM and EMM from the stream and transmits them to the attached smart card. The smart card holds in its secure memory both the secret keys and the algorithms used by the Pay TV system. It decrypts the ECM and checks whether the holder of the smart card is entitled to watch the current program. If this is the case, then the smart card returns the control word to the STB. The STB descrambles the scrambled video using the returned CW.

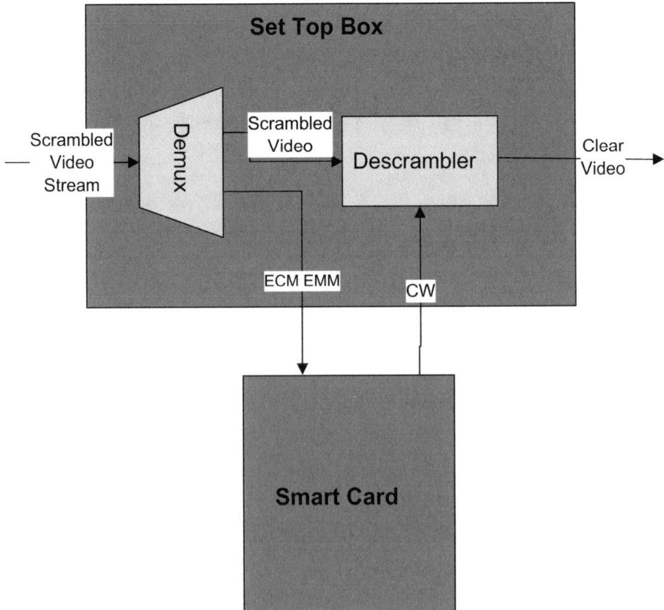

Fig. 1.4 Simplified view of a card-based Pay TV system

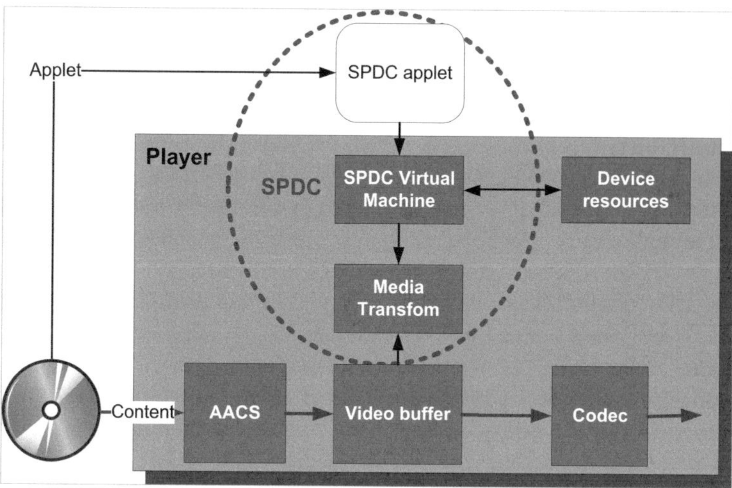

Fig. 1.5 Basic synoptic of SPDC

The initial assumption behind the trust model of modern Pay TV was that crackers would ultimately break the system and, in particular, the key management component. The assumption was that nobody would crack the scrambling method. In other words, the design assumption was that it was impossible to build a pirate decoder without first breaking the key management. Thus, the smart card implemented the entire key management and the management of entitlement rights. Changing the smart cards was a powerful way to address potential future hacks of the key management. This approach has proven to be successful.

Eurocrypt and Videocrypt survived many years. They were phased out by obsolescence of the video technology rather than by crackers. In 1995, the Digital Video Broadcasting (DVB) group standardized a way to protect MPEG2 transport streams. It employed the same smart card-based architecture. For more than 30 years, despite several successful hacks, most smart card-based CASs recovered.

The Digital Video Disc (DVD) was hacked in less than 2 years [87] after its launch. Unfortunately, its protection mechanism was not renewable. The hole could never be closed. Therefore, when the entertainment industry decided to design the Blu-ray Disc, renewability of the security was one of the main requirements of the future systems. In addition to the encryption mechanism [called Advanced Access Control System (AACS)], the Blu-ray Disc Association specified an additional layer of security, called BD+. The purpose of this layer is to enable recovery from potential security breaches occurring in the AACS. BD+ disks contain a title-specific piece of software, called the applet, which executes on a BD+ player before and during the movie playback. The program is written in self-protecting digital content (SPDC) language, a dedicated interpreted language designed by

Cryptography Research Incorporated[18] [88]. Figure 1.5 describes its main elements. SPDC is mainly a virtual machine (VM) and a media transform block. The VM is far simpler than Java's, another well-known VM. The media transform block can apply simple modifications to the compressed video. When reading a BD+ disk, the player loads the applet from the disk. Then, the player executes the applet. Usually, this applet first checks whether the host player is compliant and not revoked. Then, the applet provides configuration parameters to the media transform block. This block deterministically modifies, using the supplied parameters, the compressed content, for instance, by swapping some bytes in the stream.

In the BD+ environment, the applet is stored on the disk. This applet modifies the AACS-decrypted content before video decoding. This downloadable applet ensures the renewability of the security of BD+; if ever AACS or one of its implementations were to be hacked, the use of an applet would allow the broken AACS to heal. At the authoring stage, the essence is modified with a function f before being AACS-encrypted. Thus, on the disk, there is $AACS(f(essence))$. The applet that defines the reverse transformation, i.e., f^{-1}, is delivered within the Blu-ray Disc. When the disk is played back, the applet will apply f^{-1} to the AACS-decrypted essence, i.e., $f(essence)$. It is assumed that non-compliant players will not properly execute this applet or that the applet will detect non-compliant players. Thus, if the applet does not execute, the non-compliant player will try to render $f(essence)$ rather than the essence itself; hence, the rendering will fail. The applet may also enforce a given manufacturer's model to perform some operations that will patch a security hole.

Rule 1.3: Design Renewable Systems

Renewing security is possible in three ways.

1. Renewing the piece of software that handles the security: This is useful if the algorithm or its implementation has been broken. This either replaces the compromised algorithm by a secure one or replaces the weak implementation of the algorithm by a more secure one.
2. Renewing the cryptographic material used by the system: This is mandatory when a secret key leaked out. Of course, the transfer to the host, as well as the installation of the new cryptographic material, has to be secure. The secret data may be changed for the entire installed base in the case of a class attack or only for a set of principals if the attack is more localized.
3. Renewing the complete system, i.e., both the algorithms and the secrets: This is in the case of extremely severe attack.

In the example of Pay TV, as a smart card encompasses the execution environment, the software, and the secrets, all three kinds of renewability are extremely

[18] Cryptography Research Incorporated (CRI) is the company founded by Paul Kocher who designed the first side-channel attacks: timing analysis and power analysis. In 2013, Rambus acquired CRI.

simple to activate. In the extreme case, renewability just requires the replacement of the physical element (although this replacement may imply serious challenges of logistics). Of course, the secure download of software and data may also be used for partially renewing security in smart cards. In the case of embedded devices, such as consumer products, methods 1 and 2 are possible but require a root of trust. Usually, the secure loader builds this Root of Trust (Sect. 1.2.2). If the secure loader is compromised, then it is not possible anymore to renew the system. In the case of computers, often the third method is preferred. Replacing an entire system is rather easy in this configuration. Only the new version of the piece of software needs to be downloaded. This ease of replacement has, of course, a price: the lack of hardware Root of Trust (Sect. 4.2.5). There is no way to guarantee execution in a trusted environment. Malware may compromise the newly downloaded software. In current general computers, no trusted software or trusted hardware checks the signature of a piece of software.[19] The verification of the signature may not operate in a trusted environment and can itself be the victim of an attack.

Some people wrongly associate revocation to a mechanism of renewability. The aim of revocation is to disable a principal whose security has been compromised or who is not anymore trusted. No compliant system should interact with a revoked principal. Revocation is a major weapon in the battle against attacks, but it does not replace renewability. Indeed, once a secret revoked, there is a need to replace the revoked secret by a new valid one. This is the role of renewability. Existing standards handle revocation, such as X509 [89]. Unfortunately, there is no standard for renewability mechanisms.

1.3.2 *Design for Secure Failure*

The previous framework for in-depth defense recommends addressing any consequences of an attack. A smart way to cure is to prevent the failure from impairing the system. In other words, the system should fail securely in the case of a successful hack. When the system is hacked, the attacker should not gain undue advantages. Her benefits while gaining access should be annulled or at least minimized. Safe failure is a concept widely used in the safety of systems. Critical systems are built so that if a failure occurs, the system may still operate correctly, or, at least, its critical, vital functions should be maintained. Nuclear power units, aircraft, and automotive systems are all designed with a strong requirement in mind to fail safely, i.e., avoiding putting users in danger. There is a clear distinction between failing safely and failing securely. For instance, a fail-safe lock will be unlocked in the case of power shutdown for people to leave the room safely. Under

[19]To be precise, many modern computers have the possibility to implement such a Root of Trust because they are equipped with Trusted Platform Modules (TPMs). Unfortunately, the major operating systems do not take advantage of this feature.

the same conditions, a fail-secure lock will remain closed to prevent attackers from entering the room. Safety and security are sometimes opposites.

HSMs present such secure failing defense. HSMs are highly secure devices dedicated to cryptographic operations. They have two key features.

- Dedicated, embedded cryptographic coprocessors, which perform exponentiation, drastically accelerate public key cryptography.
- Highly secure components that, like smart cards, they include many tamper-resistant features. These devices prevent the disclosure of stored secrets.

Particular care is given to protecting the most valuable assets of the HSM, i.e., the private or secret key(s) stored in its secure memory. A typical security measure is the safe erasure of these keys whenever the HSM suspects that an attack is occurring. The range of monitored attacks includes power attacks, as when glitching the power supply to attempt fault injection; timing attacks, as when slowing down the CPU clock to better analyze the datagrams; chemical, as when depassivating the chip, for instance, with nitrofluoric acid; and physical attacks, through microprobing. Thus, even if the attacker succeeds in her attack, she will only access random data rather than the actual keys.

Securely failing is important also when designing secure protocols. One of the difficult challenges relies on the so-called atomicity of the protocol. If ever a protocol is interrupted before its normal completion, it should fail safely. An attacker should not benefit from such an interruption. In 2015, a black-box device was available to brute-force the PIN of iPhones without triggering the restriction of ten failed attempts [90]. The hack used an optical sensor that checked the screen, a relay to command the power supply, and a USB cable. The box tries a PIN number and sends it to the iPhone using the USB port. If the dialed PIN is wrong, the optical sensor detects the error message, and the box turns off immediately the power supply of the iPhone and reboots it. The box tries another potential PIN. As the phone has to reboot after each trial, each failed attempt took about 40 s. Brute forcing a four-digit PIN took about 4 days.

Figure 1.6a represents the mistake. The iPhone likely checks first the validity of the PIN. If the dialed PIN is wrong, it displays the result. Then, it updates the failed attempt counter and checks whether the counter reached the maximum number of authorized failed attempts. The black box switches off the power supply before the iPhone can update the failed attempt counter in its nonvolatile memory. When rebooting, the failed attempt has not been registered in the nonvolatile memory. Fault injections often exploit this classical known mistake in the design. Figure 1.6b represents a potential countermeasure. The initial value of the failed attempt counter is set to the maximum number of trials. The failed attempt counter should be decremented before the verification of the PIN occurs. If the PIN is correct, the failed attempt counter is set back to the top limit. In this order of operations, the protocol fails safely. If the hacker switches off the power supply after the verification of the PIN, then the counter is already decremented. She gains nothing.

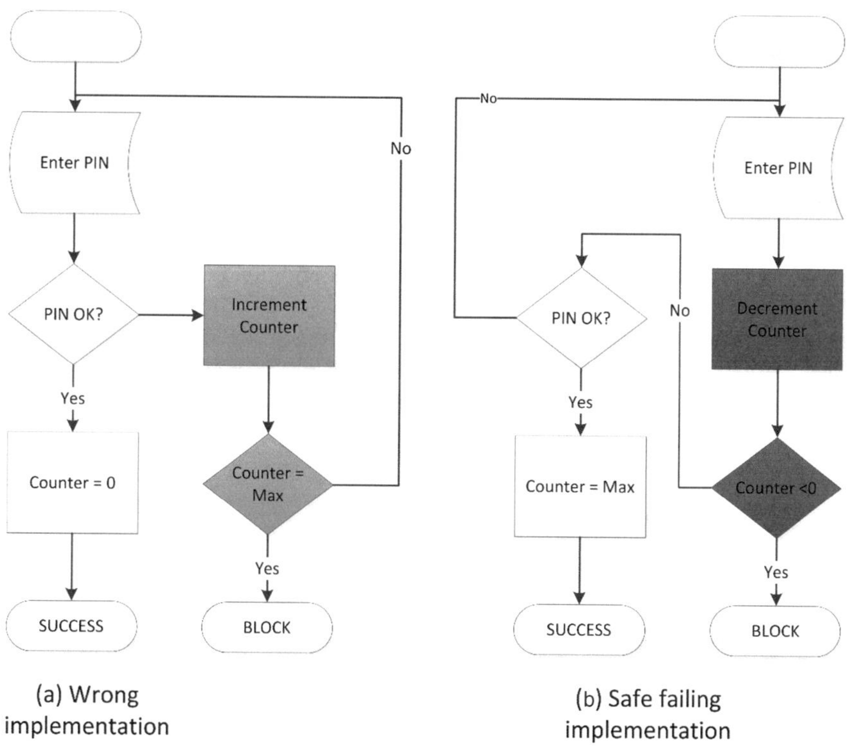

Fig. 1.6 Implementing a PIN verification

The designer should think how the attacker might derail his implementation for her advantage by injecting fault. Several techniques are possible for making the execution flow more robust against fault injection. Usually, the objective of a fault injection is to counter a Boolean test. One such technique is to split one Boolean test into a set of successive Boolean tests. For instance, rather than performing a comparison on the full data, the designer may use successive comparisons on subparts of the complete data. The attacker has to identify the occurrence of each comparison rather than one unique occurrence. The needed effort increases proportionally, and the likelihood decreases proportionally.

Securely failing is also important when writing security policies. There is no possible sound security in an organization if there are no well-documented security policies. The policy defines, at a high level, what is authorized and what is forbidden. It is essential that there be as little ambiguity as possible. Gray areas are always potential weak points in a security policy. The likelihood is extremely small that the editors have explored thoroughly and defined all the possible situations exhaustively. This likelihood may even be zero. Therefore, it is necessary to state that if a situation is not expressly permitted, it is prohibited by default [91]. This is a secure-fail condition. Good security policies ban an unknown legitimate situation rather than authorize an unknown dangerous situation [92]. Afterward, it is always possible to update the policy to cope with unforeseen legitimate situations without having to heal after the wreckage of an attack.

1.3.3 Defense in Depth

Medieval castle builders had substantive notions of security. Figure 1.7 illustrates a typical castle from the Middle Ages. Most of the fortified castles followed the same architectural core principles. A first, outer rampart protected the castle. Often a deep ditch surrounded this wall, or, even better, the castle was on top of a cliff. A portcullis protected the access to the inside of the castle. Defenders lifted the portcullis in the case of an attack. Usually, the houses and common infrastructures,

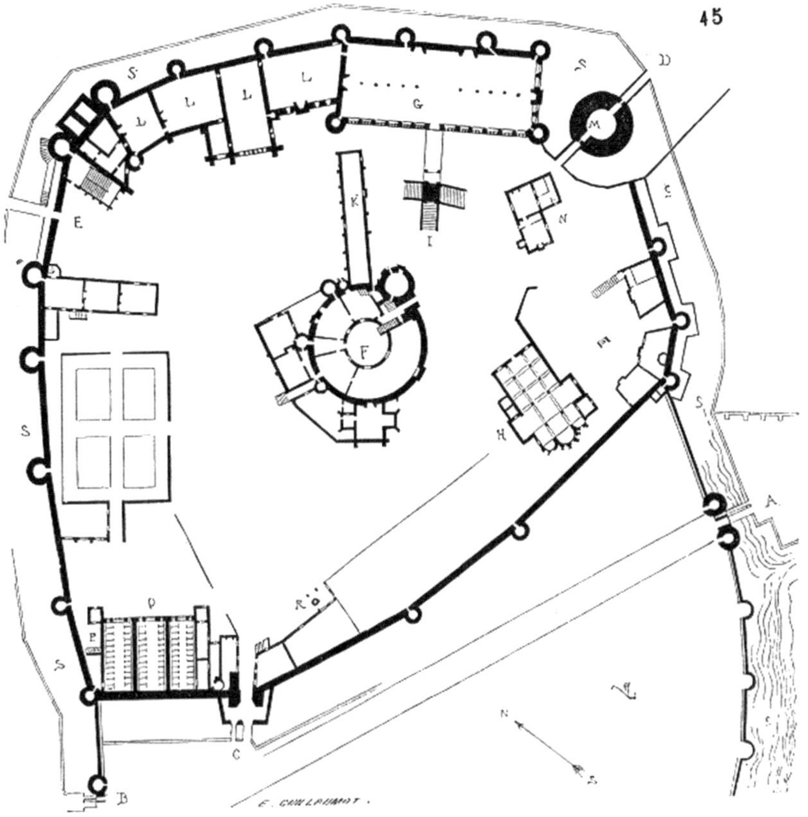

http://commons.wikimedia.org/wiki/File:Plan.chateau.Montargis.png#mediaviewer/Fichier:Plan.chateau.Montargis.png

Fig. 1.7 Medieval castle

such as the blacksmith's workshop or shops, were located behind this first outer wall. A second concentric wall, which was higher than the previous rampart, isolated from the common infrastructures the most critical infrastructures, such as the armory, the warehouses, and soldiers' barracks. The ultimate protection was the keep (location F in Fig. 1.7): a highly fortified inner tower. The noblemen and their families lived in the keep. In the case of a siege, the attackers had to defeat each successive protection before being able to reach the commander. Each successive wall was expected to be easier to defend and harder to breach than the previous one.

What are the lessons from this medieval architecture? The first lesson is that the medieval builders of castles knew about Law 1. They knew that they would inevitably lose the outer walls in the case of a siege. Therefore, they used several successive walls. Furthermore, they also knew which asset was the most precious one. In their context, this asset was the life of their Lord and his family (Chap. 2). They put this asset within the ultimate wall, i.e., the keep. Therefore, the lordship benefitted from all the successive defensive layers. This principle of successive in-depth defenses is still valid in our modern IT world. An attacker should have to breach many protections before reaching the most valuable assets.

Rule 1.4: Design In-depth Defense

The second example is about how to protect data centers, secure rooms, or vaults against physical intrusions. Physical security obeys to the rule of "security in depth." Physical security must create a set of successive zones isolated by physical barriers and gates [93]. Only the users who are the most privileged should access successive zones. The following example illustrates this rule. A facility receives visitors and handles sensitive assets in dedicated areas. Of course, most visitors should never approach these sensitive assets. Physical security may suggest a layout delimiting different zones.

- Zone 0: This zone is the external world where the security of the site does not apply. Everybody can freely access this area. The tenants of the facility have no control over this zone.
- Zone 1: Usually, the site is isolated from the external world (Zone 0) by a fence. Ideally, a clear area should be around the fence. This clear zone creates an unobstructed view to detect illegal intrusion. Furthermore, the fence drives traffic to one or a set of controlled entrances. Physical access controls the entrances to the site. Guards or receptionists monitor this access. Employees have company access badges. Visitors receive dedicated guest access badges.
- Zone 2: In this area, the visitors and employees can interact. Automatic gates control its access from Zone 1. Demonstration rooms and meeting rooms are in this area.
- Zone 3: This area is limited to employees. Visitors should not enter into this zone. Physical access controls admission to this area. Visitors' badges do not grant access to Zone 3. It is only accessible from Zone 2 area.
- Zone 4 (and above): This area is restricted to a limited set of employees. Only employees who need to access it should be granted corresponding access rights.

The access control may be more sophisticated than a badge reader. For instance, it may use two-factor authentication with biometrics. It is only accessible from Zone 3 area. Usually, vaults and data centers are located in such zones. Few people should be granted access to them.

The defense and control mechanisms between the successive layers should not be identical. The defense mechanisms of each zone should be more restrictive than the one protecting the previous zone. The mechanisms act as a sieve. Ideally, the consecutive defenses should feature increasing robustness. Indeed, the goal of the "inner" defense is to stop attackers who succeeded in defeating the "outer" defense. An attacker who manages to reach the "inner" defense already has proven that she is stronger than the "outer" defense. Were the "inner" defense to be weaker or equal to than the "outer" one, then the attacker would also easily defeat the inner defense if defeating the "outer" one. Indeed, in such unrestrained configuration, the only adequate defense would be the outer one. The builders of castles knew this design principle. The inner walls were larger and higher than the outer walls and thus stronger. Each successive defense slowed down the attackers. In network security, the use of firewalls and DMZs is a perfect illustration of in-depth defense (Sect. 8.2.1).

In-depth defense should not be limited to successive layers of increasing similar protection. It should also use a set of complementary defenses. Then, the security widens the defense perimeter. Diversity is a good security attribute.

In 2014, the Council on Cyber Security published the fifth version of its critical security controls [94]. It defines 20 security controls. Any organization should mandatorily establish these twenty control points. The variety of control points ensures that attacks are detected and remedied, guarantees that already compromised systems are identified, and prevents disrupting attacks. They constitute a comprehensive list of in-depth defenses. The ordered twenty critical security controls are as follows:

1. Inventory of authorized and unauthorized devices,
2. Inventory of authorized and unauthorized software,
3. Secure configurations for hardware and software on mobile devices, laptops, workstations, and servers,
4. Continuous vulnerability assessment and remediation,
5. Malware defenses,
6. Application software security,
7. Wireless access control,
8. Data recovery capability,
9. Security skills assessment and appropriate training to fill gaps,
10. Secure configurations for network devices such as firewalls, routers, and switches,
11. Limitation and control of network ports, protocols, and services,
12. Controlled use of administrative privileges,
13. Boundary defense,
14. Maintenance, monitoring, and analysis of audit logs,

15. Controlled access based on the need to know,
16. Account monitoring and control,
17. Data protection,
18. Incident response and management,
19. Secure network engineering,
20. Penetration tests and red team exercises.

An organization that implements all these twenty controls has both an in-depth defense and a wide defense.

Content protection within an audio/video professional postproduction environment is another example of in-depth complementary defenses. In the professional environment, there should ideally be four different types of protection—each fulfilling a different goal. The goals are as follows:

1. Control access to the asset; in this case, the video in preparation,
2. Protect the asset itself,
3. Trace the asset,
4. Limit illegal use of the asset.

Controlling Access: The first defense involves controlling access to a digital asset. This control was already in place during the analog era. It uses the principles of physical security described in the previous example, such as guards, physical access controls, and vaults. In the digital world, the type of access control is IT security. Typically, the IT department defines a perimeter that it defends against intruders using firewalls, DMZs, or virtual private networks (VPNs). Within this perimeter, IT segregates the access of corporate users to data using tools such as access control lists and role-based policies. Only authorized corporate users should be able to access the assets legitimately.

Protecting the Asset: The second defense targets direct attacks on the asset, such as theft, alteration, or replacement. The deployed tools widely use encryption and cryptographic signature. Encryption enforces the confidentiality of the asset, whereas cryptographic signature enforces its integrity.

Trace the Asset: The third barrier complements the previous one. Ultimately, any digital content has to be rendered in the analog world where the protection provided by encryption does not work anymore. Video content has to be displayed on a screen. Audio content has to be played on loudspeakers or earphones. It is prone to being captured by recording devices. Forensic marking appends information about the source that rendered the content. The protection technology used is digital watermarking [95]. A digital watermark is a message embedded into digital content that can be detected or extracted later. Watermarks are imperceptible or perceptible. Imperceptible watermarks are harder to defeat than perceptible ones.

Typical watermark information includes copyright details and the identifier of the expected recipient of the watermarked piece of content. In the case of a leak, watermark detectors can extract this information and thus trace the source of the non-legitimate disclosure. Forensic marking is useful in numerous contexts: spotting leakage in the postproduction environment, "protecting" screeners used for award selection or reviewing, or detecting the source of illegal rebroadcasting of content by identifying the infringing STB.

Thwarting Illegal Distribution: Unfortunately, as stated by Law 1, content will always leak. So, this fourth barrier attempts to limit the incurred losses. The first step is to detect illegal content. Fingerprinting is the most efficient technology for this operation. A reference database includes fingerprints that contain unique characteristics of every referenced content [96]. These characteristics may use visual hashes, distributions of color, time stamps, or points of interests. Fingerprinting may be audio-based or video-based or employ both media. Then, the system extracts the relevant fingerprints of suspicious content and compares them with those in the reference database of copyrighted content.

Once illegal content is identified, the corrective action taken depends on the context. Currently, some user-generated content (UGC) sites filter the submitted candidates. In the case of peer-to-peer (P2P) networks and file-sharing sites, a takedown notification may be sent to the sharers. In this context, fingerprinting is a superior solution than identification using cryptographic hash values because fingerprinting is robust to geometrical modifications, mashups, and camcording.

The second step is to slow down the dissemination of the stolen content. The first objective is to deter the downloaders, for instance, by delivering a bad piece of content instead of the expected one. Typical poisoning techniques spread decoys, fake content, or even encrypted content that requires payment. The second objective is to inhibit access either by slowing down the bandwidth or by routing the requester to controlled peers.

Using this second example, we can define an in-depth defensive framework as illustrated in Fig. 1.8. The proposed framework uses three complementary actions.

1. *Defend*: The assets have to be protected by any means. The employed defensive tactics should be numerous, varied, and complementary. One unique class of protection is not sufficient because it will be defeated eventually. This defeat is the inevitable consequence of Law 1. The defenses should encompass physical security, network security (such as firewalls, intrusion detection systems (IDSs), and DLPs), antivirus and anti-malware software, encryption, authentication, and reputation-based protection.
2. *Monitor*: As ultimately the attacker will win, it is crucial to know as soon as possible when she succeeded and what benefits she gained from her exploit. Acquiring this knowledge requires carefully monitoring the system (Law 9: Quis Custodiet Ipsos Custodes?). The purpose is to know which asset was affected and how (theft, alteration, destruction, or substitution). As speed is of essence, practitioners should not wait until a breach happens to trigger the monitoring; rather, monitoring should be active and continuously enabled.
3. *Cure*: This action is, unfortunately, sometimes forgotten. Limiting the impact of the exploit is paramount. Defining strategies of mitigation for every critical asset should be part of any strategy of defense. Furthermore, once the exploit is analyzed, a new set of protections should be designed and deployed, or the existing ones should be enhanced to defeat this class of attack in the future.

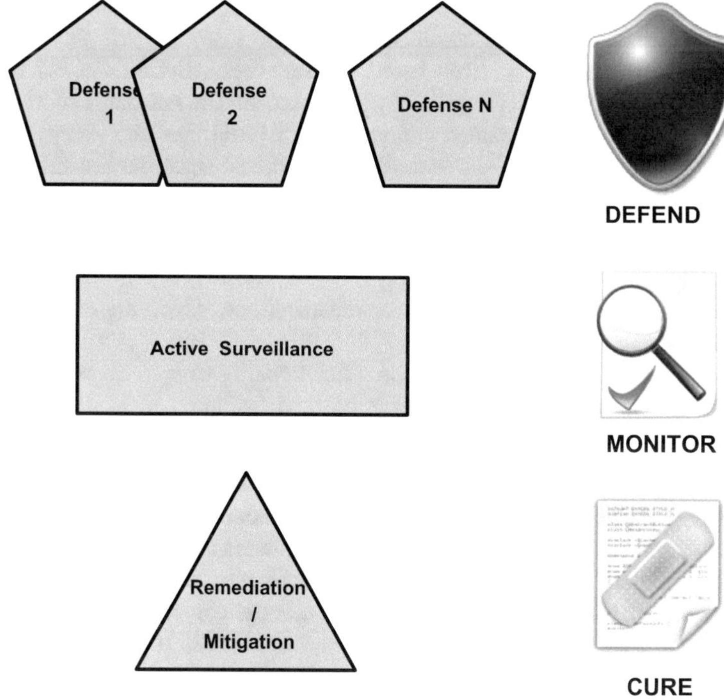

Fig. 1.8 Framework for active, in-depth defense

1.3.4 Backup

Around 2005, a new type of malware, coined ransomware, appeared. The noun ransomware is from the combination of two nouns: ransom and malware. Ransomware is a kind of malware that disables some functionality of the infected system [97]. Then, the ransomware requests payment to reactivate the blocked feature. The first generation of ransomware encrypted the files of the infected computer. Since 2009, a new generation of ransomware has also blocked the user's display[20] rather than just encrypting files [99]. More recently, a new generation of ransomware blocked the functionality of some applications such as browsers. In 2014, ransomware extended its target field from computers to specialized hardware. For instance, the ransomware Synolocker attacked Synology's Network as Storage solutions [100], encrypting remotely stored files. In 2015, ransomware extended its reach to Web sites. A new type of ransomware started progressively to encrypt the

[20] In some reported cases, ransomware blocked the screen by displaying pornographic pictures or even pedophilia [98] to shame the blackmailed person. The authors of such ransomware expected that the hacked person would not dare to call for help. It seems that this tactic was rather efficient as the ratio of paid ransoms was rather high. This is a very nice, dirty piece of social engineering.

database used by Web sites. During this encryption period, the ransomware decrypted the data so that the administrators of the infected Web sites were not aware of the ongoing attack. This lasted several weeks. Before the full backup, blackmailers removed the decryption key and asked for the ransom. Unfortunately, the incremental backup contained encrypted data [101], and the decryption was inactive. The ransomed Web site lost all data produced since the last full backup.

Typically, ransomware uses three stages:

- Seek a target: This is the usual step to finding a victim. Usually, the attacker randomly looks for an easy target to penetrate. The targets are often individuals or small companies rather than larger organizations, which are expected to be better protected.
- Enable blackmailing by disabling some functionality such as access to the files.
- Claim for the ransom to reactivate the blocked feature.

The operating mode for claiming the ransom is interesting. The first phase explains to the attacked user why he does not have access anymore to his computer, smartphone, or files. The initial strategy was to announce openly that the attacker had taken control of victim's device. Recently, ransomware used another technique by "impersonating" a legal authority [for instance, the Federal Bureau of Investigation (FBI)]. The ransomware claimed that the computer had been blocked due to illegal activities conducted from the machine [102, 103] (Fig. 1.9). In the second phase, the blackmailer proposes a method of anonymous payment. The blackmailer does not want the payment to trace back to her to avoid retaliation and potential prosecution. Anonymous payment is the trickiest part of her attack. Obviously, traditional electronic payment means such as credit cards, electronic transfers, or PayPal cannot be used. Investigators could find who received the money. The recent advent of prepaid electronic payment such as MoneyPak and WebMoney and digital currency such as BitCoin eliminates this risk. The recipient of the payment remains entirely anonymous.[21] Therefore, many instances of ransomware use these new types of payments. Once the victim has paid, he waits (often uselessly) for the blackmailer to enable again the blocked functionality. In 2014, a UK online survey demonstrated that about 40 % of the victims of CryptoLocker[22] paid the ransom [105]. In 2015, an annual report evaluated the return on investment (ROI) of ransomware at 1425 % [106]. This is an excellent lucrative business.

The Devil Is in the Detail On June 8, 2008, Kaspersky Lab, the Russian antivirus editor, detected a variant of the virus `GPcode`. `GPcode` is a ransomware that has many variants. The initial versions mainly renamed some files. They replaced the original names by random-looking names. Later,

[21]Indeed, cryptocurrencies such as Pecunix, AlertPay, PPcoin, Litecoin, Feathercoin, or Zerocoin are the payment methods used by the black market of the Darknet [104].

[22]CryptoLocker is the most well-known ransomware.

Fig. 1.9 An example of a ransomware screen

variants became more sophisticated and started to employ cryptography, often not very smartly. This newly discovered variant encrypts some data files on the victim's hard drive, renames them with extension `._CRYPT`, and adds the file `!_README_!.txt` in their folder. Then, it displays a ransom message announcing the encryption of the data and giving a contact email. The ransomware claims to use RSA-1024. Thus, it is out of the reach of brute force attacks against the private key. The pirated individual must contact the pirate, pay the ransom, and may receive in return a decryption tool.

Through reverse engineering, Kaspersky Lab extracted the public keys used by the ransomware to encrypt the files. The ransomware uses two public keys depending on the version of the OS. On June 10, 2008, Kaspersky Lab called for the help of the cryptographic community to try to crack the private key. This attempt was illusory. It would have required too much calculation power (or it would mean that RSA-1024 is not safe anymore). Moreover, there were two keys to crack!

One week later, Kaspersky Lab withdrew their challenge. Nevertheless, thanks to one "common" mistake of the ransomware's authors, there may be some hope for careless users who did not back up their data. When encrypting a file, the virus creates a new file that it renames with the expected extension and then deletes the original file. The method used was a standard deletion rather than a secure deletion. Indeed, it is common knowledge (at least in the security community) that a simple deletion does not erase the file [107]. It mainly clears the data fields of the file system's indexing tables. Consequently, the deleted files are still present on the hard drive as long as a new file has not overwritten them. Therefore, if there has not been too much activity on the hard drive since the infection, usual commercial recovery tools may be able to retrieve the "deleted" files. Kaspersky Lab proposed such a tool issued by the open source community. If the ransomware would have used secure deletion, then this fix would have been impossible. Usually, secure deletion requires overwriting several times the physical location on the hard drive of the deleted sections with random data. The US National Institute of Standards and Technology (NIST) published a guideline for media sanitization (still in draft version) [108].

Lesson: Even the bad guys make mistakes. Law 1 also applies to viruses and malware. Furthermore, deleting data is not a straightforward operation (Sect. 6.3.3).

In the previous example, it was possible to retrieve the clear files. Sometimes, it is also possible to retrieve the keys used by the ransomware. In June 2014, the FBI brought down the infrastructure of the Gameover ZeuS botnet and the initial CryptoLocker. The investigation teams seized the databases used by the malware. Among them was the database of RSA private keys that encrypt the AES encryption keys used by CryptoLocker. Using this information, security companies FireEye and Foxit provided an online tool that potentially may decrypt CryptoLocker-encrypted files [109]. Unfortunately, access to the decryption key is not available for most ransomwares. The best and unique response to ransomware is still periodic, frequent backup.

Rule 1.5: Back up Your Files

This book will never repeat this recommendation enough: Back up your data. Usually, people believe that the purpose of a backup is merely to recover from a crashed system. A backup is also the ultimate tool for recovering from a successful attack. Of course, there will be disenchantment of the affected people and loss of productivity, but at least the entire past work will not be lost. The recovery is only possible if the backup is regularly performed and with a sufficient periodicity. Furthermore, as some recent instances of ransomware can reach networked directories [110], it is safer to use both non-local physical storage and cloud-based storage. In other words, the best defense is to use an air-gapped backup. Malware cannot step over air gaps.

On June 17, 2014, the company Code Spaces suffered from a well-orchestrated distributed denial of service (DDoS) attack [111]. The blackmailer(s) requested a ransom to stop the attack. The problem was not limited to the denial of service. The attackers succeeded in getting access to the company's Amazon EC2 control panel. Thus, they owned the cloud infrastructure.[23] As the management team of Code Spaces refused to pay the ransom, the attackers started to delete large chunks of random file directories and backups. Only after Code Spaces succeeded in retrieving access to its control panel, could it assess that it had lost most of the software repositories of its customers, as well as all the backups irremediably. With this disaster, Code Spaces could never get back into business. This story highlights

[23]Amazon Elastic Compute Cloud (Amazon EC2) is a Web service that provides resizable computational capacity in a cloud.

the need for air-gapped backup. If an attacker who penetrated a system also gains access to the backups, then these backups only play their safety role but not their security role.

After a severe attack, the best solution is to rebuild the hacked system from scratch. First, install a new genuine OS, updated with the latest security patches. Then, install the usual security tools such as an antivirus and a personal firewall. After setting up a genuine, protected environment, install the applications from genuine sources and apply all applicable security patches. Once the applications are properly running, restore the data from the backup. It may be wise to sanitize the backup data with a scan by an up-to-date antivirus. Some backuped data may have been infected.

1.4 Summary

Law 1: Attackers Will Always Find Their Way
No secure system is infallible. Any secure system is doomed to fail. Attackers will always find a way to defeat it. Security designers must not deny this fact, but rather put this heuristic at the heart of their design.

Rule 1.1: Always Expect the Attacker to Push the Limits Any design operates within a set of limits defined by its initial requirements. The system should work correctly within these boundaries. An attacker may attempt to operate outside these limits to get unexpected behavior. The security designer should ensure either that these limits are out of reach or at least that the system should detect the violation of these boundaries to react accordingly.

Rule 1.2: Responsibly Publish Vulnerabilities Publishing vulnerabilities is one of the best methods to reach a safer cyber world. Not only will the solution provider close the holes, but the publication of the vulnerability will also educate the designers. Obscurity is dangerous for security (Chap. 3). Nevertheless, implementers must have a reasonable amount of time to fix the issue before the public disclosure of the vulnerability.

Rule 1.3: Design Renewable Systems As any system will be broken, the designed system must be ready to survive by the updating of its defense mechanisms. Without renewability, the system will be definitively dead.

Rule 1.4: Design In-depth Defense As any defense will fail, a secure system should implement many defenses. It should construct successive obstacles that the attacker has to cross successfully. Diversity in protection makes the exploit harder to perform.

Rule 1.5: Back up Your Files As any system will be broken one day, data may be corrupted or lost. Regular, frequent air-gapped backup of all non-constructible data is the ultimate defense.

Chapter 2
Law 2: Know the Assets to Protect

The most useless types of physical security controls are the kinds that don't protect against what you need them to and those which protect against anything for no reason
OSSTMM [112]

2.1 Examples

There is untold information about Ali Baba and the forty thieves [113]. Indeed, his cavern had two entries, as illustrated in Fig. 2.1. This fact may have changed the course of the story. What was the most valuable asset in this story? Not the treasure, but the secret passphrase "Open, Sesame." If Ali Baba had used a different protocol to demonstrate the existence of the treasure, his brother Karim could have survived. Ali Baba could have shown Karim the location of the cavern. This would have been insufficient to convince Karim. Then, Ali Baba could have proven to his brother that he knew the secret passphrase without disclosing the passphrase. This would have changed the story, but with less dramatic effect. How could have Ali Baba proven his knowledge of the passphrase without telling Karim the actual passphrase? Indeed, there is a way. Ali Baba could have entered the cavern while Karim stayed in the open. Once Ali Baba was in the cavern, Karim could have entered the tunnel and stayed at the fork. He could have asked Ali Baba to come out of the right or left exit. If Ali Baba knew the passphrase, he would have succeeded. If Ali Baba did not know the secret, then he had 50 % chance of success only. They could have repeated the challenge several times. After ten trials, if Ali Baba did not know the passphrase, he would have only about 0.001 % chance of succeeding. With this protocol, Ali Baba would have protected his most precious asset. In 1987, Uriel Feige, Amos Fiat, and Adi Shamir introduced a new concept: zero-knowledge authentication [114]. In zero-knowledge authentication, the user demonstrates his identity by proving that he knows it without actually disclosing any piece of information of this knowledge. They define a cryptographic protocol implementing

E. Diehl, *Ten Laws for Security*, DOI 10.1007/978-3-319-42641-9_2

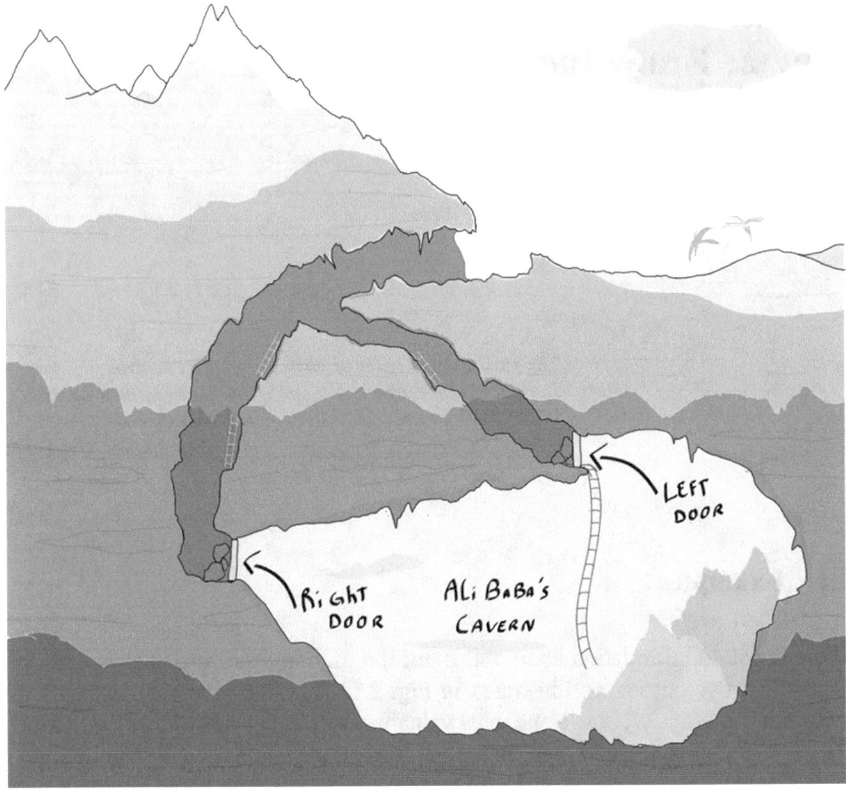

Fig. 2.1 The cavern of Ali Baba and the forty thieves

this feature. As for the case of Ali Baba, it is based on multiple successive challenges. The prover can prove to the verifier that he knows a secret without disclosing any information related to this secret.

2.2 Analysis

2.2.1 Classification of Assets

The primary goal of security is to protect. However, to protect what? "What are the assets to protect?" is the first question that every security analyst should answer before starting any design. Without a proper response, the resulting security mechanism may be inefficient. Unfortunately, answering it is tough.

The first step in answering this question is identifying all potential assets. Asset is anything that has some value to an organization or an individual and thus requires protection. Therefore, assets come in many forms. According to ISO 27002 2005, assets are any tangible or intangible things that have some value for an enterprise [1]. This definition is extremely broad and too generic to be useful. The definition of assets must be more fine-grained. Hence, many taxonomies exist. For instance, the RAND Corporation proposes classifying assets as resources, physical access, hardware, software, information, and human [115]. Other proposals classify assets as information assets, software assets, physical assets, and services [116]. Such classifications often focus mainly on information security and thus describe the world with a narrow scope. This book proposes a wider-scope taxonomy of five asset types: human, physical goods, information goods, resources, and intangible goods. This taxonomy encompasses the full range of security issues.

Human: Despite it seeming demagogic, often, the most valuable asset of any company is its staff. Depending on the context, employees can be under many threats, natural and malevolent. People need protection in the case of natural catastrophes such as fires, floods, and earthquakes. Traveling employees are also an important issue and not only when they travel to risky or politically unstable countries. Malevolent actions against employees are serious risks for every organization. For instance, in many countries, employees are under the threat of criminal organizations. For example, they may be blackmailed or threatened to collaborate in extortion or fraud. In some countries, employees are under the menace of being captured by a terrorist or criminal organization for ransom from their employers [117, 118]. Any decent corporate security policy should have a section dedicated to the protection of all of the organization. Another interesting threat targeting humans is the competitors that may attempt to attract their talent. A risk analysis should also take this issue into account. Corporate risk management should address it. This complex issue of managing human-related risk is out of the scope of this book.

Physical goods are often the assets that first come to mind when listing what to protect. This category encompasses many items ranging from real estate to device. The following example based on physical goods pinpoints a typical issue: identifying what the real physical asset to protect is. Let us do an exercise using a jewelry store. What are the assets to protect in this shop? The evident and right answer is the jewels. Jewels are the primary assets. As noted earlier, employees and customers are also assets to protect. The owner of the shop does not want them to be killed or injured during a holdup. They are substantial indirect assets as the threat (robbery) does not directly target them. Nevertheless, they may become unfortunate, collateral damages. A common mistake is to identify vaults as assets to protect. In this example, vaults are not the assets to protect. Of course, they must present a maximal defense but only because they are the ultimate line of defense against the burglars. Vaults are tools that protect the valuable assets. Per se, they are not assets that need protection. Were there never to be any jewels within the vault, it would be a waste to protect the vault itself. The value of the safety box is small compared to the value of the items stored in the vault.

Information goods are the assets that the practitioners most often focus on. Information goods encompass all digital information created, processed, stored, rendered, and backed up in the digital world. They are digital files of all kinds:

documents, databases, lists of contacts, personally identifiable information, and multimedia content.

Resources are assets that provide some service or functionality. They encompass data availability, network bandwidth, calculation capabilities, printing and rendering capabilities, and storage capacities. In many cases, one major attribute of these assets is their availability. For instance, the bandwidth and the availability of the access to the network are critical assets for an e-commerce Web site. Denied access to the Web site means a potential loss of revenue for the owner. Thus, DoS attacks that attempt to block access to the Web site are dangerous. Therefore, network availability may be an asset to protect.

Intangible goods are the last category of assets. These assets are difficult to identify and even harder to evaluate. They are the intangible concepts that define the value of a company. They encompass notions such as brand, reputation, trust, fame, reliability, intellectual property, and knowledge. For instance, in the case of a business selling security solutions, reputation and trust are critical assets. A severe successful attack would inevitably tarnish the reputation of this security company. The loss of reputation would mean a loss of revenue as customers might question the actual skill set and trustworthiness of this security company. For instance, in February 2011, Anonymous successfully hacked the servers of HBGary Federal. HBGary Federal provided security-related services and tools to the US government for background checking of individuals who might require a security clearance. After the hack, HBGary Federal closed in 2012, as trust in it vanished [119]. The same argument applies to a financial institution that is successfully breached. Preserving the reputation of a company seems to be one of the major arguments for justifying cybersecurity expenditures [120].

The second step needed to answer the question is the classification of the assets. Once all assets are identified, it is mandatory to rank or cluster them according to their value. Not all assets will have the same value, else everything would need equal protection. The more the number of assets to protect, the more elaborate and expensive the defense will be. Only the most valuable assets will require complicated and potentially expensive protections. Usually, the cost of protection should not exceed the value of the protected asset. Only rare cases, generally linked to national security, may this rule be not valid.

Physical goods are easy to rank. They have an associated purchase cost that may be balanced against some depletion factors. Classification by value becomes a rational, objective process. The assessor defines a set of ranges with minimum and maximum values. Each asset is clustered in the proper corresponding category, with regard to its actual value.

Information goods are probably the type of asset most studied in the security-related literature. Even before the IT, physical documents were classified. The initial requesters of such classification were military organizations and government agencies. The seminal work in this field is the Bell-LaPadula model [121].

The corporate environment often uses the following typical multi-level classification of documents (electronic or physical), inherited from the defense environment, in order of decreasing value:

- *Restricted* means that only persons within the enterprise in a defined nominal list should have access to the information. A signed Non-Disclosure Agreement (NDA) should govern access to restricted information by an individual or an organization external to the enterprise. This category of information represents the most valuable digital goods of the company. Involuntary disclosure of restricted information could generate an enormous potential loss for the enterprise.
- *Confidential* means that access to the information is restricted to a set of persons within the company, for instance, a team or a division of the organization. A signed NDA should govern access to confidential information by an individual external to the enterprise. The main difference with restricted data is that the authorized people are not nominally listed, but defined as belonging to the organizational structure. As such, the value of confidential data is lower than the value of restricted data. Nevertheless, the involuntary disclosure of confidential information may result in severe losses for the enterprise.
- *Regulated* means that access is governed by government regulations or by industry rules. The Sarbanes–Oxley Act (SOX), the Gramm–Leach–Bliley Act, the US Health Insurance Portability and Accountability Act (HIPAA), the UK Data Protection Act, and the US data breach laws are examples of governmental regulations. The Payment Card Industry Data Security Standard (PCI-DSS) that defines the security of credit card processing is an instance of industry rules. Such regulations increasingly protect data of personal information such Social Security Numbers (SSNs) and phone numbers.
- *Private* means that access to the information is limited to the usage within the company. Related information should not leave the enterprise.
- *Public* means that access to the information is unrestricted.

To illustrate the different types of information, we will use the family of Alice and Bob Doe, who have two children. We will classify some of the information related to the Doe family. The incomes of Alice and Bob are restricted information. Even their children should not know their wages to avoid any risk of disclosure. Alice and Bob do not want other members of the Doe family to know this information. As Bob and his brother compete for the success of their children, Bob does not want his brother to be aware that his son's results for this quarter are poor. Therefore, this information is confidential and is only shared with the grandparents. All members of the Doe family know that Aunt Gaby's mink fur coat is indeed a synthetic fur coat. Nevertheless, no member of the Doe family would like that the neighbors be aware of it. This information is private. Alice is the chairperson of the white hacker association of her town. This information is public.

Multi-level classification is useful only if at least two conditions are present. The first condition is that users correctly classify information. It is of the uppermost importance that the actual classification be accurate. The risks associated with a too-lax classification are evident. Precious sensitive information may "legitimately" leak out. Indeed, were a confidential document not to be marked as such, an employee could disclose it without any mischievous purpose. Employees often understand well this risk. Therefore, the natural temptation is often to overrate documents, i.e., classify private (or even public) data as confidential. The extreme strategy could be to classify by default all information as confidential. This lazy approach also has serious risks and even operational issues. Unfortunately, these risks are often not properly perceived. The first problem is that this may deplete the actual value of the information. The purpose of documentation is to transfer and to exploit (or maintain) knowledge. If access to the information were to be too restrictive, it would drastically limit the availability of information. Then, this rough categorization would annihilate the actual purpose of creating this information. By construction, classifying a document means restricting its access to fewer people. Thus, the best heuristic for deciding which category to use is to grant access on a need-to-know basis.

Rule 2.1: Give Access Only to Those Who Need to Know

The need-to-know basis means that individuals get only the facts they need to know, and only at the time, they need to know them and nothing more. This policy makes it harder for unauthorized people to gain access to classified information. It is also a way to mitigate risk from insiders. Insiders have more difficulty acquiring more confidential information than that they really need. Furthermore, splitting confidential information into different subparts with distinct access groups limits the risk of secret disclosure. To get complete information, individuals from separate groups have to collude. The rule also applies to the type of access that is granted. Principals should be granted read-only access to information if they just need to read the information. As these principals do not modify data, they should not be granted write access. Fewer principals should be given write access to the same information—indeed, only the principals who have reason to modify the information. This restriction limits risk in the case of error and in the event of voluntary misuse.

Too strict policies lead to operational issues, but the second problem is more related to human nature. In the minds of people, the value of a good (physical or digital) is associated with scarcity. The scarcer a good is, the higher its perceived value will be [122]. For instance, were private data to be classified as confidential data, the perceived value of real confidential data would be watered down because people would see too much confidential information. In the end, people would handle confidential data in their mind more as private data rather than as confidential data. Employees would become laxer about them. The notion of perceived value is paramount. The higher the perceived value of something is, the more careful the people will be with it. Unfortunately, the perceived value is purely subjective. Proper classification helps enforce the alignment of perceived value with actual value.

Confidentiality is not the unique security attribute of information goods. Another attribute is its integrity. It is sometimes important to ensure that the information good is genuine and has not been altered. There are two types of protection of information. Tamper evidence detects whenever digital information has been modified without the knowledge of its owner. Digital signature is one example of a tamper-evident technique. Nevertheless, tamper evidence does not prevent tampering. Tamper-resistant protection attempts to prevent the attacker from changing the information.

Evaluating the value of resources becomes even harder than classifying physical or information goods. Of course, the investment and the operational cost could be indicators. Nevertheless, these indicators are rather insufficient. The assessor must also evaluate the impact of a shortage or blackout of the analyzed resource. For instance, if a hospital has an electricity blackout, this may endanger the lives of patients. The lives of patients are invaluable. Therefore, in the context of a hospital, the value of reliable power supply is high, relative to its actual reasonable production cost. Therefore, hospitals have auxiliary electricity generators to provide power in the case of a general blackout. For example, in the context of surgery, availability of electricity is invaluable. The same resource may have minuscule value in another context. In a domestic context, interruption of electricity is a mere annoyance rather than a critical issue. In that context, the power supply has less value. In the broadcasting environment, on-air time is a precious resource. A TV channel should always display something and, ideally, the expected program. A black screen is not acceptable, even for one second. On-air time is a high-value asset, despite its having no intrinsic "direct" cost. Thus, broadcast facilities have sophisticated redundant architectures to compensate for any failure or hack of even one piece of equipment. Similarly, the availability of an e-commerce site is critical for a merchant. While the Web site is down, the merchant may miss sales. Therefore, availability is paramount for the merchant. The primary resource to ensure this feature is the availability of the network bandwidth. A typical attack is to flood the Web site with requests, consuming more bandwidth than available. This attack is a DoS attack. The value of one resource may be approximated by the loss incurred if it were to be blocked or unavailable.

Evaluating the value of intangible assets is even harder as it is most often subjective. For instance, how does one assess the value of the reputation of an organization? Depending on the business, some events may be far more detrimental than others for its reputation.

2.2.2 *Classification of Attackers*

After this phase of analysis, the security analyst should have an ordered list of assets to protect from the most valuable to the least valuable ones. These assets may be under the threat of attackers. The attackers may attempt to steal them, to replace them, to impair them, to corrupt them, or even to destroy them. The strategy will depend on the objectives of the attackers. The second important question to the analyst is, who are the attackers? The answer to "Who may launch the attacks against your system?" is the second piece of information that every security analyst should seek. Without knowing who the attackers will be, any defense may be suboptimal or even pointless.

Rule 2.2: Know Your Opponent

This rule is ancient. In his famous treaty "The Art of War" [123], the Chinese general Sun Tzu stated this rule:

> If you know the enemy and know yourself, you need not fear the result of a hundred battles. If you know yourself but not the enemy, for every victory gained you will also suffer a defeat. If you know neither the enemy nor yourself, you will succumb in every battle.

How can the security practitioner learn about the enemy? The purpose is not to know the actual identity of the enemy but rather to know her techniques, her tactics, and her procedures. Collecting this knowledge requires an active stance. As in any intelligence operation, the analyst must gather as much information as possible. He will have to dive deeper than the Surface Web. He will have to visit and review sites and forums dedicated to hacking. He will have to read the corresponding fanzines to understand this underground culture better and to grasp its tactics. He will have to go to the Darknet [124] and the infamous Deep Web [125]. He will have to follow the hacking groups and attend conferences such as Black Hat, CCC, and Defcon. This task is long and arduous. Nevertheless, it is paramount to gain useful knowledge. The attackers are keen to learn new techniques and to exchange their expertise with peers. The security practitioner has to be equally knowledgeable and use these sources of knowledge.

Once the security analyst knows his potential enemies, he will have to classify them. In the 1990s, IBM researchers introduced a classification of attackers [126]. This simple classification is extremely interesting and can be widely used. IBM defines three types of attackers.

- Class I: Clever outsiders who are often brilliant people. They do not have a profound knowledge of the attacked system. They have no funds and, thus, will only use widely available tools. They will often exploit published weaknesses and known attacks rather than creating new ones. They often act as copycats of more skilled attackers. An example of such class I attackers are script kiddies. Script kiddies are hackers who exclusively use tools developed by others and attack random weak targets. Their motivation is mostly amusement, sometimes vindictiveness, or even the unrealistic expectation of fame. Usually, they will not fight hard and, at the first resistance, will look for easier prey.
- Class II: Knowledgeable insiders who have specialized education. They have good, although potentially incomplete, knowledge of the targeted systems. They are likely to use specialized tools and have some financial resources. Examples of specialized tools are debuggers, logic analyzers, emulators, memory scanners, and memory dumpers. Interestingly, in the second-hand market, very sophisticated tools are available at a price that passionate hobbyists can afford. Their motivation is the intellectual challenge either in succeeding to break into the system or trick the system into doing something it is not supposed to do. In

some cases, it may be financial reward. Another kind of motivation is currently on the rise: political or ideological protest. In this case, the attackers are called hacktivists or whistle-blowers. For instance, the *Anonymous* hacker team is one of the most famous groups of hacktivists [127].

- Class III: Funded organizations that can recruit teams of complementary world-class experts. They have access to professional, expensive tools. They may even design their own custom hacking tools. Moreover, they have "unlimited" financial resources. The attackers performing APT are an example of such Class III attackers. With the rise of cyberwarfare, the press is often focusing much on them [128]. Their motivation is either economic (when driven by a criminal organization) or political (when operated by governmental agencies). Obviously, these attackers are the most dangerous and skilled ones.

There are many other taxonomies. For example, Annex C of ISO 27005 introduces another segmentation: hacker, computer criminal, terrorist, industrial spy, and insider. This classification is more motivation-driven than IBM's classification, which is resource-driven. The Merdan Group defines an interesting five-scale classification of attackers.

- Simple manipulation: The attacker uses only what is readily available through the OS and applications. The attacker has no technical knowledge. The motivation is to extend the use of a product beyond its expected features. These attackers are not even part of the IBM Class I category.
- Casual hacking: The attacker uses utilities and tools commercially available. The attacker has some technical knowledge and is an OS power user. The motivation is to extend the use of a product beyond its expected features. Casual hackers correspond to the IBM Class I category.
- Sophisticated hacking: The attacker has access to sophisticated equipment such as computational resources, disassemblers, simulators, logic analyzers, or hardware prototyping tools. Often these resources come from laboratories, e.g., those of major universities. Sophisticated hackers have significant knowledge of programming and internal details of operating systems. They may design their own tools. The investment would never exceed $10,000. They may team up with other hackers. The motivation is intellectual challenge and recognition by peers. In some cases, the motivation may be to earn additional revenue. Sophisticated hackers correspond to the IBM Class II category.
- University challenge: The attacker is similar to the sophisticated hacker but with extended resources. A university may provide access to expensive tools such as E-Beam. The capital investment could reach $100,000. The motivation is an intellectual challenge and the recognition by peers. In some cases, the motivation may be to earn additional revenue. University challengers correspond to the IBM Class II category.
- Criminal enterprise: The university challengers' capabilities are extended to capital investment that exceeds $100,000. The motivation is to generate

financial benefits or to steal classified information in case of state-sponsored hacking. Criminal enterprise hackers correspond to the IBM Class III category.

2.2.3 Threats

Once the assets have been listed, and the attackers have been identified, the next step is to determine the potential threats. This identification is the purpose of a threat analysis. Threats may be natural or may originate from humans. Threats may be accidental or deliberate. Annex C of ISO 27005 introduces several types of threats: physical damage, natural event, loss of essential services, disturbance due to radiation, compromise of information, technical failure, unauthorized action, and compromise of functions. These threats are all related to information systems.

There are several methods for representing the threats. The easiest model to understand and to implement is the attack tree model [129] introduced by Bruce Schneier. This model has several advantages. It is easy to comprehend, very graphical, and usable in many ways. Many commercial tools are available to undertake this type of analysis.

The first step is to define the target of the attackers. It represents either the objective of the attacker or the asset to protect by the defender. A system may have several targets for attackers. One specific tree represents each target. Then, the analyst tries to list all the potential methods to succeed in reaching this target. For this, the analyst has to think like an attacker. For each method, he tries to list all the possible attacks. As such, the analyst builds a tree of attacks. Each branch of a node is a potential attack (logical OR branch, the default type); or, in some cases, the node requires all the branches to succeed (logical AND branch). The expected level of detail of the attack defines the depth of the tree. We will use a simplified example to illustrate the attack tree model. Let us try to find the password of Bob.

At the root of the tree, the attacker either may brute-force the password, may steal the password using a keylogger, or may induce Bob to disclose his password using social engineering. Figure 2.2 illustrates this root level.

To install a keylogger, the attacker will have to find one possible vulnerability on Bob's computer and install the keylogger. As the attack requires two actions to succeed, the corresponding branches in Fig. 2.3 are marked with AND. Using social engineering, the attacker either may read it from a Post-it note that Bob may have stuck on his computer's screen or view it as he types by shoulder surfing.

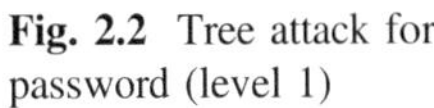

Fig. 2.2 Tree attack for password (level 1)

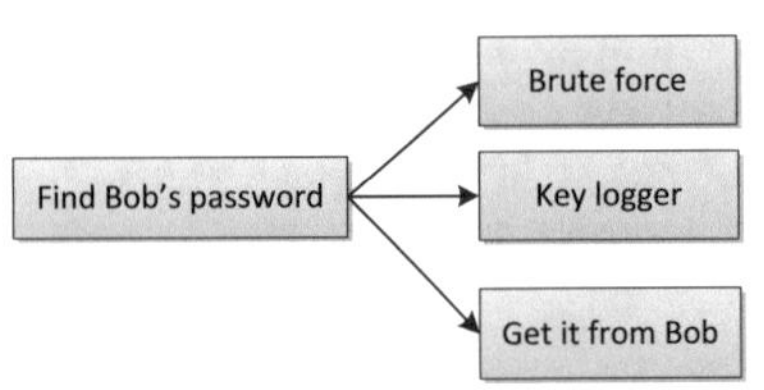

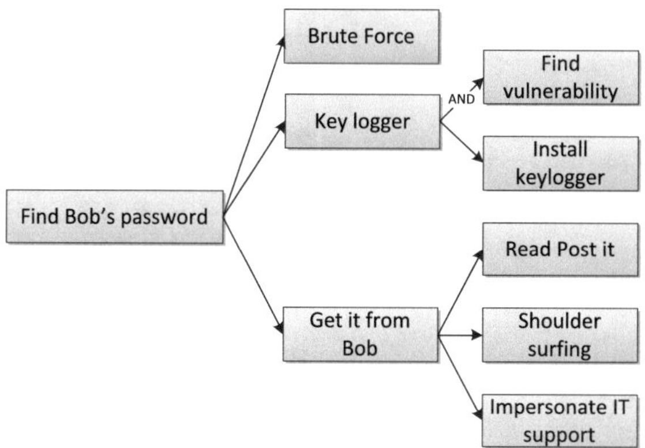

Fig. 2.3 Tree attack (level 2)

Using social engineering, the attacker may also try to impersonate the IT support team and convince Bob to communicate his password to the attacker.

Of course, to succeed in reading the Post-it note or to shoulder surf, the attacker will have to enter the building. These attacks require physical presence in Bob's office. To impersonate IT support, the attacker has to collect enough information about Bob's company's IT support. Figure 2.4 represents this third level of attack. Of course, this analysis is oversimplified. The real world is far more complex than the theoretical view. Figure 2.5 shows an extract of a more sophisticated analysis of an actual example.

Attack tree analysis is not limited to the graphical representation of the potential attacks. Starting from this initial graphical representation, an interesting enhancement is to add more information and to exploit it. Usually, the first appended information is the likelihood of the attack succeeding. It is possible to associate with each attack a probability of its success. Then by combining the probability of

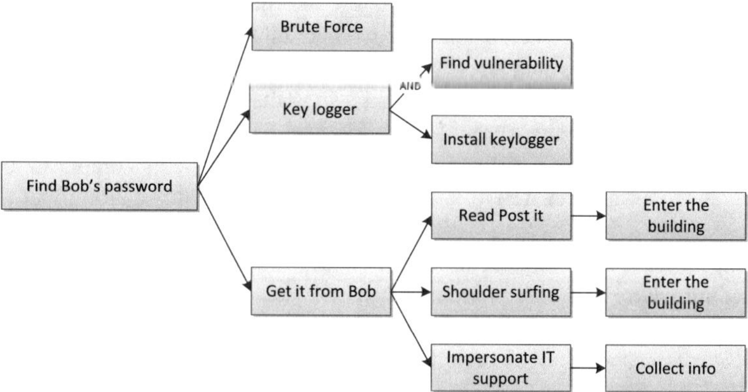

Fig. 2.4 Tree attack (level 3)

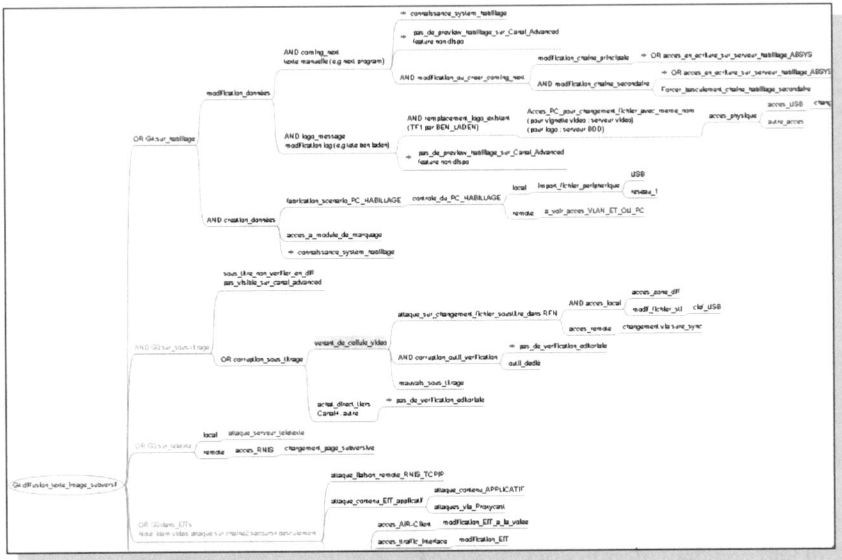

Fig. 2.5 An example of tree analysis

successive successful attacks, the analyst can identify the most probable attack branch. The implementation of countermeasures may focus on the branches displaying the highest probabilities. The tree attack scheme provides a good indication of where to apply countermeasures most efficiently. The nearer the countermeasure is from the root of the tree, the more efficient will its impact be, as the countermeasure covers more attack paths. Of course, several countermeasures should be spread all along the critical paths in the application of Rule 1.4: Design In-depth Defense.

The second useful information to add is the cost of the attack for the attacker. Summing the cost of the successive attacks necessary to fulfill the target gives a good indication of the effort that the attacker will have to make to succeed in her exploit. As the first phase of the analysis evaluated the value of the assets to protect, it is possible to decide which attacks are worth preventing. Usually, if the cost of the entire exploit exceeds the value of the asset, then the corresponding branch can be discarded, as a rational attacker will be looking for a reward. This assumption is not true if the attacker has a different motivation than financial compensation. An example is the case of an APT that steals critical national information (Sect. 1.2.4). Other defensive strategies are possible. For example, the defender may first prevent the cheapest attacks if the attacker looks for the best ROI.

Deciding where to invest in security is a complex topic. Usually, when deciding whether an investment will be worthwhile, the standard practice is to analyze its ROI. ROI defines after what period the benefits generated will exceed the money invested. Unfortunately, when investing in security, the organization does not make a profit but rather prevents loss. Sometimes, this parameter is called Return on

Non-loss. As the avoided loss is only potential and not well defined, it is hard to justify a security investment from a quantitative perspective [130]. Several metrics may be used. Equation 2.1 gives the first metric. It defines the relative value (RV) of each countermeasure.

$$\text{RV} = \text{Probability compromise} \times \frac{\text{SLE}}{\text{cost of solution}}, \tag{2.1}$$

where

- *Probability compromise* is the absolute probability of occurrence of the corresponding compromise. *Probability compromise* $\in [0, 1]$.
- Single loss expectancy (SLE) represents the expected amount of monetary loss that will be incurred if the risk materializes. It should encompass the total cost of the incident, including the remediation process, productivity loss, and support time.
- *Cost of solution* represents the cost of the countermeasure, including deployment cost and additional operational costs.

This formula is simple to apply and is a first indicator. Nevertheless, a countermeasure will not necessarily eliminate the risk. It will just reduce it. Therefore, the Return on Security Investment (ROSI) indicator is a more accurate metric. Equation 2.2 defines the ROSI calculation for a given risk.

$$\text{ROSI} = \frac{\text{ARO} \times \text{SLE} \times \text{mitigation ratio} - \text{cost of solution}}{\text{cost of solution}}, \tag{2.2}$$

where

- Annual rate of occurrence (ARO) is the probability of occurrence of the risk in the coming year; *ARO* $\in [0, 1]$.
- *Mitigation ratio* evaluates the reduction of loss or probability that the new countermeasure will generate. It measures the effectiveness of the countermeasure in reducing the risk. *Mitigation ratio* $\in [0, 1]$.

Is there an optimal strategy to invest in countermeasures?

The Devil Is in the Detail In 2002, two economists, Lawrence Gordon and Martin Loeb, attempted to answer this complicated question [131]. Their conclusions were surprising. The problem is to define how much a firm should invest in a given type of security solution for better security. They characterize each asset with three parameters: λ, the loss induced by a successful attack (*SLE* in previous equation); t, the probability of the successful threat; and v, the vulnerability. In their model, the vulnerability is the probability that once the threat realized, the attack will be successful. λ is a finite monetary value. As they are probabilities, $t \in [0, 1]$ and $v \in [0, 1]$. Therefore, the expected loss for the firm without any additional investment is $\lambda t v$.

Gordon and Loeb make the simplified assumption that for a given asset, the loss and the threat are constant. They call this fixed value for the asset the potential loss $L = \lambda t$. Furthermore, a monetary investment z in security countermeasures will impact the vulnerability v. They define the security breach probability function, $S(z, v)$, which establishes the link between z and v, i.e., how much the investment will impact the vulnerability.

The expected benefit for investment in security z is defined as follows:

$$\text{Benefit}\,(z) = (v - S(z, v))L$$

Consequently, the expected net benefit or ROI is as follows:

$$\text{ROI}\,(z) = (v - S\,(z, v))L - z.$$

Thus, when looking for the optimal investment z_o, the partial derivative of the net expected benefit respective to z has to be null, i.e., $\frac{\partial \text{ROI}\,(z_o)}{\partial z} = 0$. This gives the following equation: $\frac{\partial S\,(z_o, v)}{\partial z} = -\frac{1}{L}$. This underscores the fact that the optimal investment depends on the breach probability security function, $S(z, v)$.

Gordon and Loeb analyzed two classes of vulnerabilities. The first class, denoted by $S^{\text{I}}\,(z, v)$, was defined as $S^{\text{I}}\,(z, v) = \frac{v}{(\alpha z + 1)^{\beta}}$, where the parameters $\alpha > 0$ and $\beta > 1$ measure the productivity of the security system. For instance, if there is no security investment, i.e., $z = 0$, then the vulnerability v remains constant. They demonstrated that for this class of vulnerabilities, it is useless to invest as long as $v < 1/\alpha\beta L$. Furthermore, the higher the investment is, the higher the benefit will be for this class of vulnerability. In other words, it is more efficient to invest in higher loss vulnerabilities.

They considered a second class denoted, $S^{\text{II}}\,(z, v)$, defined as $S^{\text{II}}\,(z, v) = v^{\alpha z + 1}$, where the parameter $\alpha > 0$ measures the productivity of the security system. Once more, if there is no security investment, i.e., $z = 0$, then the vulnerability, v, remains constant. They showed that for this class of vulnerabilities, the conditions of investment are ruled by $1/L > \alpha v \ln\,(v)$. This investment is optimal for $v = 1/e \approx 0.3679$. They concluded that for these two classes of vulnerabilities, the investment is best lower than 37 % of the potential loss due to the corresponding threat.

Unfortunately, no solution vendor provides a mathematical model of the security breach probability function $S(z, v)$. Furthermore, no practical work has proven that these classes are significant and that there are no other important classes.

Lesson: Defining the right value of the investment to get a proper protection is a complex exercise. Unfortunately, no economic model offers a real answer yet. Currently, the security practitioner's expertise drives this decision.

In 2006, Stephano Bistarelli, Fabio Fioravanti, and Paolo Peretti enhanced the concept of attack trees by introducing the notion of defense trees [132]. In this new model, a tree of countermeasures, the so-called defense tree, enriches each leaf node. Defense trees offer a global view of the attack space as well as the available range of defenses. Having both of them in the same representation facilitates the analysis of the economics of defense. In most cases, if the cost of the protection exceeds the potential loss, then the corresponding countermeasure should not be implemented. Nevertheless, figures must not drive the final decision. They are only some of the parameters to analyze when deciding upon the countermeasures. Furthermore, the deciders should be aware that the evaluations are subjective estimates with unknown margins of error.

There are many other threat analysis methodologies. For instance, Verizon introduced a framework to describe incidents: Vocabulary for Event Recording and Incident Sharing (VERIS) [133]. VERIS is now an open source community project. This framework uses four elements to describe incidents.

- Asset, the asset impacted by the described incident.
- Actors, who caused the incident. There are four categories of actors:
  - External actors are not members of the "attacked" organization. They span the full range of attackers, from script kiddies to organized, funded teams. VERIS proposes several subcategories, including activist, auditor, competitor, customer, force majeure (nature and chance), former employee, nation-state, organized crime, an acquaintance of employee, state-affiliated, terrorist, and unaffiliated person(s).
  - Internal actors, who are members of the "attacked" organization. They may be of two types.

    Active actors, who deliberately contribute to the attack. Usually, we refer to them as insiders. Disgruntled employees who are vindictive or blackmailed employees are examples of active internal agents.
    Passive actors, who do not deliberately contribute to the attack; nevertheless, they are actors of the attack. Victims of social engineering are an excellent example of internal passive agents.
  - Partner actors, who are not members of the "attacked" organization. The attack uses their infrastructure or members. Usually, their participation in the attack is not voluntary.
  - Unknown actors, attackers whose type is not yet identified.
- Action which affected the asset. The classification is interesting.
  - Malware, any piece of software that has been specifically designed to compromise the asset. This definition is broader than the usual definition of malware.
  - Hacking, any set of actions performed to get access to or impair information.
  - Physical, using physical devices to perform the attack.

  - Misuse, using trusted resources or privileges to gain unintended access
  - Social, meaning social engineering
  - Error, or anything done incorrectly or inadvertently; often this is the action of internal passive actors or partner actors.
  - Environmental, the consequence of a natural catastrophe often referred to as an act of God.

- Attribute, which describes the impact on the asset from the point of view of information security. Attributes are confidentiality, integrity, and availability.

With this formalism, it is possible to describe any incident. An incident can be a succession of events expressed by the previously described elements. A better understanding of incidents is essential for a pertinent threat analysis. Past incidents educate the analysts on the operating modes of the attacker. They represent real threats that have been activated. A repository of the past incidents is a precious tool for the practitioners. It helps back up the threat analysis with real field data. The VERIS community provides a public database of past incidents that can be used for education and research purposes.

Once the threats are known, the next step is to define whether a response is needed and what type of response is required. Ronald Westrum proposes a taxonomy of three kinds of threat [134].

- Regular threats occur often. Thus, it is possible to develop a standard response. Their likelihood of occurrence is high. Nevertheless, they are rather predictable. The expected response should be well documented.
- Irregular threats occur randomly. Addressing them is more challenging. Nevertheless, their impact may be severe, and they thus need to be addressed. These threats will be the most difficult ones and will challenge the security analysts. The reaction may need to be ad-hoc.
- Unexampled threats are exceptional and unexpected. It may appear impossible that such attacks can happen. For instance, before September 2001, an aircraft deliberately being flown into a building would have been unthinkable. By construction, these types of threat cannot be foreseen. Addressing them requires ad hoc reaction and improvisation at the time of their occurrence. Due to their unexpected aspect, no planning is possible.

Regular threats have to be treated. The security team should be prepared and trained to answer them. The list of foreseen regular threats should be as exhaustive as possible. As much as possible, the response to irregular threats should be documented. Of course, the security team may have to make opportunistic adaptations. The only possible answer to unexampled threats is a highly skilled and trained security team.

2.3 Takeaway

2.3.1 Overprotecting Can Be Bad

Accurately identifying the target of the attack is paramount. Sometimes, the wrong valuation of the valuable asset to protect may overprotect the system and create unforeseen side effects. This was the case for Sony and its PS3. Section 1.2.2 analyzed in detail the hack of PS3. The execution of software in the initial versions of PlayStations, PS1 and PS2, and the first release of PS3, was not protected. Executing arbitrary code on the game console was possible. Furthermore, it used Linux as the operating system. Linux is the favorite OS of geeks. Soon hobbyists used these platforms to develop new homebrew applications. Furthermore, researchers used these powerful platforms for heavy calculations. Indeed, PS3 used a powerful, expensive processor: the IBM Cell. The cost of one PS3 console was less than that of IBM's Cell development board! These PS3 consoles often were used in parallel architectures to create a cluster of parallel computers for massive calculations. This clustering created a powerful computer with low investment cost. For instance, Marc Stevens and six colleagues were able to forge X509 certificates with colliding MD5 signatures [135]. This forgery would enable impersonating another entity's certificate. For this, they used a network of 200 PS3s on the EPFL PlayStation cluster.

Incidentally, while every competing game console was being hacked in a matter of months, PlayStation was never even attacked. One potential explanation is that as Sony was friendly to the Linux community, this community had no incentive to hack Sony. History shows that whenever the Linux community has been involved, hacks have occurred faster [136]. For instance, DVD John created the first hack of DVD content protection in 1998 [87]. Allegedly, because there was no official support for DVD playback for Linux, DVD John decided to write the decryption tool to watch DVD movies on Linux workstations. He took only a few months to reverse engineer and write the DeCSS hack. Similarly, as soon as Sony decided to forbid the execution of homebrew applications on their newest game console, the Linux hacker community concentrated on jailbreaking the PS3. As reported in the previous chapter, they succeeded.

In the context of game consoles, the actual asset to protect is the control of the distribution of official games. The aim is to prevent the game console from executing illegally distributed or replicated titles. Hence, Sony decided to control not only the applications coming from discs but also every application executed on the platform. The protected asset was larger than the actual valuable asset. This restriction provoked the Linux community; we know the results. This overprotection doomed the security of the console as indeed it attracted more attackers.

The following insert studies another use case issued from the game industry: Microsoft Kinect.

The Devil Is in the Detail One of the main success factors of Nintendo Wii was its innovative remote control using gyroscopes. Wii could measure movements of the players. Motion became a new interactive means for gaming. To compete with this innovation, Microsoft introduced its Kinect for the Xbox 360 game console. The Kinect is a device incorporating a set of cameras. Through image processing, the Kinect generates a depth map of its field of vision. A depth map provides an image with the 3D coordinate of each pixel. Initially, Kinect's API was not public. Nevertheless, as in the case of PlayStation's Cell, this solution piqued the interest of hobbyists. Furthermore, a US company, Adafruit, offered a $3000 bounty to the first developer who would provide a library that connects to the Kinect [137].

The initial reaction of Microsoft to Adafruit's challenge was to threaten a lawsuit charging hacking. "*With Kinect, Microsoft built in numerous hardware and software safeguards designed to reduce the chances of product tampering. Microsoft will continue to make advances in these types of safeguards and work closely with law enforcement and product safety groups to keep Kinect tamper-resistant.*" [138].

One week after the publication of the challenge, Hector Martin won the prize. His library gave access to the RGB data from the camera together with the depth map. He published his library as open source under the name `LibFreenect` [139]. Indeed, the first person who reported being able to connect to Kinect was alexP from the open source group Natural User Interface (NUI). Nevertheless, he did not publish his drivers [140]. Later, all these works evolved into a common open source project: the OpenKinect project. Currently, drivers are available for Windows, Linux, and Mac OS for connecting with Kinect.

Meanwhile, Microsoft softened its position. It did not claim anymore that this library was a hack. Stricto senso, this may be true. The new official Microsoft position was:

> Kinect for Xbox 360 has not been hacked – in any way – as the software and hardware that are part of Kinect for Xbox 360 have not been modified. What has happened is someone has created drivers that allow other devices to interface with the Kinect for Xbox 360. The creation of these drivers and the use of Kinect for Xbox 360 with other devices is unsupported. We strongly encourage customers to use Kinect for Xbox 360 with their Xbox 360 to get the best experience possible [141].

This turnaround of Microsoft was very smart. In no way did this library harm Microsoft's business. Soon, hobbyists used the Kinect and created innovative applications extending its use of Kinect beyond just gaming [142]. They probably found some ideas that Microsoft's engineers or game designers would exploit later. This initiative was even a good advertisement for Kinect. The asset to protect is not Kinect's API. The actual asset to protect is the control of the distribution of official games. That a device other than Xbox 360 can access Kinect does not enable the playing of pirated games on

Xbox 360. Furthermore, it may even increase sales of Kinect. Microsoft acknowledged that the API of Kinect was not an asset to protect.
Lesson: Protect only assets that are valuable.

The previous example highlights the next rule.

Rule 2.3: The Cost of Protection Should Not Exceed the Potential Loss

In most cases, the total cost of protection should not be higher than the possible loss of the hacked entity. Evaluating the cost of the protection should be simple. It encompasses the cost of designing or selecting the protection, implementing the protection, exploiting or managing the protection, and maintaining the protection to be up to date and state-of-the-art. Evaluating the potential loss is harder. The overall loss is composed of three types of costs [143].

- The direct, easily quantifiable costs: Recovery of the system and data and replacement of credit cards (if such information is leaked) and the direct financial loss due to fraud or theft. The insurance premium will increase. In some cases, there may be some regulatory fines (for instance, for privacy breaches). The breach may end up with lawsuits and settlements.
- The direct impact on business: There may be some business downtime. Stolen or leaked information may affect the business and marketing strategy, requiring adjustments.
- The indirect impact on intangible assets: The reputation may be tarnished, which may end up in a loss of competitive advantage. Often, severe breaches end with the firing of executive staff, who will need to be replaced, leading to recruitment costs and temporary loss of leadership.

In most cases, Rule 2.3 should never be forgotten. A constant risk for designers is to ignore this rule and overprotect an asset because it has to be defended. This rule is not applicable in some rare cases, such as top governmental and national secrets. It could be considered that these secrets are invaluable.

2.3.2 Know Your Enemy

As it is mandatory to identify what asset to protect, it is also paramount to determine the kind of attacker who will attempt to access the asset. Knowing the adversary gives a hint about her skill set, her preferred operating modes, and her resources. The IBM classification is an excellent starting point [126]. As the attacks are contextual, often it is wise to delve deeper into the description of the different types of attackers to get a finer picture of the potential attackers.

For instance, one of the major technology providers of disk anti-ripping technologies, Rovi, introduces the pyramid of determination. The pyramid defines five categories of users who will attempt to duplicate protected discs [144].

- Regular users will try to reproduce the disk using the usual commands of their operating system or commercial, illegal ripping software. Nevertheless, after the first failure, they will abandon the ripping. They correspond to IBM's Class I category. Regular users correspond to 60 % of the users.
- Advanced users use the same illegal ripping pieces of software. If they fail, they will consult forums to check whether somebody had the same issue and found a solution. They also correspond to IBM's Class I category. They may devote more time trying to defeat the protection using solutions designed by others. Advanced users correspond to 25 % of the users.
- Power users will experiment to find solutions in the case of failure. For instance, they may combine several ripping tools. They may devote a serious amount of time to overcoming this challenge. They may sometimes consult specialized forums. They correspond to IBM's Class II category. Power users correspond to 10 % of the users.
- Committed users will beta test new ripping solutions. Frequently, they will answer the questions of other users in specialized forums. Committed users still belong to IBM's Class II category. They correspond to 5 % of the users.
- The companies design and sell the illegal ripping pieces of software. They correspond to IBM's Class III category.

Rovi asserts that its solutions will not defeat committed users. Rovi assumes that its solutions will deter 95 % of the users. This level of protection may be sufficient for most content owners. Rovi's classification can be easily extrapolated to other tools or fields.

In another domain of content protection, one of the major conditional access system providers, Irdeto, provides an interesting description of the landscape of the Pay TV systems. Irdeto splits content piracy into six categories [145].

- *Consumers*, who only access legal content.
- *Confused consumers*, who use online video sites but do not know whether the Web site is legitimate. The current situation is that it is difficult for non-experts to distinguish between legitimate Web sites and Web sites managed by pirates. Some illegal Web sites are even better looking than the legal ones and propose attractive useful features. The proper answer is to promote better legal offerings. Confused consumers are not pirates and do not enter into IBM's classification. They do not voluntarily try to hack. The illegal offerings just mislead them.
- *Frustrated consumers*, whose goal is to access the content even if it is not legally available. For instance, US movies or TV shows are often not yet available in foreign countries because they are either not yet dubbed or subtitled or due to commercial distribution agreements. Frustrated consumers will look for the coveted piece of content on file sharing sites such as peer-to-peer networks, cyberlockers, or streaming sites. In the long run, frustrated consumers

may turn into casual pirates. The proper answer is to increase the availability of the legal offering. Frustrated consumers correspond to IBM's Class I category.

- *Casual pirates*, whose goal is to access occasionally illegal content. They know where to find it and how to download it. They will not use solutions to break content protection systems; rather, they will consume illegal content that has been posted by other people. The proper response is to remove illegal pieces of content or make the access to them more difficult. Casual pirates still correspond to IBM's Class I category.
- *Hackers*, whose goal is to break as fast as possible the content protection scheme. The objective is not the theft of content, but rather the intellectual satisfaction of having overcome a hard challenge. Hackers may collaborate to break the system. For notoriety, they will distribute the tools or methods they designed. The previous categories of users will benefit from these tools. The proper response is to increase the security of the content protection schemes. Although hackers do not seek money, their exploits compromise revenues of content owners. Hackers correspond to IBM's Class II category.
- *Criminals*, whose goal is to make money by trading illegal content. They are well organized and well funded. They will not share the techniques used to steal pieces of content. As they can break any content protection scheme, the only valid response is legal action. Unfortunately, the legal response is sometimes complicated and challenging [146]. Criminals correspond to IBM's class III category.

At CloudSec 2015, the FBI Supervisory Special Agent Timothy Wallach disclosed a different gradation system of cybercriminals, with six types.

- *Hacktivism* exploits computer to advance certain political or social causes. One of the most famous groups of cyber hacktivists is "Anonymous."
- *Crime* is the theft of personal and financial information to extort financial gains from its victims, which may be individuals or companies.
- *Insider* steals information from a corporation or organization she belongs to, mainly for financial gain or, sometimes, ideological reasons.
- *Espionage* is primarily organized by state actors for stealing state secrets or, when targeting private companies, strategic business intelligence.
- *Terrorism* targets critical infrastructure for sabotage.
- *Warfare* is conducted by state actors who target critical civilian infrastructure or military systems for sabotage. Its main difference with terrorism is that its objective to gain an advantage in the event of war.

This FBI classification focuses on the motivation of the attacker rather than on her skill set and available resources. This classification covers IBM's categories II and III, with strong bias toward the IBM category III. This taxonomy does not encompass casual hackers or unskilled script kiddies. As such, it cannot be used for any threat analysis. VERIS proposes a slightly different grid of motives: espionage, fear, financial or personal gain, fun, grudge, ideology, or convenience. The convenience motivation is an important one within an enterprise (Sect. 7.3.2).

As there are many available classifications of attackers, the choice of the taxonomy used is important for the analyst. The taxonomy should be adapted to the security context of the system. Once the taxonomy is selected, his knowledge of the adversary allows him to fine-tune the response to her attacks. This knowledge helps him understand her motivation and her skill set. Knowing her motivation, the defender can try to thwart her attacks or divert them to other targets. Knowing her skills, the attacker has a better sense of the optimal level of defense to apply.

2.4 Summary

Law 2: Know the Assets to Protect
Identifying the assets to protect is at the heart of any security analysis. The identification of the valuable assets enables defining the ideal and most efficient security systems. The identification should specify the attributes of the asset that needs protection (confidentiality, integrity, anti-theft, availability, and so on).

Rule 2.1: Give Access Only to Those Who Need to Know A best practice in security is to restrict access to assets only to the principals who need access. This restriction is the first defense line against insiders and intruders. It also mitigates the risk of leakage due to human mistakes.

Rule 2.2: Know Your Opponent It is not sufficient to know what to protect to design proper defense. The knowledge of the enemies and their abilities is paramount to any successful security system. This knowledge can be collected by surveying the Darknet and attending security conferences.

Rule 2.3: The Cost of Protection Should Not Exceed the Potential Loss Usually, a defense is sufficient if the cost of a successful attack is equivalent to or higher than the potential gain for the attacker. Similarly, a defense is adequate if its expense is equal to or greater than the possible loss in the case of a successful attack.

Chapter 3
Law 3: No Security Through Obscurity

3.1 Examples

This law is often regarded as the most famous law of security. It inherits from the early work of a Dutch cryptographer, Auguste Kerckhoffs, who taught in France in the late nineteenth century. In his famous treatise, entitled "La cryptographie militaire," he stated[1]: "A [cryptographic system] must require secrecy, and should not create problems should it fall into the enemy's hands" [147]. In other words, the robustness of a cryptographic system should rely on the secrecy of its key rather than on the secrecy of its algorithm. As with Law 1, history has demonstrated the truth of Law 3 repeatedly. This section explores some related examples. They illustrate specific cases that did not take this law into account and thus may have miserably failed.

The Greek historian Herodotus told an interesting story. In 499 BC, the tyrant Histiaeus wanted to command his nephew Aristogoras to start a revolt against the Persians at Miletus. To send his command, Histiaeus shaved the head of his most trusted slave. He tattooed the message and waited for the slave's hair to grow back. He sent his slave to Aristogoras with the instruction to ask Aristogoras to shave his scalp. Aristogoras could then read the hidden message. This is a perfect example of security by obscurity. Anybody knowing the ruse could have recovered the message. The only advantage of this message is that eavesdropping would have been detected.

The second example comes from our day-to-day life: shopping carts. Shopping carts are common elements of the urban landscape. Every supermarket provides them to its customers free of charge. They are part of the essential services of a parking lot. Customers expect shopping carts to be available to them in shops or a mall as well as in the parking lot. Unfortunately, sometimes, indelicate customers remove them from the parking lot and abandon them outside the mall. This leads to

[1]Military Cryptography (in French).

E. Diehl, *Ten Laws for Security*, DOI 10.1007/978-3-319-42641-9_3

many issues. Stealing shopping carts induces a cost for the owner of the mall, who has to replace the stolen shopping carts. The mall's employees have to scout the neighborhood to locate the stolen carts and to retrieve potentially lost shopping carts, which induces additional costs. Stolen shopping carts left unattended on the street represent a serious risk to car safety. Furthermore, stolen shopping carts may deprive new customers of them and alienate them.

In the USA, it is common to see painted, thick, yellow lines that delimit car parking lots, as illustrated in Fig. 3.1. Whenever customers attempt to cross over this yellow line with a shopping cart, the cart brakes. In other words, the malls have installed an anti-theft system for shopping carts. There are two commercial systems deployed: Carttronics's CAPS and Gatekeeper Systems' GS2. They both employ a similar principle [148]. Inside one wheel of the cart, a mechanism can lock or unlock the brakes. The radio frequency (RF) receiver in the wheel decodes the signal received from the antenna buried in the ground. When decoding a specific set of radio signal characteristics, the system locks the brake, thus immobilizing the shopping cart. When decoding another specific set of radio signal characteristics, the system unlocks the brake, thus releasing the shopping cart. A buried wire beneath the yellow paint generates the RF signal that locks the brake. When crossing this buried, emitting antenna, the wheel's receiver captures the blocking signal. This wire delimits the boundary of the protected area. The mall's staff has a gadget that emits the unlocking signal. For these systems, the security depends on the secrecy of the specific signals.

In 2000, Nolan Blender analyzed the signals used by CAPS [149]. Using a digital oscilloscope, he found out that the locking signal was an 8-kHz signal, with a 50 % duty cycle and a cycle time of 30 ms. He discovered that the unlocking signal was a pure 8-kHz signal. With this information, it was trivial to design

Fig. 3.1 Cart anti-theft system

circuits that could lock and unlock CAPS-protected shopping carts. In 2007, a hacker by the nickname of "*orthonormal*" performed a similar reverse engineering operation for GS2 [150]. The locking signal was a set of different length pulses at 7.8 kHz. Once more, it was easy to design the corresponding hacking circuitry which leads to a DoS attack. The weakness of CAPS and GS2 is that the "secret key" is readily observable. Attackers just need to eavesdrop on it. Indeed, the "secret key" cannot be protected.

Is this an issue for these anti-theft systems? No. The objective of these systems is not to prevent the theft of any single shopping cart. The goal of these systems is to make trespassing the yellow line difficult for most customers. Therefore, they succeed in fulfilling their objective. Consequently, it would not be cost-efficient to design stealthier or more complex systems. This example is another illustration of Rule 2.3: The Cost of Protection Should Not Exceed the Potential Loss. The loss of a few shopping carts is affordable compared to investment in a more complicated and more expensive security system.

When first introduced in the market, the digital video disc (DVD) was already protected against illegal duplication. The protection mechanism was called the content scramble system (CSS) and was designed by Japanese manufacturer Matsushita in 1996. It was strictly confidential. The access to its specification is still contingent upon the signature of an NDA, implementation of strict security policies, and acceptance of a licensing agreement. This specification defines the method used to protect the content of DVDs.

CSS uses three 40-bit keys.

- The title keys protect the actual content. The content is encrypted with a proprietary symmetric cryptographic algorithm using the title key.
- The disk key protects the title keys. The disk key is unique for a disk, i.e., for one given movie title, there is one single disk key. This disk key encrypts the title keys. The disk holds the encrypted title keys as well as the encrypted content.
- Player keys are unique for each manufacturer of DVD players. Every player of a given manufacturer holds the player key of this manufacturer. CSS supports up to 409 manufacturers. A data structure called secured disk key data contains a table with the disk key encrypted under all 409-player keys. In other words, the secured disk key data includes 409 encrypted disk keys.

The details of the encryption and decryption algorithms were secret. They were neither standard nor published algorithms. The scheme may seem complex enough to deter attackers lacking the specifications.

The first DVD players appeared on the Japanese market in 1996. Three years later, in October 1999, Jon Lech Johansen, nicknamed DVD Jon, published a piece of software that circumvented the CSS protection. How did he succeed? A subsidiary of RealNetworks, XING Technology, did not adequately protect its player key in its software-based DVD player. Through reverse engineering, DVD Jon managed to extract Xing's player key. With Xing's valid player key, he could write an application that implemented the secret specifications [151]. In fact, DVD Jon reverse-engineered the complete software to rebuild the decryption process. Furthermore, using Xing's player key, he and his team reverse-engineered the player keys of 170 other manufacturers. In fact, with the Xing player key, DVD Jon had access to the pair {encrypted disk key, clear disk key} for each manufacturer key. The manufacturer key was 40 bits long. At this time, 40-bit keys were already far too easy to break by brute force. DVD Jon could brute-force all manufacturer keys. His software, DeCSS, eventually implemented all keys. Once DVD John published the first version of DeCSS on the 2600.org forum, the hacker community was extremely creative in disseminating this piece of software. First, it was modified, enhanced, and ported to many languages such as PERL and Java. The best examples are most probably the PERL implementations, which were rather tiny. Second, a surprising variety of dissemination supports were used to spread the hack. For instance, the CSS descrambler source code was printed on T-shirts and ties used to compose a dramatic reading on the radio and was even presented as a Haiku poem! There was no possible way to stop its dissemination. CSS was irremediably doomed.

3.2 Analysis

3.2.1 Designing a Secure Encryption Algorithm

Designing a secure cryptographic algorithm or a secure protocol is a complex task that involves many actors. Today, it is probably impossible for a single individual to develop such secure solution alone. As Bruce Schneier stated in 1988, it is easy for a designer to create an algorithm that he cannot break. It is far harder to design such an algorithm that no one else can break [152]. This statement is sometimes quoted as Bruce Schneier's law. Thus, for many years, the cryptographic community has defined a design procedure that complies with Kerckhoffs's law. We will use the selection of the Advanced Encryption Standard (AES) to illustrate this design methodology.

In November 1976, the US National Bureau of Standards (NBS) published the Data Encryption Standard (DES) algorithm [153]. It was a modified version of *Lucifer*, an algorithm designed by IBM in 1973. IBM's designers enhanced DES following the recommendations of the US government's NSA.[2] DES is a 64-bit block cipher using a 56-bit key. Soon afterward, DES became the standard

[2]From its inception, DES has resisted to an advanced cryptanalysis method called differential cryptanalysis. IBM's Lucifer was not resistant to this attack. The modifications made DES resistant to differential cryptanalysis. Interestingly, the academic community did not find out about differential cryptanalysis until the late 1980s [154]. In 1994, Don Coppersmith, who was part of the design team of DES, confirmed that the NSA already knew about differential cryptanalysis in 1974 and that they had helped to make DES resistant to this "unknown" attack [155].

algorithm used by most contemporary systems. The use of DES became compulsory for US governmental agencies. In the 1990s, the length of the DES key became insufficient to resist brute force attacks. In 1998, Deep Crack, a $250,000 custom-built machine, could break a DES key in about 1 week [156]. Triple DES, a more sophisticated version of DES, extended the size of the key to 112 bits. This size is still considered out of the reach of current brute force attacks [157]. Nevertheless, the NIST, which is the successor of the NBS, decided to define a new encryption algorithm to phase out the aging DES.

On January 1997, the NIST launched an international contest for selecting a new symmetric encryption algorithm. The minimal requirements were that:

- It should be a 128-bit block symmetric cipher.
- The size of its key should be at least 128, 192, or 256 bits.
- Its implementation should be efficient both in hardware and in software.[3]
- It should be secure against all known attacks.
- It should be made public and royalty-free.

On June 1998, the NIST received fifteen submissions. The contenders were CAST-256 from Entrust Technologies; CRYPTON from Future Systems; DEAL by Richard Outerbridge and Lars Knudsen; DFC from the French CNRS, i.e., the National Centre for Scientific Research; E2 from NTT; FROG from TecApro Internacional; HPC by Rich Schroeppel; LOKI97 by Lawrie Brown, Josef Pieprzyk, and Jennifer Seberry; MAGENTA from Deutsche Telekom; Mars from IBM; RC6 from RSA; Rijndael by Joan Daemen and Vincent Rijmen; Safer+ from Cylink; Serpent by Ross Anderson, Eli Biham, and Lars Knudsen; and Twofish by Bruce Schneier, John Kelsey, Doug Whiting, David Wagner, Chris Hall, and Niels Ferguson. The submissions had to provide the detailed algorithm and a reference software implementation in C and Java. Once the submissions were received, the NIST requested public comments on these fifteen algorithms. The community of cryptanalysts attempted to break all of them. Security was not the unique criterion of selection. The computational efficiency of the implementation (i.e., performance) and complexity of the design of the algorithm were other key criteria for the selection. The NIST organized several public workshops to discuss the identified attacks and compare the respective performances.

In August 1999, the NIST short-listed five contenders: Mars, RC6, Rijndael, Serpent, and Twofish. Table 3.1 provides the results of the evaluation of the five candidates for the three top criteria: the complexity of design, the speed of execution, and the security margin. Apparently, no candidate offered the ideal configuration, i.e., a simple, fast algorithm with high-security margin.

Reference hardware designs, written in VHSIC Hardware Description Language (VHDL), were published for each remaining algorithm. Once more, public

---

[3]DES was not optimized for efficient software implementations. Hardware implementations were straightforward and could be fast, but software implementations were laborious (for instance, due to the use of modulo 32 operations).

Table 3.1 Evaluation of the five short-listed AES candidates

Candidate	Design	Speed	Security margin
MARS	Complex	Fast	High
RC6	Simple	Fast	Low
Rijndael	Clean	Fast	Good
Serpent	Clean	Slow	High
Twofish	Complex	Fast	High

comments were collected. The selection criteria evolved. Of course, security was the major factor. Also, the efficiency of hardware and software implementations was analyzed and compared. Possible attacks on the implementations, such as side-channel attacks, were also studied. On October 2, 2000, the NIST announced that Rijndael was the winner of the contest. Rijndael would become the Advanced Encryption Standard. On November 26, 2001, AES became a public standard, namely FIPS-197. Since then, AES has become the established de facto standard for symmetric encryption, quickly phasing out DES.

In 2007, the NIST initiated a similar process to define the successor of the most widely used hash algorithm, SHA-1. In 2005, Xiaoyun Wang, Yiqun Lisa Yin, and Hongbo Yu had disclosed a method to generate collisions on SHA-1 [158]. Once the attack was confirmed, SHA-1 could no longer be considered secure enough. In October 2008, the NIST received 51 candidates for its hash function competition. The contest followed the same process of public examination. In October 2012, KECCAK was designated as the winner and officially became SHA-3.[4]

The process that the NIST employs to define new cryptographic standards is entirely compliant with Kerckhoffs's law. The security of the algorithms is based on and demonstrated by theoretical analysis by the best brains in the academic and industrial communities. It is a perfect example of security by transparency.

3.2.2 Kerckhoffs' Law Does Not Mean Publish Everything

A very common misconception about Kerckhoffs' law is that this law mandates publishing everything, including the source code of the implementation. However, a full publication has two consequences that may each lower the security of the system in a real-world deployment, depending on the context.

Full publication is the strategy behind the open source movement. In an open source project, the entire source code is available for everybody to review. Furthermore, any developer may modify it for his own use.[5] For security-related

[4]One of the designers of the Keccak algorithm, Joan Daemen, is also one of the two designers of the AES.

[5]Depending on the licensing model, there may be some obligation to publish the modifications. For instance, an open source project licensed under GPL3 requires the publication of all derivative works.

open source projects, such as OpenSSL, it is widely believed that this strategy results in a more secure implementation of standard algorithms than a proprietary one. The alleged explanation is derived from Linus's law: "given enough eyeballs, all bugs are shallow." With open source, more people may review the source code, and thus, the likelihood of finding existing vulnerabilities increases. Unfortunately, this law is not based on actual facts and is often challenged [159]. For instance, the optimal number of reviewers seems to be between two and four. Regardless whether the belief that open source is more secure is actually true, open source security has one severe limitation, described hereafter.

It turns out that all security-related open source projects operate under a given, often unstated, trust model, i.e., that the implementation operates in a trusted environment except for the communication channel. A trusted environment means that the execution environment is trusted, i.e., that the code will always execute as expected. No adverse piece of software can interact with its context. No adverse piece of software can impair the software or access its volatile memory. A trusted environment also implies that the operator or the owner of the execution environment is not an adversary. Not all security solutions can operate in such a trusted environment. Content protection is an example of a non-trusted environment. Usually, the content distributor cannot trust all the viewers of the content to act lawfully. Thus, he will have to protect the secret keys protecting its content from being accessed by the watcher while decrypting the protected content. If the content protection software would use an open source cryptographic library, then the owner of the platform watching the content would know where the decryption keys are located within the memory. Thus, she could simply dump them. Thus, Digital Rights Management (DRM) cannot use open source cryptographic libraries. DRMs are not the only ones. Some devices such as Automated Teller Machines (ATMs) or Points of Sale (POS) also operate under such hostile trust models. The most "paranoid" industry of security is the smart card and HSM industry (Sect. 6.2.2). Unfortunately, more and more critical applications are executing on open platforms such as laptops or mobile devices. Malware can entirely control these platforms and can, therefore, extract secrets. Thus, critical applications such as banking applications should not use open source cryptographic libraries. If the environment does not meet these trust conditions, then the security of the system may be in danger as the trust model is no longer valid.

The second potential issue of open source is the publication of the bug history. As is the case for every properly managed software project, an open source project maintains a database of bugs. This database lists all the declared bugs, the open bugs, the bugs under correction, and the closed bugs. Users may log newly found bugs. Developers may fix them once they are acknowledged. Project managers may close them once the fix has been tested. This database is publicly available. In some cases, this database may become a weakness, as illustrated by the following example.

The Devil Is in the Detail For many years, TV manufacturers have introduced the concept of Smart TV or connected TV. A Smart TV does not simply display broadcast video or video rendered by other devices such as an STB or a DVD player. Modern Smart TVs can execute downloadable applications, run a browser, and play any media available either on the LAN or supplied by a USB drive. In 2014, Benjamin Michele and Andrew Karpow presented a scary proof of concept using Samsung Smart TVs [160]. They exploited the integrated media player of the Smart TV set. They discovered that this TV set used a 2011 version of the open source FFMPEG `libavformat` library. FFMPEG is one of the major open source frameworks to handle multiple video formats. Many devices and pieces of software use it. The `libavformat` library identifies the type of content to be played. Once the type is identified, the library demuxes the recognized format before transmitting the elementary streams to Samsung's proprietary media player. The `libavformat` library supports many containers. It is a complicated piece of software, and as such, it has many newly discovered bugs. By scanning the bug-tracking database of this open source library, the two security researchers selected one vulnerability that was not patched in the older version of FFMPEG used by the TV set. This vulnerability allowed them to execute arbitrary code when playing a piece of forged content. As the Samsung proprietary player executed in a root shell, the forged payload carried an arbitrary code that would also execute in a root shell. Thus, the malicious payload had full access to the platform. As the Smart TV had an integrated camera and microphone, they wrote an exploit that captured the video of the camera and the sound from the integrated microphone. The obtained information could then be sent to a remote server. Of course, the payload could do other potentially nefarious things. As the payload is encapsulated in a real movie, the consumer is not aware that his TV set is being infected and that he is being spied upon. The movie will seamlessly play while stealthily infecting the TV set. Furthermore, the researchers found a way to flash the persistent memory of the Smart TV set and, therefore, to make the infection permanent.

Of course, the problem is not in the open source management of bugs. The issue is that the firmware of the TV set was not patched to fix the latest vulnerabilities. Section 6.3.2 tackles the problem of patching with embedded devices such as connected TV sets.

Nevertheless, the attackers found the information they needed to build their exploit in the database of bugs.

Lesson: Exploiting vulnerabilities in an open source project is easier for the attacker than exploiting vulnerabilities in proprietary software. The attacker has access to two interesting resources: the bug-tracking database and the corresponding source code. With such information, she can write the optimal exploit.

In an open source project, the attacker may act as a white-box penetration tester. In a proprietary project, the attacker usually has to act as a black-box penetration tester. The first one is more powerful and likely to find vulnerabilities.

Section 1.2.1 showed that vulnerabilities should be published rather than kept secret. Open source projects should not, and cannot, deviate from this rule. Unfortunately, the problem of the right timing of disclosure is exacerbated with open source projects. To fix a vulnerability, this vulnerability must first be referenced in the bug-tracking database. In the case of open source software, this database is publicly available. This means that malicious hackers may be aware of the existence of the vulnerability as soon as it is registered into the bug database, i.e., before the vulnerability has been fixed. Of course, for large open source projects that have dedicated full-time developers and a strong project management, it is always possible in case of severe vulnerabilities to set up a dream team and to hide the vulnerability until the dream team released a patch. Usually, bug-tracking databases such as Bugzilla can classify bugs or vulnerabilities as private. These private bugs are only accessible to a set of trusted developers. Unfortunately, sometimes, these bug-tracking databases are riddled with vulnerabilities that allow circumventing the private classification [161]. Furthermore, the majority of open source projects do not have enough resources available to properly handle critical vulnerabilities.

This section may seem a charge against the use of open source in security. It is not the case. The author strongly believes in open source software. Open source projects like OpenSSL have made the cyber world more secure than would be the case if a non-security-aware developer were to have developed his own cryptographic library. Nevertheless, sometimes the evaluation of the actual security of these open source projects is too idealistic. Security-related open source libraries are not fit for all purposes. The trust model of the systems that expect to employ such open source libraries must be adequate.

One of the consequences of Kerckhoffs' law is the necessity to publish the algorithms and protocols used by the system. Cryptanalysts can evaluate the actual strength of these algorithms. Kerckhoffs' law does not mandate publishing the details of the implementation. If the trust model indicates that this publication would not provide an advantage to the attackers, then the publication of the source code could strengthen the security by adding more code reviewers. If the trust model indicates that this publication would give an advantage to the attackers, then the publication of the source code should be banned. In the latter case, the details of the implementation should be considered as confidential.

3.3 Takeaway

As illustrated in previous sections, designing a cryptographic algorithm or a security protocol that is robust against all known attacks is an extremely challenging task. It requires a very specialized skill set both to design a system and to assess its robustness. For many decades, the cryptographic community has developed a process for designing suitable algorithms that is based on peer review. The cryptographic community continues to search for new attacks on the approved algorithms. Many specialized academic conferences, such as Crypto, Eurocrypt, or Fast Software Encryption (FSE), are the venues for proposing new schemes, disclosing progress on the attacks of established algorithms, or publishing new theoretical concepts. For each kind of security goal, several algorithms are available which the community thoroughly scrutinizes. They are considered to be secure. Their strength at any time is continuously assessed. These published algorithms should be the preferred ones when designing a new system.

Rule 3.1: Prefer Known Published Algorithms

Usually, there is no good reason to prefer a proprietary algorithm to an established published algorithm.[6] Kerckhoffs' law teaches us that the publication of an algorithm should not reduce the security of a system. If it were the case, then the choice of the algorithm might be questionable.

Furthermore, with awareness in security becoming more widespread, using a known secure algorithm is a good way to increase the trust in the system. It is easier for the corporate customer to trust a system that uses widely used and approved algorithms. Furthermore, many governments bless the use of such algorithms. For instance, the NIST publishes many recommendations through its FIPS publications. FIPS-197 defines AES for symmetric encryption. FIPS-202 defines SHA-3 for hashing. NIST SP800-90 defines the required characteristics of a true random number generator.

The second lesson of Kerckhoffs is that nothing is more important than keeping the keys secret. Choosing the proper algorithm is not a difficult task. Protecting the secrets that the algorithm will use is the most critical mission of the security designer. This protection is referred to as key management. It is the cornerstone of every secure system.

Rule 3.2: Protect the Key

The keys need to be protected both at rest and while in use. Protection at rest is the role of the digital key store. Secure hardware is harder and more expensive to break than secure software. Thus, hardware-based key storage is preferable. If the secure, hardware memory is not able to store all keys, then, at least, a master key should be stored in the secure memory. The master key should encrypt and authenticate all the

[6]Designers of military applications may have a different opinion.

secondary keys. Therefore, the encrypted and authenticated secondary keys may be stored in insecure memory. As long as the attacker cannot access the master key, she cannot decrypt the keys stored in insecure memory. Obviously, once the secondary keys are decrypted, they are again exposed and vulnerable. Thus, they also need protection while in use.

Protection of the key while in use is the role of the secure implementation. This protection is not always necessary. If the trust model assumes that the key management operates in a trusted environment, typically the secret or private key should not need any additional protection. The trusted environment implies that its internal data are safe from eavesdropping and tampering. Nevertheless, side-channel attacks may be an issue (Sect. 6.2.2). In that case, the implementation has to be carefully designed to prevent this type of attack. If the trust model assumes that the key operates in a hostile environment, then special care is needed to prevent the attacker from discovering the key stored in the volatile memory. Examples of such precautions are never to transfer or handle an unencrypted key as one block of contiguous bytes and never to store the entire key in adjacent memory locations. In the first case, the attacker may observe the exchanged data on the memory bus or within the disassembled binary code when the code handles the key [162]. In the second case, the attacker may scrutinize the RAM and try systematically the contiguous blocks as candidate keys. Such a weak configuration of the memory reduces the key space to explore drastically. In 1998, Adi Shamir disclosed a simple method to detect potential locations of keys in a memory space [163]. Binary code is extremely structured. Similarly, data often has semantic meaning. Thus, the binary representation of data is structured. Both binary code and data present low entropy. On the contrary, good keys are random. Therefore, random keys show high entropy. It is possible to identify areas that may store keys by analyzing the entropy of different areas of the memory. The keys will be located in the high-entropy areas.

It is not sufficient to hide the keys inside the implementation of the algorithm. It is also important to hide some known identifiable binary patterns. For instance, AES uses substitution tables (S-boxes) with known data. Looking for such data pattern provides a good hint of where the encryption is happening within the software. The reverse engineer needs just to do a simple static analysis of the binary code. Many algorithms present such identifiable patterns if the code is unobfuscated. IDA Pro, the ultimate reverse engineering disassembler, offers an automatic method called cross-reference analysis [164]. Cross-references show how a particular piece of information is being accessed. The information may be either a piece of code or data. The cross-reference describes where and when the piece of information was accessed and who accessed it. IDA Pro can even display it graphically with all the accesses to many cross-references. Such power tools help us identify and understand the handling of the keys.

The ideal implementation of a cryptographic algorithm would be such that even if the attacker were to have the source code and entirely control the platform, she would not be able to retrieve the secret key. In 2002, Stanley Chow and his colleagues proposed a new concept called the white-box cryptography [165]. The threat model of white-box attack assumes that the attacker has full access to the encryption software and entirely controls the execution platform. White-box cryptography attempts to protect the keys even under such a hostile threat model. The main idea is to create a functionally equivalent implementation of the encryption or decryption algorithm that uses only lookup tables [166]. Corresponding lookup tables, with the corresponding hard-coded secret key, replace the S-boxes, Feistel boxes,[7] and XOR functions usually employed by symmetric cryptography. Then, the lookup tables are further randomized. In theory, the randomization hides the hard-coded key. White-box cryptography is a difficult challenge for skilled reverse engineers. Unfortunately, white-box cryptography has a high penalty in terms of both size and performance. Usually, the footprint of a white-box implementation is at least 50 times larger than the reference implementation. In addition, a given implementation works only for the specific hard-coded key it was designed for. If another key needs to be used, a new particular implementation is developed. Abundant cryptographic analysis has demonstrated that these constructions are not theoretically secure. Nevertheless, well-crafted real implementations seem to resist reverse engineering. Many vendors offer such white-box cryptography for AES.

Security through obscurity is not fundamentally bad when used as an additional layer of security and as a measure of defense in depth. Code obfuscation is one example of such an additional layer. Obscurity will ultimately vanish. Nevertheless, obscurity will have slowed down a determined attacker and hopefully discouraged many casual attackers. A general solution to reducing the attack surface is to publish only the minimal required information about the implementation. The less information the attacker has, the more difficult her work will become. There is no valid reason for a defender to facilitate the attacks. Any method that may frustrate the attackers is a valid method.[8]

3.4 Summary

Law 3: No Security Through Obscurity

The robustness of a cryptographic system should rely on the secrecy of its key rather than on the secrecy of its algorithm. As such, a strong assumption is that if an attacker knows the algorithm used, she should gain only a minimal advantage.

Security should never rely exclusively on obscurity. Nevertheless, obscurity may be one component of the defense. However, whenever the secret or ruse obfuscated by the obscurity is disclosed, the system should remain secure. In other words, obscurity should not be the centerpiece of the security of any system.

[7]For instance, for the DES algorithm.

[8]Of course, assuming that the method is legal.

Rule 3.1: Prefer Known Published Algorithms A design should only use known, published, cryptographic algorithms and protocols and avoid proprietary solutions.

Rule 3.2: Protect the Key Keys are the most valuable assets in cryptography. Key management is the most critical and complicated task when designing a secure system. The implementation details are dependent on the trust model. In a hostile environment, the implementation should use obfuscation techniques or white-box cryptography and avoid open source.

Chapter 4
Law 4: Trust No One

Honesty is the best policy
(old saying)

4.1 Examples

There are many definitions of trust. Trust has many aspects. Trust encompasses different fields such as psychology, sociology, economics, and computer science. Trust is essential when the environment is uncertain and risky. Trust allows taking some decisions based on prior knowledge and experience. However, trust is subjective and evolves with time. This chapter will use Roger Clarke's definition [167]:

> Trust is confident reliance by one party on the behavior of other parties.

In other words, trust is the belief that the other parties will be reliable and that they are worthy of trust.

History is full of examples of wrongly placed trust that resulted in the doom of a king or an empire. Following is a list of some famous traitors. In 480 BC, the Greek shepherd Ephialtes of Trachis directed the Persian army of Xerxes to a hidden path that bypassed the Thermopiles pass. This betrayal was the doom of the 300 Spartans led by Leonidas, who defended this pass. In 44 BC, the adopted son Brutus partook the assassination of his father: Roman Ceasar Julius. The stabbed emperor was supposed to have claimed: "Tu quoque mi fili?"[1] In 778, Frankish knight Ganelon betrayed Karolus Magnus at Ronceveaux.[2] Ganelon plotted with the Saracens an ambush at the pass of Ronceveaux. Karolus Magnus's champion knight, Roland, died there defending the retreating army against the Saracens driven by King Marsile. During the American Revolutionary War (eighteenth century), Benedict

[1]"You also my son." Brutus was a proponent of the republic that Julius Caesar replaced by the empire.

[2]Ganelon is indeed a fictitious character who was created three centuries later in the "Chanson de Roland."

E. Diehl, *Ten Laws for Security*, DOI 10.1007/978-3-319-42641-9_4

Arnold was a successful American commander. Nevertheless, in 1780, he proposed to the British adversaries to sell them the Fort Clinton at West Point, which he was commanding. Once the plot was exposed, Arnold fled and joined the British army. In the nineteenth century, Karl Schulmeister began his career as a spy for the Austrian Empire. Later, French emperor Napoleon recruited him to spy on his enemies. The information that Schulmeister gathered as a double agent led to the French victory at the battle of Austerlitz. These historical examples highlight the importance of trust. Whenever trust is misplaced, defeat is never far.

Trust is the foundation of any security system. If the foundations of a building are weak, then the building may collapse. If the foundations of a building are firm, then the building will be robust and will resist many incidents. The same goes for trust. Without a sound trust model, the security system will inevitably collapse.

Knowing who and what the security designer can trust is of utmost importance. Even the hackers know that they cannot trust anyone and especially not their peers. We will start with a few examples coming from the Scene.

In August 2010, the first "commercial" hack for Sony's PS3 was available. Its name was PSjailbreak. This USB drive enabled the execution of copied games. The hackers who designed it did not trust other hackers. They feared that soon other hackers would reproduce their device and sell counterfeited hacks of their hacking tool. Indeed, they were right because soon the reverse engineering of PSjailbreak started [168]. Soon, clones of PSjailbreak such as PS3stinger, PS3key, and X3JailBreak appeared on the black market. Therefore, the distributor of PSjailbreak explicitly warned about imitators and created a logo to authenticate their official suppliers (Fig. 4.1). The sticker used an expensive hologram to prevent cheap duplication of this logo. The hackers used the same techniques that software distributors employ to thwart piracy. The hackers attempted to create a trust relationship with their customers against hackers hacking the hackers.

SubSeven is a well-known Trojan malware. Once SubSeven is installed on a computer, it is possible to take control remotely by this machine. Of course, as with other Trojans such as Back Orifice or SubNet, it should operate stealthily. Usually, a password, defined by the attacker who installed the Trojan, protects access to SubSeven. This password prevents another hacker from taking control of an already infected computer. In 2000, Defiler reverse engineered the current version of SubSeven. He found out that the author of SubSeven introduced a non-documented master password in the software. With this master password, it was possible to connect to any installed SubSeven without knowing the password set by the first hacker [169]. A hacker could steal the hacked computers from the first hacker.

Unfortunately, sometimes seemingly robust security assumptions are wrong. Air gapping two systems is supposed to be the ultimate solution for isolation. An air-gapped, or air-walled, system is physically isolated from any insecure network such as the Internet or even a corporate LAN. The air-gapped system should not have any means to leak out information. If any information is to be sent out through a digital connection, then a DLP tool most probably monitors the output data to

Fig. 4.1 A logo to authenticate a hacking device of a game console

detect any illegitimate data transfer. Unfortunately, sneaker net[3] may be an efficient method to cross this air gap isolation. The systems have to sanitize any data that are coming via sneaker net to avoid contamination. Some researchers have attempted to create communication channels that cross the air gap isolation. Mordechai Guri and his three colleagues disclosed an attack that used thermal measurements to transfer digital information over an air gap [170]. On the emitting computer, they modulated its temperature by increasing or decreasing the use of power-consuming components such as graphical processing units (GPUs). On the receiving computer, they used the integrated thermal sensors to measure the variations in temperature. They achieved an effective bit rate of 1–8 bits per hour. The information crossed the air gap. Such a communication channel is called a covert channel. Researchers studied many other covert channels using different physical phenomena as carriers. Audio is probably the most studied one. For instance, in 2005, Madhavapeddy Anil and his four colleagues proposed an audio network between disconnected devices [171]. Thus, under some circumstances, even air-gapped isolation may be not trusted.

[3]Sneaker net is the vocable that describes the use of a physical detachable storage to exchange data between two systems.

4.2 Analysis

4.2.1 Supply Chain Attack

Can you trust the devices that you use every day? It would be comforting to believe so. Unfortunately, the answer is negative.

In July 2005, 3700 Zen Neeon mp3 players from the Japanese brand Creative were shipped with the virus "`W32.Wullik.B@mm`" [172]. The virus was loaded at the manufacturing site and not in the field. In August 2006, McDonald's and Coca-Cola launched a contest whose prizes were 10,000 branded mp3 players preloaded with songs. Unfortunately, these players were also infected with a variant of the virus "`QQPass`" [173]. When connecting to a host computer, the virus stole the stored passwords and personal information. The virus emailed back to its author the collected data. McDonald's reaction was exemplary. It set up a dedicated hotline, recalled the infected mp3 players, and replaced them with new mp3 players [174]. In September 2006, Apple announced that it had shipped some Video iPods that were infected with the virus "`RavMon.`" This virus could only contaminate Windows-based systems. It infected fewer than 1 % of the shipped devices [175]. In January 2007, the Dutch company TomTom had to admit that Trojan viruses infected some of its GO910 satellite navigation systems V6.51 [176]. When the device connected to a host Windows computer, the virus "`Win32:Perlovga`" attempted to infect the host. A security-savvy customer was alerted by his anti-virus software when connecting his brand new TomTom GO910 to his computer [177]. Unfortunately, TomTom did not take this infection as seriously as it should have. First, it minimized the potential effects of the viruses, quoting it as a very low risk; then, it did not provide sufficient guidance for removing the virus. In February 2008, Best Buy sold Insignia digital picture frames NS-DPF-10A. This device was initially infected with the virus "`Mocmex`" at the factory. This virus was much more complicated to remove than the previous ones [178]. Once installed, Mocmex deactivated the anti-virus that was protecting the host computer. Then `Mocmex` gathered passwords for online gaming, especially targeting World of Warcraft. Then, it loaded other viruses such as "`W32Rajump,`" which sends back the infected Internet Protocol (IP) addresses [179]. A Trojan opened a backdoor and displayed pop-up ads. In November 2008, Lenovo distributed its Trust Key software for Windows XP infected with the Trojan dropper "`Win32/Meredrop`" [180]. In March 2014, the security company Marble Security announced that it had discovered that a malware disguised as a legitimate Netflix application was pre-installed on Android phones and tablets. It affected, at least, four manufacturers: Asus, LG, Motorola, and Samsung [181]. The malware application captured passwords and credit card information and sent them to Russian servers.

The list is never ending and could be updated with more recent similar events. What is the common denominator of all these incidents? The infection of the shipped devices occurred during the manufacturing phase. The supply chain was infected and could not be trusted anymore. It is not always clear if the infection was due to negligence or due to malevolence. In the case of negligence, the infection could have come from a production computer already infected in the manufacturing plant. These incidents highlight that the factory's computers were not regularly sanitized and that the pieces of software to be loaded onto the devices did not pass through an anti-virus or, at least, an up-to-date anti-virus. In the case of malevolence, an insider would voluntarily install the malware in the original product. The next example will illustrate a case where attackers deliberately compromised the supply chain.

In October 2008, Joel Brenner, the US National Counterintelligence Executive, disclosed that European law enforcement teams detected an ongoing attack. For several months, hundreds of European banking card readers had sent account details, together with the corresponding PINs, to a Pakistani and Chinese fraud ring [182]. The attackers modified the credit card POS for the British, Irish, Dutch, Danish, and Belgium markets. The modifications occurred either during the manufacturing phase of the card readers in China or at the shipping stage. The attackers modified the POS with the addition of a metallic box containing three electronic boards. The first board collected information from the inserted credit card and the keyboard. The second board encrypted the collected information and stored it in one memory buffer. The third board used the terminal's phone line to send the stored skimmed data to Pakistan [183]. Thus, a non-intrusive method to detect the hack was to check whether the weight of the POS exceeded its original value. The hack was sophisticated and could even filter out the type of credit card to skim. With all the information at hand, the hackers could clone cards. The thieves were smart, as, depending on the shop where the skimming occurred, they stole different amounts. If the credit card was used for small purchases, then the cloned card would be used only for small purchases; whereas if the credit card was utilized in a luxury shop, then the cloned card would be used to steal a larger amount of money. Furthermore, the attackers continuously modified the pattern of theft. This strategy prevented easy triggering by the banks' fraud pattern detection. The estimated losses were in the range of 50–100 million euros [184].

In 2008, the US government discovered that thousands of CISCO routers being used by their military services, the FBI, defense contractors, American universities, and financial institutions were not genuine CISCO routers. They were cheap counterfeited routers. The government bidding system indirectly helped this infection. Genuine CISCO 1721 routers cost about \$1375, whereas the counterfeited routers cost about \$234 [185]. Therefore, the counterfeited routers won the governmental bids. Of course, these counterfeited products raised some issues. The counterfeited routers had a higher failure rate, thus being less reliable than the genuine routers. Furthermore, they presented a risk of duplicate MAC address, which could generate network shutdowns.[4] The unanswered question is whether

[4]Network protocols assume that every MAC address is unique. If, in a network, two devices present the same MAC address, the network may have some failures. Furthermore, if the conflicting devices are routers, then the network will collapse.

this counterfeiting was driven purely to gain higher profit or whether the Chinese government sponsored it. The FBI identified some threats related to a foreign government initiating the counterfeiting of routers. The first goal of a state-sponsored attack may be to cause premature or frequent network failures. The second goal could be to implement backdoors that would enable stealth access to secure or critical systems. A third goal could be to weaken the cryptographic tools, allowing easier eavesdropping. Naturally, CISCO experts thoroughly analyzed some counterfeited products. They claimed that there were no trapdoors in the investigated counterfeited routers. The fact that they did not find any traps does not mean that these routers were benign. The attackers may have been cleverer than the CISCO experts.

In 2012, the US House of Representatives' Permanent Select Committee on Intelligence warned the government and American companies to ban Chinese Huawei and ZTE network equipment [186]. The committee assessed that the likelihood that the Chinese government would have tampered with these pieces of equipment for eavesdropping was too high. The Chinese manufacturers could not provide convincing evidence that this threat was not existing. Likewise, the NSA allegedly implements spyware and trapdoors in routers and network equipment manufactured by American companies for export to foreign markets. Once installed, these tampered devices attempt to connect back to the NSA surveillance network [187].

"Interdiction" attacks are a variant of the supply chain attacks. The attacker intercepts shipped physical goods and replaces them with an infected version. This type of attack was detected several times in the wild (CD-ROM of proceedings of a scientific conference, Oracle installation disks, and so on) [188].

During the last quarter of 2014, Chinese manufacturer Lenovo knowingly installed on its shipped PCs a bloatware from Superfish. Superfish is a California company that provides a solution for performing a visual search. Superfish designed Visual Discovery, a piece of software that displayed context-driven advertisements during Web browsing. To perform this hijacking of the browsers, Visual Discovery used a software stack designed by Komodia: SSL Digestor. According to the site of Komodia, its advanced SSL hijacker SDK allows access to data encrypted using SSL and performs on-the-fly SSL decryption. It enables easy access to the data and the ability to modify, redirect, block, and record the data without triggering the target browser's certification warning. On 19 February 2015, The CERT disclosed the solution used to fool browsers [189]. Komodia Redirector with SSL Digestor installs non-unique root CA certificates and private keys, making systems broadly vulnerable to HTTPS spoofing. Komodia installs its own root certificate within the browsers' CA repository stealthily. The software stack holds its private key. This allows it to "self-sign" certificates to forge SSL connections. The software then generates a typical man-in-the-middle attack. Although the private keys were encrypted, it was possible to extract some corresponding private keys. As long as browsers do not erase the root public key from their repository, an attacker may use the corresponding private key. She may sign malware that would be accepted by the machine or generate phony certificates for phishing. Lenovo caused a supply chain attack by voluntarily infecting its products.

All these examples illustrate the diversity and the complexity of cybersupply chain security [190]. The incidents can be classified into three categories.

- Accidental attacks due to lack of or lax control; the first examples are part of this first category.
- Attacks that try to access some assets, organized criminal organizations often drive these attacks. The example of the compromised credit card terminal belongs to this second category.
- Cyberwar attacks when state-funded organizations attempt to penetrate critical infrastructure and steal confidential information. Business intelligence may also be one strong motivation.

Supply chain attacks will most probably become more prevalent in the coming years. The reasons for this increasing risk are multiple. The first reason is that today rarely one single team designs an entire secure system. The design always relies on components or subelements that other teams have created. Even military systems use off-the-shelf microprocessors and other electronic components. No country can afford anymore to have a manufacturer that would be able to design general-purpose processors as powerful as the commercial ones just for military applications. They use the same processors as consumer products.[5] Thus, they have to trust that the other designers did a proper job. Unfortunately, this assumption is sometimes wrong. Studies have shown that it is possible to implement stealthy functions inside semiconductors [191]. Thus, the designer of the complete system has to trust the combination of several sub-systems. It is easier for an attacker to infect the system, especially at the boundaries between the sub-systems or within a sub-system.

The second reason is the globalization of the industry. Most of the large companies are now transnational. Most of their production is located in other countries, especially in the Far East. Therefore, the industry has to trust that the production chain and the supply chain are properly managed and that quality control is of a high standard. International legislations may sometimes even impair the enforcement of such verification. This evolution of the IT world is unavoidable and inexorable. Thus, the corresponding risk of a supply chain attack is here to stay and will most probably increase.

[5]The main difference with consumer products is the resistance to environmental conditions. Military grade components must operate on a larger range of temperature and higher humidity ratio. For instance, military grade components operate from −65 to 175 °C, whereas commercial grade components operate from 0 to 70°C. The mean time before failure should also be higher for military grade components than for consumer grade ones. Usually, the military grade is not reached by a different design but rather by a stricter final qualification test during manufacturing.

4.2.2 Who Can You Trust?

Security blogger Technion disclosed that HP StoreOnce had an undocumented backdoor installed in its product line [192]. When connecting via SSH with the identity `HPSupport` and the password with SHA-1 hash of value 78a7ecf065324604540ad3c41c3bb8fe1d084c50, the connection opens an administrator account. In previous versions of the product, a button on the server allowed one to reset the server to factory defaults. This button was useful whenever the administrator forgot his password. As it was a potential vector of attack for an insider seeking to gain administrator access, HP removed this button. HP replaced it with an undocumented administrator account that used the same password for all servers. HP support could remotely reset an administrator password by entering the undocumented administrator account. Unfortunately, SHA-1 hashes cannot resist brute force attacks (Sect. 7.2.2). Technion claimed that the password had been cracked. The initial version of the software hard-coded the credentials of the administrator account, i.e., customers could not disable it. A few months later, HP issued a patch that solved this problem [193].

In accordance with law 1, the reader should be now aware of never blindly trusting any security solution. What if the technology provider of these security solutions cannot be trusted anymore? This misadventure happened to Bit9 in February 2013. Bit9 is a company that offers security solution controlling which applications are authorized to execute on a platform. Rather than relying on detecting malicious applications (using signatures or blacklists), Bit9 uses an engine that only permits trusted applications belonging to a white list, combined with a cloud-based reputation service. Every application that does not belong to the white list is by default considered suspect and denied access. Of course, the Bit9 engine considers every application issued by Bit9 as trusted. A Bit9 private key signs the applications published by Bit9. The control verifies whether the Bit9 private root key correctly signed the piece of software. On 8 February 2013, security consultant Brian Krebs announced that a malware signed by Bit9 affected some companies [194]. Later on the same day, Bit9 Chief Executive Officer (CEO), Patrick Morley, acknowledged the problem [195]. The company's solution did not protect some of the Bit9 servers. Among these, unprotected servers were the servers used to sign digital applications. Using a Structured Query Language (SQL) injection attack, professional hackers from the Hidden Lynx team penetrated the company's network. Once inside, they succeeded in having thirty-two malicious pieces of code signed by Bit9. Thus, any Bit9 engine would accept these signed pieces of malware as trusted applications. Once the breach was detected, Bit9 announced that they had started to fix the issue. They applied their proprietary solution to their entire infrastructure. They revoked the digital certificate used to sign the malware and informed their customers.

According to Bit9, only three undisclosed customers were affected. It is likely that the attack was part of a broader APT: the VOHO campaign [196]. This campaign targeted, via waterhole attacks, some targeted companies. As in the case of the RSA attack (Sect. 1.2.4), penetrating a system via one of its protection tools could be an excellent, efficient vector of attack. Ironically, a few hours before the disclosure of this attack, Bit9 blogged that the unique answer to APTs should be "to deploy a trust-based security solution with policy-driven application control and whitelisting" [197]. Bit9 claimed that there was no need for other defense mechanisms than their own solution. Due to marketing considerations, Bit9 had to follow their own ingenious approach. Thus, Bit9 forgot the Rule 1.4: Design In-depth Defense. Oversight and perhaps overconfidence affected trust.

In 2013, Edward Snowden shook the world of intelligence and security. Snowden was a system administrator at the NSA. In a series of successive revelations, he informed the world that his former employer widely spied on all electronic communications with systems called PRISM and MUSCULAR. He backed up his revelations by publishing classified documents that proved he told the truth. It should not have been a surprise that the NSA was spying even on its alleged allies. Spying is at the heart of any intelligence agency. The surprise was the scale of the deployment and the methods used. For instance, some documents proved that the NSA approached large corporations such as Microsoft and RSA Ltd. [198]. These enterprises are suspected to have implemented backdoors in their cryptographic libraries. For instance, the NSA requested RSA to weaken the random generator for the elliptic curve cryptosystem of its iconic cryptographic library: BSAFE [199]. Consequently, the NSA could decrypt any encrypted message with this library via these backdoors. Furthermore, the NSA subverted some certification authorities to be able to mount stealthy man-in-the-middle eavesdropping. These revelations put a large veil of suspicion on all US originated products and US originated services [200]. For instance, the French administration banned the use of commercial off-the-shelf smartphones [201], as all operating systems used by smartphones come from US enterprises.[6] Often, governments are suspicious of other countries. For instance, the Japanese government banned the use of a popular free Input Method Editor, which was widely used in Japan. This editor helped to write Japanese hiragana and katakana characters and translate them into Kanji characters. The software sent all dialed characters to the Chinese Baidu[7] Cloud, even when the device was not connected to the cloud [202].

Two-factor authentication is spreading and replacing the usual password-based authentication. This is a good trend and provides a more secure authentication. Nevertheless, the assumption underlying the trust model is that the attacker is unlikely to have access to both factors. For instance, many financial institutions enable money transfer when the account owner can log into his account and has access to his mobile phone. The strongest assumption is that nobody other than the user has access to a phone with the same phone number. In October 2010, New Dehli's police arrested two hoodlums [203]. Even though two-factor authentication

[6] Apple produces iOS; Google produces Android; Microsoft produces Windows RT and Windows 10.

[7] Baidu is the Chinese equivalent of Google.

protected online financial transactions, the attackers stole many accounts of Indian people and orchestrated fraudulent transactions. Furthermore, the account owners did not lose their registered mobile phones. The criminals purchased banking account information and phone numbers from Nigerian scammers. They asked the mobile companies mentioned in the bank accounts of their victims for a new SIM card. Thus, they got a duplicate SIM card issued. After activation of the duplicate SIM card, they accessed the account of the victim on the Internet using the purchased details. They initiated a money transfer from their victim's account to straw man accounts. With their phones deactivated, the victims could not react in time. The protocol used, like many protocols using a mobile phone, assumes that only the legitimate owner of a mobile phone receives messages. The assumption may be wrong if the mobile carrier does not properly manage the renewal of SIM cards.

4.2.3 Is This Certificate Yours?

When publishing their seminal paper "New directions in cryptography" [204] in 1976, Whitfield Diffie and Martin Hellman revolutionized the world of security. They proposed the foundations of public cryptosystems (Appendix A). Prior to Diffie and Hellman's work, cryptography required Alice and Bob to share a common secret key. They lifted this constraint. Alice could securely send a message to Bob by encrypting the message with Bob's public key, which he published. Bob would decrypt it using his private key. Provided nobody other than Bob knows his private key, the message sent by Alice should be safe as only Bob can decrypt it. This paradigm is at the heart of the security of Internet, for instance, via Secure Sockets Layer (SSL) and Transparent Layer Security (TLS). Thus, we explore its trust model thoroughly.

The trust model of public cryptosystems is that an attacker cannot access, calculate, or guess the private keys. Of course, this model has a salient consequence: The private key needs protection. It may be stored in a secure memory, for instance, in a smart card or an HSM. Smart cards provide efficient protection against software attacks and also against many physical attacks. The HSM provides even more defenses. For instance, most HSMs will erase the stored secrets whenever they detect an attempt to intrude. If a private key is to be stored in a non-secure memory, then usually it is encrypted with a secret symmetric key only known to Bob. Access to Bob's private key will require first getting the secret key from Bob and then decrypting his private key using the secret key. Pretty Good Privacy (PGP) uses this mechanism to protect its private keys. PGP users have to define a long passphrase (the longer the passphrase, the more secure it will be) that will serve as a secret key. In the case of PGP, the secret key is never stored. The secret key remains only in the consumer's mind: a safe place, at least in theory. Passphrases, although longer than passwords, are supposed to be easier to remember.

The private key must be kept secret but also securely backed up. If ever the private key is lost or destroyed, then all encrypted messages and data are also lost.

In 1999, when testing the first deployment of the German eHealth smart card, the engineers experienced a serious problem. They could not add new smart cards or revoke smart cards. Following a voltage drop, the HSM handling the Certification Authority's (CA) private root key assumed it was under attack and thus erased the private key. Unfortunately, there was no backup of this HSM [205]. Do never forget Rule 1.5: Back up Your Files.

Unfortunately, protecting the private key from the attackers is not sufficient. Then, a man came in the middle of the communication. Let us assume that Mallory can intercept messages exchanged between Alice and Bob and that she can modify them. Mallory sends her public key to Alice pretending that it is Bob's public key. When sending the message to Bob, Alice encrypts the message with Mallory's public key believing that she uses Bob's public key. Mallory decrypts the encrypted message with her private key. She encrypts the intercepted message using Bob's public key and forwards the encrypted message to Bob. If Mallory does not modify the message, Alice and Bob cannot discover that Mallory is indeed eavesdropping on them. This type of attack is called the man-in-the-middle attack. The weakness is that the public key is just a set of random bits that are not linked to its owner. Mallory can easily claim her public key as being Bob's. Random bits have no meaning for humans.

Therefore, it was necessary to link the public key tightly to the identity of the owner of the corresponding private key. This is the purpose of the digital certificate. A digital certificate is a data structure that contains, at least, the following information:

- The public key,
- The identity of the principal owning the key; this identity must be in a human-readable format so that Alice can easily verify that the provided certificate is indeed Bob's certificate.
- A digital signature of previous information by a trusted third party hereafter called the signing CA. The digital signature prevents Mallory from modifying previous information without Alice being aware.
- The identity of the signing CA; once more, this identity must be in a human-readable format so that Alice can easily verify who the signing CA was. It is mandatory that Alice trusts the signing CA. Without the identity of the signing CA, Mallory could use her own CA infrastructure.

When Alice receives Bob's alleged certificate, she should:

1. Verify that the identity on the certificate is Bob's;
2. Check whether the signature is valid; and
3. Verify that she trusts the signing CA.

Only once these three conditions are fulfilled can Alice believe that the provided certificate is indeed Bob's certificate. Ideally, there will be some additional information to verify. A certificate may have an expiration date. Furthermore, the CA should publish a list of revoked certificates. For instance, if Bob's private key leaks out, he will not want to use it anymore because the initial security assumption that Mallory cannot access or guess his private key is not valid anymore. Thus, Bob requests the CA to add his revoked public key to the Certification Revocation List (CRL). In addition to the three previous conditions, Alice checks:

- Whether the certificate is still active, i.e., the current date does not exceed its expiration date.
- Whether the certificate is not revoked, i.e., the certificate does not belong to the CRL.

An international standard, X509, defines all these parameters [89]. Currently, most of the issued digital certificates are X509-compliant.

Hence, the second trust assumption of any public key cryptosystem is that the certificate contains signed, valid information from a CA that is trusted. In March 2001, Microsoft issued a security bulletin disclosing that VeriSign issued two signed certificates that were wrongly assigned to the Microsoft Corporation [206]. The recipient of these digital certificates and associated private keys could sign any arbitrary code, for instance, a piece of malware, on behalf of Microsoft. Users would wrongly believe that the signed piece of software was a piece of legitimate Microsoft software and thus trustworthy. VeriSign revoked the problematic certificates. Unfortunately, the illegally generated certificates did not contain a CRL distribution point, i.e., the location where to seek the latest CRL. Therefore, Microsoft had to design a patch that blacklisted these certificates without relying on the CRL mechanism. The mistake was that VeriSign delivered the certificates to a person claiming to be a Microsoft employee without verifying that he was actually a Microsoft employee entitled to issue such a request. The usual process of VeriSign involves also sending one email to the requesting company to check whether the company has sought the issuance of certificates. Unfortunately, it took several weeks to detect the illegal publication. A certificate that identifies a principal is valid only if the CA has properly checked this identity. This task is not simple with digital identities. Sometimes, this verification might be delegated to an external registration authority. A registration authority is an entity that handles the identification and authentication of principals but does not sign or issue certificates.

In 2007, the importance of this verification triggered a new flavor of X509 SSL certificates: Extended Validation SSL (EV SSL). Certification authorities that are authorized to issue EV SSL certificates have a strictly controlled documented process to validate the requester of a domain-based certificate as the actual owner of the domain. Browsers provide a distinctively enhanced display for sites using EV SSL certificates. Usually, they show in the address bar the name of the entity owning the certificate, the name of the issuing CA, a locked padlock, and a distinctive color (usually green). Web sites using EV SSL certificates should be more trustworthy than sites using standard X509 certificates.

It is interesting to analyze the robustness of the second security assumption. Closed proprietary systems usually employ one single CA. The certificates use a hierarchical structure. A hierarchical structure is defined as a typical tree structure.

One unique public/private key pair is allocated to each node. The private key of the root signs the certificate of the public key of each direct subordinate leaf. Subsequently, the private key of each leaf signs the certificate of the public key of each direct subordinate leaf. The root of the hierarchy is unique. The second security assumption is simple to satisfy by adequately protecting the private root key and by using well-designed processes to deliver signed certificates and to manage the CRL correctly.[8] Thus, in this configuration, the security assumption is valid and robust. Unfortunately, in the real world, systems are more and more open. Furthermore, they need to interoperate and collaborate. Using one unique source of trust may become unpractical. Therefore, several models define architectures enabling the use of multiple Roots of Trust [207].

The first one is the cross-certified CAs or bridged trust model. In this model, two CAs agree to share their trust. Each CA uses its own pair of root keys and issues certificates signed by its own private root key. Nevertheless, principals using this system will accept the certificates issued by both CAs. Indeed, each trusted CA signs the root key certificate of the other CA also. The security assumption slightly weakens, as now we must trust two entities rather than one unique entity. This model can be extended to more than two CA roots. There has to be a certified mesh that allows navigating among the different CA hierarchies. This mesh of certificates is implementable when the system implies only a few root CAs. It becomes inadequate when too many CAs participate in the system. The second model is adapted to this scenario (Fig. 4.2).

The second model is the trust listed CA model. In this model, the system manages a list of CAs that it will trust. Indeed, the list contains the self-signed certificate of the public key of each trusted root CA. A self-signed certificate is the certificate of the public key that is signed by its private key. The CAs are independent. This model is weaker than the bridged CA model. The Internet and browsers use this model. For many years, Secure Hyper Text Transfer Protocol (HTTPS) and SSL/TSL have become widely deployed. Currently, millions of certificates are generated and distributed. They "protect" Web sites (using HTTPS) and computers. For logistical, economic, and political reasons, it was impossible to have only a handful of root CAs. Currently, several hundred CAs issue certificates for the Internet. Browsers have to access all legitimate Web sites to offer a satisfactory user experience. Thus, browsers have to trust most existing CAs. Therefore, browsers come with a large bunch of preinstalled public root keys of CAs. Internet Explorer hosts about twenty trusted root CAs. Mozilla recognizes more than 120 trusted root CAs. Each trusted root CA signs other CAs' certificates. Furthermore, the supported trusted root CAs are not necessarily the same across the different browsers. The current Internet has more than 1500 CA that the browsers trust [208]! In this context, the second security assumption becomes frail. Only one wrongdoing

[8] If the attacker is able to request revocation of a valid certificate, she can create a nice, efficient DoS attack, especially if the revoked certificate is rather high in the certificate hierarchy. Indeed, if the certificate of one leaf of the hierarchy is revoked, then all the certificates of its branches are also revoked.

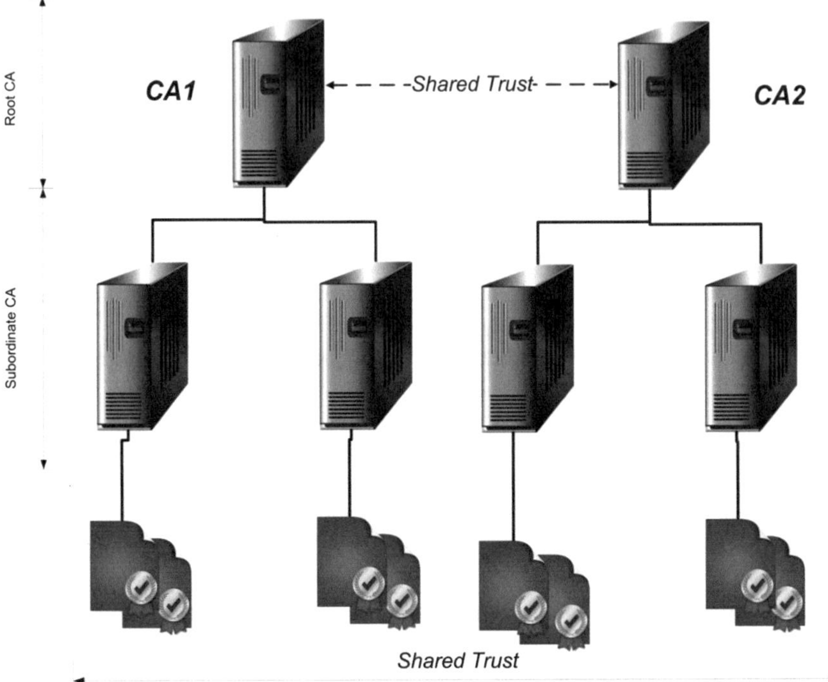

Fig. 4.2 Bridged CA model

CA among these 1500 CAs is needed to hack the Web. Trusting 1500 entities is, of course, a fragile assumption that can easily fail. The next example will illustrate this fact.

The Devil Is in the Detail In 2012, a hacker compromised the Dutch CA DigiNotar. An Iranian citizen, under the nickname ComodoHacker, claimed to have authored this attack [209]. This hack resulted in the creation of rogue digital certificates that granted authentication to malicious sites impersonating the name of well-known and trusted domains, like Google.com. As a valid trusted CA issued these certificates, every browser trusted them.

For DigiNotar, the breach marked the death of the company because all the major browsers revoked DigiNotar's root public key. DigiNotar did not discover the intrusion of ComodoHacker when the attack occurred. DigiNotar discovered it only once the rogue certificates were widely deployed. DigiNotar reacted only ten days after the issuing of the first rogue certificate (Fig. 4.3). It revoked some malicious certificates. The critical certificate linked to the domain **.google.com* was revoked 1 week later. DigiNotar could not provide an exhaustive list of all issued malicious certificates. In fact, some rogue certificates had no record in the CA database of

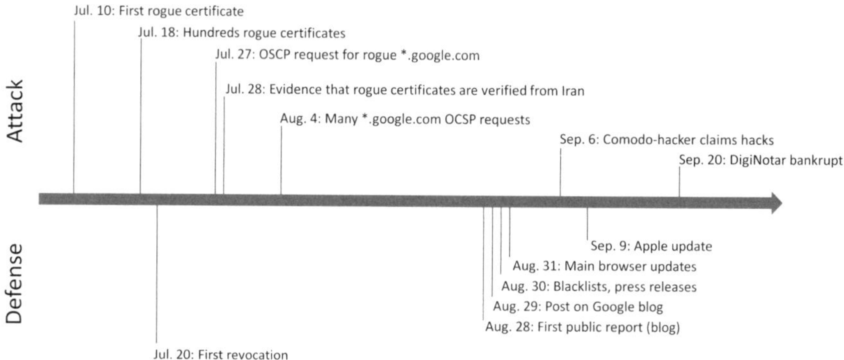

Fig. 4.3 DigiNotar case: timeline

DigiNotar-issued certificates. The problem got national attention when Google found evidence that a **.google.com* certificate was issued. The Dutch government enrolled a security consulting company [210] to perform an assessment of DigiNotar. After this audit, Vasco closed its subsidiary DigiNotar by filing a voluntary bankruptcy application [211].

In the meantime, major browsers prepared new versions revoking the DigiNotar public root key CA. More than 6 weeks had passed before these browsers issued the patched versions. Interestingly, users of Google's Chrome browser were not impacted by this rogue **.google.com* certificate. Chrome trusts only a small list of CAs as issuers for its own domain name. DigiNotar did not belong to this limited list.

The security audit report [210] showed that the intrusion mainly used well-known, off-the-shelf tools. Furthermore, the hacker also developed his own tools. He left a signature in some of the tools: *Janam Fadaye Rahbar*, meaning, in Persian, "*I will sacrifice my soul for my leader*." The audit concluded that DigiNotar had poorly managed its security. DigiNotar did not comply with the industry's best practices. For instance, one single username/password controlled all DigiNotar servers. This account was the administrator of a Windows domain. This password was too easy to brute-force. Frontal Web servers were not protected by an anti-virus, which could have easily detected some of the hacking tools used. Nor were these servers patched with the latest security fixes. An Intrusion Prevention System was in place but failed to block attacks. Finally, no central logging system was in place. Basic security practices were not in place, and an intrusion was inevitable eventually. Obviously, there was no monitoring and accurate tracking of the generated certificates.

Lesson: Trust is serious matter. We have the right to expect exemplary behavior and practices from companies that manage our trust. Unfortunately, this is not necessarily the case.

Will similar events happen again? Yes. Because—quoting Bruce Schneier [212]—"there are too many single points of trust." Many CAs are trusted by default without the user even knowing it. Furthermore, not all CAs are aligned to the best practices of public key infrastructure management.

The previous example illustrates that many current security models are brittle. They rely on a set of security assumptions that users cannot evaluate. Why should users trust one CA rather than another CA? These assumptions are not often documented. If the documentation is available, only experts can assess their efficiency and reliability. Furthermore, only an audit of the actual implementation can determine the real effectiveness. Most people have no rational way to make such a decision. Ordinary users cannot evaluate the security methods used by the company. They do not understand the mathematics used in cryptography. They cannot assess the quality of implementation of cryptographic algorithms. They cannot evaluate the policies defined and enforced by these CAs. They have to trust CAs blindly. This is true for every security system. The public has no other choice than to trust what is proposed.

Another unstated assumption with digital certificate chaining is that the system verifies the entire chain. Unfortunately, this is not always true. At the 2014 Black Hat conference, Jeff Forristal disclosed an Android vulnerability called Android Fake ID [213]. The Android package installer did not verify the authenticity of the certificate chain. It only verified that the signature of the package complied with the provided final public certificate and not the full chain of certificates. In other words, the packager acted as if they were self-signed applications. Unfortunately, some CAs, such as Adobe Systems, have special privileges in the Android operating system. A malware self-signed with a false Adobe Systems identity would gain system privileges. This vulnerability could create malwares, allowing access to Near-Field Contact (NFC) payment, for instance. This vulnerability was present since Android 2.1 (January 2010). The corresponding patch has been available since April 2014.

The Onion Router (TOR) is a system designed to guarantee the anonymity of the Web user and to prevent traffic analysis (Sect. 8.3.3). The promise of TOR is that by using it while navigating the Web, there is no possibility of tracing back the origin, nor of finding any information about the viewing habits of the user. In 2013, in the wake of Snowden's revelations, the presentation in "TOR stinks" [214], an NSA confidential document, shed some light on the actual anonymity of this protocol. The document is a 2012 analysis of TOR. The first reassuring conclusion is that it is impossible to deanonymize all TOR users all the time. Nevertheless, the NSA claimed to be able to deanonymize a tiny fraction of TOR users. First, the NSA and the British intelligence agency Government Communications Headquarters (GCHQ)[9] run many TOR proxies. Nevertheless, their number is

[9]The British GCHQ is the equivalent to the US NSA.

negligible compared to the number of TOR routers. Under some circumstances, some cookies seem to survive TOR sessions. The intelligence agencies attempted to exploit them to identify users by looking for these particular "quantum" cookies. Furthermore, they exploited a zero-day vulnerability of Firefox present in many old versions of the Tor Browser. The exploit, named *Egotistical Giraffe* [215], enabled transmitting users' MAC and IP address to the attacker. The FBI seems to have used this attack for legal investigations for child pornography cases. Many other attacks were mounted against TOR and countered by successive updates.

The advent of social networks, the systematic logging of electronic transactions, and the new promises of big data have created an intense interest in a new, security-related characteristic: privacy. How anonymous can individuals stay in this new digital world? How can individuals allow the collection of private data to offer some benefits without sacrificing our privacy? Research of anonymous schemes in big data processing has become an extremely active academic field. The proposed schemes, which, unfortunately, still offer meager privacy, employ complex, sophisticated mathematics [216]. There is no way to build a trust relationship with customers using the sanity of the mathematics as a convincing argument. People will not or cannot understand it. As people cannot understand the underlying mathematical foundations, they will look for trusted sources as in their real-world social life [217]. They will blindly trust services they already use for other purposes. As long as there is no severe scandal with the sources they trust, the trust relationship will continue. This type of trust relationship is usual for most users of security systems. As most users cannot base their trust on rational criteria, they will fall back on social and reputation-based criteria. This fallback strategy highlights why reputation is often a valuable non-tangible asset (Chap. 2).

4.2.4 Is the Cloud Trustworthy?

There are several architectures of clouds. The NIST defines four types of architecture [218].

- *Private cloud* is for exclusive use by a single organization. The organization or a third party may own, manage, and operate the infrastructure. Only entities belonging to the organization share the resources. The infrastructure may be on or off the premises of the organization. For instance, a private cloud can be designed and hosted by the organization's IT team or can be provisioned from a cloud provider.
- *Community cloud* is for the exclusive use of a community of organizations that share a common concerns or goals. One or more organizations of the community or a third party may own, manage, and operate the infrastructure. Only entities belonging to the community share the resources. The infrastructure may be on or off the premises of the members of the community.

- *Public cloud* is for open use by the general public. A third-party cloud provider owns, manages, and operates the infrastructure. Entities may share the resources. The infrastructure is on the premises of the cloud provider.
- *Hybrid cloud* is the composition of two of the previous architectures.

It is essential to understand the trust assumptions for the different architectures. Architectures rely on many trust assumptions. This book proposes categorizing them into four large clusters.

- *Trusting the hardware*: this category encompasses all assumptions about the physical context. Usual assumptions are as follows:
  - That the hardware is reliable and genuine; genuine means that it was not the target of a successful supply chain attack or was not maliciously modified on site.
  - That the hardware is located in a secure area whose physical access is controlled. Physical security often requires fences, human guards, surveillance cameras, and access control. Only trusted individuals are granted access to the installed hardware.
  - That the hardware is at a known geographical location; the actual geographical location may be essential for compliance with some regulations, such as the European Directive on personal data.
- *Trusting the operating system (OS) and virtual machines*: it is key that the OS and VM be genuine, properly patched, and not infected. In the cloud environment, a particular focus is on the behavior of the VM. One of the most important features is the perfect isolation between two VMs executing on the same machine.

  Once this trust established, it is also important to trust the different components executing in the VM as well as the final applications.
- *Trusting the administrators*: acomputer is only as secure as its administrator is trustworthy [219]. Malicious insiders are a severe threat. Administrators are potentially dangerous due to their extended privileges.
- *Trusting the logical access*: only trusted, authenticated principals should get logical access to the system. Principals include individuals, services, and computers. This frontier is the usual assumption of perimetric defense. The system executes behind the secure walls of the IT defense.

Table 4.1 provides a subjective evaluation of the strength of these four trust assumptions for different operating environments. The rating is subjective and not absolute. For each environment, it is assumed that security has been correctly implemented and with sufficient resources. In other words, we have an ideal theoretical situation. Unfortunately, this assumption is not always accurate, and a table describing some real operational conditions will be different and may present a different assessment.

The data center is the traditional environment. The system runs from a secure data center that the organization owns, managed by the organization's

Table 4.1 Trust dilution

	Data center	Private cloud	Community cloud	Public cloud
Trust hardware	***	***	**	**
Trust OS/VM	***	***	**	**
Trust admin	***	***	**	*
Trust access	***	***	**	*

administrators, and access is only granted to principals approved by the organization. This configuration is optimal, at least from the point of view of security. Unfortunately, it misses some key cloud features, such as elasticity.

The private cloud turns the traditional data center into a cloud architecture. The private cloud runs within secure data centers that are owned (or at least controlled) by the organization and managed by the organization's administrators, and access is only granted to principals approved by the organization.[10] The trust model is similar to the model of the data center.

With community cloud, trust starts to dilute. For the sake of simplicity, we assume that companies A and B share a community cloud. Company A manages the cloud infrastructure. Company B has to trust that company A does a proper job of the management of the hardware and OS. Company B also assumes that company A's administrators are trustworthy. Each company has to trust that the principals authorized by the other company as trustworthy. Usually, the purpose of community cloud is a shared goal. Therefore, these trust assumptions may be reasonably reliable.

With public cloud, trust dilution is even greater. For hardware trust and OS trust, provided that the cloud provider is serious, the strength of these assumptions is equivalent to that of the community cloud.[11] One issue may be the geographical location of the processing units and the storage units. Customers may not feel comfortable that their sensitive data are stored in some foreign countries. The administration is under the full control of the cloud provider. Most of them do proper screening when hiring their administrators. Nevertheless, the cloud customer has no means to assess the trustworthiness of the administrators. The access is clearly not anymore under the control of the users. Unknown principals may share the same CPU, the same physical memory, and hard drives. This is referred as multi-tenancy.

When using a public or community cloud, the customer delegates part of its security responsibility to the cloud provider. As such, the security assumptions theoretically become more fragile because the customer loses some control. A hybrid cloud may reinforce some of these assumptions by carefully crafting the

[10]This section does not consider Virtual Private Cloud (VPC) as a private cloud. VPCs are public clouds that have more sophisticated secure intra- and extra-communication means.

[11]At least, this is true for company B. In the case of company A, the assumptions are weaker as it does not anymore control the hardware and OS.

distribution of tasks between private cloud with strong assumptions and public cloud with weaker assumptions.

A cautionary note is needed. Table 4.1 depicts an ideal theoretical situation. It assumes that the implied organizations operate their security models efficiently. In the real world, it may be different. For instance, large cloud providers may have a stronger trust hardware and trust OS assumption than a small or medium organization. Public cloud providers have larger security teams and potentially better-trained staff. Thus, the security analyst should assess the strength of the four types of trust before taking any decision on the best architecture from the point of view of security.

4.2.5 Hardware Root of Trust

Hardware Root of Trust (HRoT) attempts to solve the following challenge: How can we be sure that a piece of software executes on a platform that has not been compromised? In the current state of the art, this is impossible with a purely software-based implementation. In 1997, Microsoft launched an initiative with code name Palladium that was supposed to solve this problem. The proposed architecture separated the applications into two categories: non-trusted applications running in a non-trusted mode and trusted applications executing in a trusted mode. The architecture implied integrity measurement and device-specific encryption. In 2002, Microsoft disclosed the project under the name Next-Generation Secure Computing Base. This approach received a very negative response from public associations such as the Electronic Free Frontier. The critics focused on DRM mainly, completely ignoring the potential future advantages such a secure environment would provide against malware and benefits for secure storage of personal data and privacy. In 2005, the project was abandoned and was not integrated with the next generation of Windows OS: Vista. Nevertheless, many of these concepts were reused in the new initiative of the Trusted Computing Group (TCG), which gave birth to the Trusted Platform Module (TPM).

The first question to answer is what the definition of a Root of Trust is? According to the Global Platform Association, a Root of Trust is as follows:

> Generally the smallest distinguishable set of hardware, firmware, and/or software that must be inherently trusted and which is closely tied to the logic and environment on which it performs its trusted actions.

The NIST proposes an interesting taxonomy [220]. It defines five types of Root of Trust. They are logical entities and may encompass hardware and software solutions.

- The Root of Trust for Storage provides a protected repository and interface that store and manage cryptographic keys and sensitive data. It also encompasses a cryptographic engine to perform cryptographic operations without releasing the clear keys outside of the Root of Trust for Storage.

- The Root of Trust for Verification provides a protected engine and interface that verify digital signatures associated with software/firmware and create assertions based on the results. An assertion is information that proves to an external element that a given software has been validated successfully. The verification of the signature uses a public key. This public key must be immutable to anchor the trust. Therefore, the Root of Trust for Verification either has a secure internal storage for this public key, or it relies on the Root of Trust for Storage to maintain the integrity of the public key.
- The Root of Trust for Integrity provides protected storage, integrity protection, and a protected interface to store and manage assertions. The assertions are usually stored in tamper-evident locations. A controlled API limits direct access to the assertions by the external world. Only specific entities or methods can modify these assertions. Any entity or method can read these assertions.
- The Root of Trust for Reporting provides a protected environment and interface that manage identities and sign assertions. It binds the information to a given device cryptographically. The aim is to prove the integrity of information as well as provide non-repudiability of the information.
- The Root of Trust for Measurement performs measurement used by assertions protected via the Root of Trust for Integration and attested to by the Root of Trust for Reporting.

As they are at the heart of the trust of a system, Roots of Trust need to be secure by design. They cannot rely on other principals for their trust model. Therefore, HRoTs are preferred over software Roots of Trust due to their immutability, smaller attack surface, and more reliable behavior. Immutability is hard to reach through software. Furthermore, it is easier and cheaper to reverse engineer an obfuscated piece of software than to reverse engineer a tamperproof piece of hardware.

The TPM is the perfect example of an HRoT. The TCG defines, maintains, and publishes the TPM's specification [221]. As these specifications are public, we can use it to illustrate how HRoT works. The TPM is a simple, cheap component to be soldered on the printed circuit board (PCB) of the motherboard of a device. The goals of the TPM are as follows:

- Verify whether a system is in a known, defined state.
- Prove this information to a third party that requires knowing whether the system is valid.
- Grant access to some cryptographic resources once the system has been proven to be in a known, defined state.

Figure 4.4 describes the different elements that compose a TPM. The TPM has three main functional components.

- The cryptographic processor that includes the following hardware components:
  - A true random number generator; this is vital for proper operation of the cryptographic protocols.

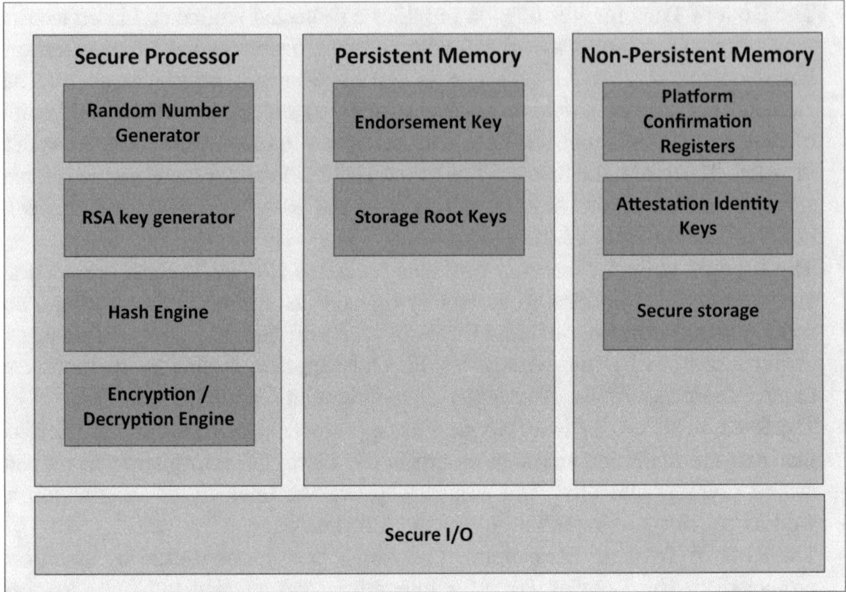

Fig. 4.4 Architecture of TPM

  - An RSA key generator that calculates the proper RSA key using the integrated RNG.
  - An accelerated hash generator to compute a cryptographic hash on a dataset.
  - A hardware engine for calculating asymmetric encryption, decryption, signature generation, and signature verification. TPM 2.0 must support RSA with at least a 2048-bit-long key.

- A secure, persistent memory that holds two public/private key pairs that never leave the TPM.
  - The endorsement key (EK), at least the private EK never leaves the TPM. This key pair is unique for a given TPM. Usually, the manufacturer initiates its creation. In this case, the TPM entity signs the corresponding public key certificate. The certificate assesses whether the private key belongs to a genuine certified TPM. The EK may be erased and assigned a new value when installed. For privacy reasons, the EK is reserved for encryption and decryption and never used for signing.
  - The Storage Root Key (SRK)

- A secure volatile memory that holds different sets of keys.
  - At least 16 Persistent Configuration Registers (PCRs) that store integrity measures. A PCR is 160 bits long.

- Attestation Identity Keys (AIKs): An AIK is an RSA key pair generated by the TPM. The public key of the AIK and the public certificate of EK are sent to a privacy CA that verifies whether the TPM is whose it claims to be. If it is the case, then the privacy CA signs the certificate of the AIK. Thus, an AIK can prove that a given TPM generated it without disclosing the identity of the TPM. The AIK acts as an alias of the EK. This ensures privacy while using a TPM as long as users trust the privacy CA. Indeed, the privacy CA can remove the anonymity of an AIK.[12]
- Generic storage keys.

As with every secure processor, a software interface with a limited set of instructions strictly controls the access to any internal resource of the TPM. The external world has no direct access to the internal memories and registers. As for HSM and smart cards, the hardware is protected from tampering and observation. Nevertheless, the tamper resistance of TPM is lower than the robustness of HSM or smart cards.[13]

The TPM supports three Roots of Trust, reflecting the three goals of the TPM.

- The Root of Trust for Measurement: The TPM uses the PCR to record the static state of the system it measures. A PCR holds the hash of the code to be verified. When updating this value, the PCR is extended. The extended PCR value is defined by the following formula for a new version of code: $PCR = hash(PCR \parallel hash(newcode))$.
  The typical flow of integrity control is shown in Fig. 4.5. The Core Root of Trust for Measurement element initiates the process. This element is a trusted element that cannot be verified at boot time. It is the first element to boot. It initiates the measurement (arrow 1 in the figure) of the integrity of the BIOS,[14] i.e., calculation of a cryptographic hash of the BIOS. Once the integrity is confirmed, the Core Root of Trust Measurement launches the BIOS (arrow 2 in the figure). The BIOS initiates the measurement of the integrity of the OS (arrow 3 in the figure). If the integrity is confirmed, then the BIOS starts the OS (arrow 4 in the figure). The OS initiates the measurement of the integrity of the application (arrow 5 in the figure). If the integrity is confirmed, then the OS launches the application (arrow 6 in the figure). The OS may launch other applications by repeating steps 5 and 6 of the figure. Each integrity measurement may use more than one PCR.

[12]There is another mode of certification of AIK, called Direct Anonymous Attestation, that does need a private CA. It uses a zero-knowledge protocol.

[13]TPM is cheaper and dedicated to consumer applications. They do not need to sustain the same attacks as smart cards or HSMs.

[14]For systems using specialized chips such as SoC, BIOS may be replaced by a dedicated secure mode of this specialized chip.

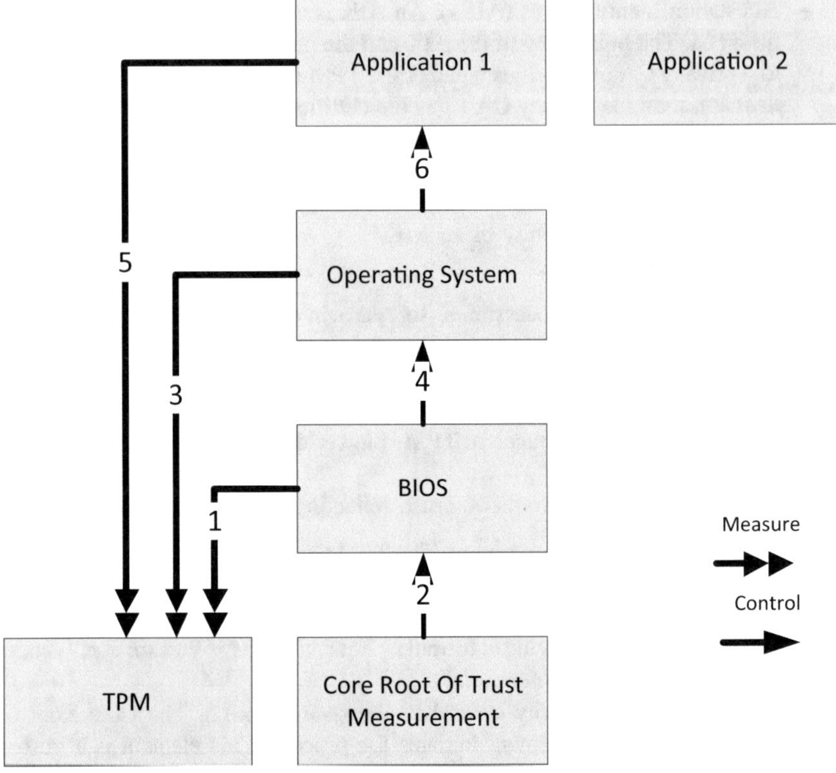

Fig. 4.5 Secure HRoT boot chain with TPM

It should be noted that the TPM does not control the flow of execution, nor authorize or ban an application. Once a piece of software has been verified, it actually controls the execution of its followers within the chain of trust execution.

- The Root of Trust for Reporting: The TPM reports accurately and correctly the state of the measured system. It uses the PCR and AIK. An external entity may request the state of the TPM (called "*quote*" in TPM linguo). The quote is the concatenation of all the PCRs signed by an AIK.
- The Root of Trust for Storage: The TPM securely stores information and grants access to it once the system is in a known approved state. The TPM shields the secured information from the external world. The SRK is used to encrypt a hierarchy of keys that will encrypt other secrets stored externally. The use of the SRK requires presenting a shared secret the *SRK authorization data*. The external application sends the SRK authorization data encrypted with the EK of the TPM.

  Secure encrypted data may be bound or sealed. Bound data are encrypted with a key handled by a given TPM. Only this given TPM can access the bound data.

Sealed data are encrypted with a key controlled by a given TPM within a particular state. An expected value of a PCR qualifies the particular state. Only this given TPM while in the given state can access the sealed data. Naturally, all the cryptographic operations are performed inside the TPM to prevent the private keys of EK and SRK from leaving the secure environment.

There are other forms of HRoT than TPM. Usually, they all have at least the following elements when embedded in a SoC.

- At least one unique secret that never leaves the chip and which is used as the root of storage encryption (equivalent to the RSK) to build the Root of Trust for Storage. The key may be symmetric.
- A mean to calculate the hash of data and securely store it, i.e., the Root of Trust for Measurement.
- An execution flow that controls the integrity of the successive elements of a chain of execution based on the Root of Trust for Integrity.

Many SoCs use their own implementation. In that case, the Core Root of Trust for Measurement is hardwired in the SoC, offering an even higher security than the TPM-based architecture. In 2013, Intel introduced a new extension of its architecture: Software Guard Extensions (SGX). Intel SGX architecture creates the concept of enclaves [222]. An enclave is an isolated region of code and data within an application's address space. Only code executing within the enclave can access data within the same enclave. Instructions that attempt to access, read to, and write from the memory of an enclave fail. Intel SGX offers hardware isolation, hardware-enforced integrity check, measurement, attestation, and data sealing for a given enclave. Enclaves can verify the attestation of a running enclave and establish a secure authenticated channel between enclaves.

4.3 Takeaway

4.3.1 Define Your Trust Model

The trust model is the linchpin of any secure system. The trust model identifies all the assumptions on which the security system is built. No sane architect would build a house without a sound foundation. Similarly, no security architect should design a system without sound security assumptions.

The first consequence is that a trust model should exist for any secure system. The trust model should be entirely defined and be documented thoroughly. It should be the initial piece of work of any security design. Also, the trust model should be available for examination. In my professional life, I always ask any potential security supplier about the trust model of the system he would like me to purchase. If the vendor is not able to describe it, that sends a strong warning message during the selection period.

Documenting the trust model is also essential all along the life cycle of the system. Whenever an evolution is applied to an already designed system, it is mandatory to verify that this development does not invalidate or weaken one of the security assumptions of this system. If this will be the case, then the new trust model should be upgraded. The impact on the design has to be defined, and potential countermeasures implemented. Without this precaution, the evolution may introduce weaknesses that would enable attackers to breach an evolving system that was secure in its previous instances.

New attacks may also affect the trust model. For example, until 1996, the assumption was that secret keys stored on smart cards were secure. This assumption was robust and reliable, at least under the belief that the attackers would not invest an enormous amount of money in hacking the cards.[15] For years, experts had considered the world of smart cards as secure. Then, Paul Kocher and his team published their first works on side-channel attacks [36]. It became possible to extract secret keys with standard, cheap laboratory equipment. The trust model was not valid anymore. Currently, when selecting a smart card supplier, the security designer has to verify whether the smart card provider has implemented countermeasures to defeat the latest side-channel attacks. If it is not the case, then the designer must change this security assumption or choose another supplier.

Rule 4.1: Know Your Security Hypotheses

Understanding the implications of the trust model is also mandatory for foreseeing the issues that may occur in the event of a hack. OpenSSL offers an excellent example. The open source library OpenSSL was initially designed to implement the security protocol SSL. As such, the initial security goals were to protect the communication between two hosts. If ever a private key would leak out, then a simple revocation would be sufficient. Unfortunately, OpenSSL is the victim of its success. Often, its cryptographic library is used as a general-purpose cryptographic library. The consequences became apparent with the Heartbleed bug [223]. The bug, present in OpenSSL versions 1.0.1–1.0.1f, allows an attacker to capture the contents of the host's memory. In the Heartbeat extension of OpenSSL, the library does not check whether the user is authorized to access the memory and how much information is returned. Therefore, by sending multiple requests, an attacker can stealthily explore the complete Random-Access Memory (RAM) of the remote server. With the dumped RAM, the attacker can find the used private key or any private or confidential information currently stored in the clear RAM. As the private key may be used for broader usage than securing communication, its disclosure may have a strong impact:

- The primary victim is communication. With the leaked key, an attacker may decrypt all communication that used it. This leak may affect even past

[15]It had always been clear for the smart card community that with enough time, plenty of money, the right skill set, and access to expensive specialized equipment such as electron beam microscopes or focus ion beam etchers, it would be possible to break into any smart card.

communication. If the attacker had recorded exchanged information before the system was patched, then she could recover the eavesdropped communication. Revocation will block the attacker from eavesdropping on new communication. Nevertheless, previous communication may be definitively compromised.
- The secondary victim is users' credentials. As usually logins use the HTTPS protocol, the attacker may have stealthily eavesdropped on the login credentials. Users will have to change their passwords, which were protected by HTTPS. This is a variant of the previous issue.
- The third victim is protected content. Some stored content may have been encrypted using a symmetric key. Often, a private key protects this symmetric key using the OpenSSL cryptographic library. A new public/private key pair has to protect the potentially compromised content.

This new usage of OpenSSL as a general-purpose cryptographic library modifies the current trust model of OpenSSL. For instance, this new usage extends the attack surface.

4.3.2 Minimize Attack Surface Area

Let us play a role game. You have a gun, and opponents are trying to shoot you with fire weapons. Which option would you prefer?

1. Be in the open to freely shoot back?
2. Be behind a wall with a large window from where you can shoot back?
3. Be behind the same wall but with an arrow slit or a balistraria to shoot back?

Architects of medieval castles preferred the third option. They designed small, thin apertures in the walls that allowed observing at besiegers and the firing at them arrows with bows (Fig. 4.6). The smaller the aperture was, the harder for an attacker it was to hurt the defender with a lucky arrow strike. With the advent of the first cannons, the shape of the arrow slits evolved by the introduction of a round aperture for the cannon to fire outside. Nevertheless, the rule remained of keeping the opening as small as possible.[16] In other words, medieval architects tried to keep the attack surface as minimal as possible.

Rule 4.2: Minimize the Attack Surface

The next example comes from the IT world. A firewall is a device or application that prohibits or permits incoming and outgoing communications of a network.[17] When configuring a firewall (Sect. 8.2.1), two approaches are possible [91].

---

[16]The advent of serious artillery made obsolete the advantages of balistraria. The artillery blew down the entire wall. Therefore, Renaissance castles used large windows and privileged aesthetics over security as balistrarias were not anymore an efficient defense mechanism.

[17]Sometimes, a firewall may protect one unique appliance rather than a set of connected principals.

Fig. 4.6 A balistraria

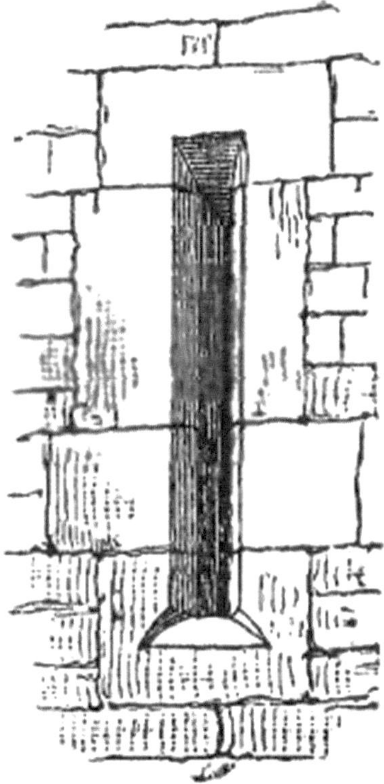

Balistraria.

http://commons.wikimedia.org/wiki/File:
Chambers_1908_Balistratia.png

- That which is not expressly prohibited is permitted;
- That which is not expressly permitted is prohibited.

The second approach is the most secure one. Unfortunately, it puts the burden on the firewall administrator to be reactive. In this approach, a best security practice for configuring a firewall is to open only the ports in use. A properly configured firewall should block all other ports. This usage is a clear illustration of the previous rule. Each open port is a hole in the defensive perimeter that the attacker may exploit. Therefore, closing the unused ports reduces the possibilities for an attacker to penetrate the system. It reduces the attack surface. Furthermore, the number of implemented rules should be kept minimal. Each new rule increases the likelihood that the administrator will misconfigure the firewall; that the new firewall rule contradicts a previously established rule; that older firewall rules become obsolete; or even that the new firewall rule becomes dangerous if the context changed. Hence, each new firewall rule increases the attack surface [224].

Unfortunately, respecting Rule 4.2: Minimize the Attack Surface is not easy, especially if the context does not facilitate its strict application. In the past, the perimetric defense was sufficient to achieve proper corporate security. From the IT point of view, the world was split into two spaces: inside the company's network and outside. The inside was secure and trusted. The outside was potentially hostile. The attack surface was clear and well documented. Unfortunately, this attack surface grew and became less precise. First, the corporate system had to communicate with other computer systems that were not part of the corporate network, such as suppliers' systems. Then, remote workers or traveling workers gained access to the company's network via a VPN. Some company's computers were now residing outside of the traditional perimetric defense. Fortunately, these computers were still under the enterprise's control. The corporate IT team defined, sourced, configured, and managed all these computers. Lately, the new paradigm of Bring Your Own Device (BYOD) has drastically increased the attack surface. Now, the company does not control anymore its employees' computers. Furthermore, some computers within the traditional corporate perimeter are not anymore trustworthy. With the advent of the cloud, once more the attack surface grows. Corporate applications execute outside the traditional perimeter. The computers on which they execute are fully under the control of third parties. In conclusion, every day, more employees, contractors, and business partners have access to the corporate network. Every day, services approved by the company are now running on external servers. More and more legitimate "devices" have access to the corporate network. These legitimate access points drastically stretch the corporate attack surface. Not only is the concept of corporate perimeter defense dead, but the corporate defenders have a more brittle trust model on which to rely and have more attacks to prevent.

4.3.3 Principle of Least Privilege

The principle of least privilege states that any principal should have access only to the data, the resources, and the services that it needs to perform its tasks successfully. Let us use one real-world metaphor to illustrate the rationale behind this principle [225]. If Alice has a pet that has to stay at home while she is going on a holiday, she most probably will give her house keys to Bob, her trusted friend or neighbor to feed the pet. If she does not have a pet and expects her friend just to take care of her postal mailbox, would she give Bob the keys of her house or would she give him only the key to the mailbox? Most probably, she will opt for the second option, regardless of the level of trust she puts on her friend. This example illustrates the principle of least privilege.

Rule 4.3: Provide Minimal Access

This rule means that each principal should be granted access only to the resources that it needs. Let us start with a non-IT-based example. Luxury cars often come with two types of keys: regular keys and valet keys. The regular key opens all the doors of the car, the trunk, and the glove compartment and turns on the ignition. The valet key does not open the trunk and the glove compartment. The valet key opens only the doors and turns on the ignition. The valet who parks the luxury car can access only what he needs, i.e., the doors and the ignition. Normally, the valet does not need to open the trunk or the glove compartment. Similarly, a user, once logged into the corporate network, should only get access to the directories that he needs. There is no valid reason a researcher should have access to the financial data of the company or a human resource employee to the source code of its software. Restricting the access to the needed directories limits the risk of transversal navigation if a hacker compromises a user's account.

Similarly, restricting the ports and services of a server to the strict minimum limits the doors that a hacker may use for breaking into the system. A device should have all its ports closed, excluding the few it needs open. Only the necessary services should be installed on the servers. The non-necessary services should be uninstalled rather than not being launched. Of course, the ports corresponding to the uninstalled services should be closed. `Telnet` is part of the distribution of Linux and Windows. This client–server protocol was widely used to connect to remote servers a few decades ago. Although `telnet` uses a password login session, it does not encrypt the communication channel. Today, there is no valid reason `telnet` should be available on a computer or an embedded device.[18] Not being secure, `telnet` is a weakness. Whenever a remote communication tool is needed, `SSH` should be preferred to `telnet`. `SSH` uses an encrypted communication channel. Therefore, `telnet` should be removed from the computer.

The principle of least privilege attempts to limit the attack surface to its minimal, strictly necessary, unavoidable size. Each violation of this rule may unnecessarily extend the attack surface, potentially weakening the security.

Implementing the principle of least privilege in IT raises several issues from the organizational point of view. The first problem is that it requires more management effort. Any evolution of the needs of the users may end up with the modification of their rights by the managing IT team. Tight control may have some impact on the reactivity of the enterprise. Urgent requests may perhaps not be served fast enough. If requests are not responded to quickly enough, users may seek alternative solutions that may be dangerous. Furthermore, any evolution of a position has to be accompanied by a possible reattribution of access rights. The new job may require new access rights and may not anymore need previous access rights. The second problem is that end users may interpret this control as a lack of trust. Education should mitigate the risk of misunderstanding. Using examples from the day-to-day life, such as the ones used to introduce this section, may facilitate better acceptance of strict control from end users. This issue is especially sensitive to so-called power users. Fortunately, power users are savvy enough to be educated on the topic.

---

[18]The only valid scenario is that the computer has to communicate with a legacy device that only supports `telnet` and not `SSH`. This scenario should be banned where possible.

In any case, more control results in more work for the IT team. No tool can make the critical decisions for wisely granting access.

4.3.4 Simplicity

Complexity is one of the enemies of security. The following rule should always stay in the mind of security designers.

Rule 4.4: Keep It Simple

The more complex a trust model is, the more brittle it will be. The number of assumptions of a trust model is a good indicator. Of course, if there are too many assumptions, then the likelihood that at least one of these assumptions will fail or will be violated is greater. Thus, the designer must attempt to reduce the number of security assumptions of his trust model.

Many vulnerabilities in a piece of software are the result of a software bug. For instance, in February 2014, Apple announced that all its most recent versions of operating systems had a critical vulnerability in their SSL implementation. The vulnerability was due to a coding bug in the SSL certificate verification routine. The following is the problematic piece of software [226].

```
static OSStatus
SSLVerifySignedServerKeyExchange(SSLContext  *ctx,  bool  isRsa,  SSLBuffer
signedParams,uint8_t *signature, UInt16 signatureLen)
{
     OSStatus err;
     ...
Heart
     if ((err = SSLHashSHA1.update(&hashCtx, &serverRandom)) != 0)
            goto fail;
     if ((err = SSLHashSHA1.update(&hashCtx, &signedParams)) != 0)
            goto fail;
            goto fail;
     if ((err = SSLHashSHA1.final(&hashCtx, &hashOut)) != 0)
            goto fail;
     ...

fail:
     SSLFreeBuffer(&signedHashes);
     SSLFreeBuffer(&hashCtx);
     return err;
}
```

The highlighted line should not be present in the source code. Due to this repetition, “goto fail” is executed in every case, regardless of the tested conditions. In other words, the code execution systematically jumps to label fail. Hence, this jump bypasses all subsequent tests of the SSL certificate. Thus, an attacker could easily forge a certificate that would be accepted as valid regardless of its actual validity. The execution of the program needs just to reach this test. With this bug, she could mount a man-in-the-middle attack. This bug was present in the source code for many years. Multiple non-regression tests never spotted the bug.

Unfortunately, current security protocols such as SSL and TLS have become more and more complex. Each new revision adds a new layer of requirements or new algorithms or increases the size of the keys. Often, the revision maintains backward compatibility with older versions. All these modifications increase the complexity of the protocol. Unfortunately, these modifications may carry their own vulnerabilities. These vulnerabilities or modifications may partially reduce the additional security provided by the new version or reveal older non-detected vulnerabilities. Implementations such as OpenSSL are riddled with vulnerabilities. Over a period of 2 years (2014–2015), many serious vulnerabilities were disclosed: Heartbleed [223], Poodle, FREAK, and Logjam (Sect. 6.2.3). These vulnerabilities are within the implementation of the protocol or in misfit use of the protocol. They are not due to software bugs in the implementation.

The more complex a system is, the more complicated its implementation in software will be. The number of bugs in a piece of software is usually proportional to its complexity and its number of lines of code. The more complicated a piece of software is, the harder it is to attain a reasonable test coverage. Attaining 100 % test coverage for the entire piece of software is illusory as it is extremely costly if not impossible. Full test coverage should be reserved for the most critical parts of the software. The likelihood of a vulnerability due to an unknown bug increases with the complexity of the software.

Complexity is also synonymous with more sophisticated design. This complexity is another source of vulnerabilities. A more elaborate design means more complex interactions between modules and more complex data structures. All these factors increase the likelihood of errors in the design and thus also the risk of vulnerabilities. Interfaces between systems or modules are among the hackers' favorite points of attack. Furthermore, complexity increases the attack surface. For instance, Yakov Haimes and his four colleagues demonstrated that cloud computing technologies are at greater risk than non-cloud-based technologies [227]. This higher risk is due to cloud computing being a set of complex, interconnected systems of systems. Using a simplified model of cloud computing, the researchers built a streamlined fault tree analysis (Sect. 2.3.2). They compared it to the fault tree of the same system using non-cloud-based computing. The failure probability of the non-cloud computing system was 8.8×10^{-8} compared to 4.1×10^{-6} for cloud computing system. In other words, for the same functionality, the likelihood of failure of the cloud-based implementation of a solution was 50 times higher than that of the implementation that was not cloud-based.

As always, there is an exception to the rule: software tamper resistance. The purpose of tamper resistance is to protect software against reverse engineering. Strength is achieved by obfuscating the source code, i.e., introducing maximum in complexity to software [228]. Obfuscating a program means generating a new program using a set of transformations while preserving the behavior of the original program. The obscurity of the obfuscated program is maximized so that reverse engineering the obfuscated code is more time-consuming than reverse engineering and understanding the original code. Many methods of obfuscation are available, such as the use of opaque predicates, the insertion of irrelevant pieces of code, the concurrency of opaque predicates, and the flattening of the control flow graphs. These techniques make the task harder for the reverse engineer. The expected result is a piece of software that is incredibly complex.

4.3.5 Insiders

On February 18, 2010, the Web site WikiLeaks, which specializes in the publication of leaked secret documents, disclosed a large set of classified, American military documents. The author of this leakage was Bradley Manning, a US intelligence analyst based in Iraqi. He collected thousands of confidential documents that he had access to due to his role of analyst. He handed them to WikiLeaks for publishing. His claimed motivation was his disagreement with some exactions of the American army in Iraqi. In August 2013, he was sentenced to 35 years in prison.

On June 5, 2013, a series of revelations about the NSA and its surveillance programs such as PRISM and MUSCULAR and its spying activities profoundly shook the world of intelligence and the security community as well as the public. Once more, these revelations were based on leaked, top-secret, classified NSA documents. The source of this enormous leakage was a US computer specialist who was a former Central Intelligence Agency (CIA) employee and a former NSA contractor: Edward Snowden. At the NSA, like any contractor, Snowden had access only to the limited privileges of a contractor: an identification card with credentials to operate the system, ssh keys for accessing the resources to which he had been granted access, and access to a basic computer console [229]. Nevertheless, he succeeded in accessing classified documents that he was not supposed to. It seems that he used a typical three-step approach. The first step was the analysis. He located the targeted documents he was interested in and identified the access rights he would need to obtain them. The second step was infiltration and lateral movement. He extended his current access rights by using the usernames and passwords he collected from dozens of colleagues. These credentials granted him access to resources that he was not authorized to access. From there, he forged self-signed keys that would extend his access rights. Once he acquired the proper access rights, the system granted him access to the confidential documents. He stored the stolen documents in a Microsoft SharePoint system. The third step was exfiltration. He had to exfiltrate the stolen information. Of course, he could not use a USB drive or recordable media. NSA terminals would not permit that. Thus, Snowden used typical command and control (C&C)[19] servers using encrypted channels as used by APTs (Sect. 1.2.4). He exfiltrated up to 1.7 million top classified documents. Probably, the weakest point of the security of the NSA was Snowden's colleagues trusting the contractor and handing him their credentials. Colleagues trusted him although he was not a regular employee of the NSA and thus should never have been trusted. Military accreditations are designed to firewall the different levels of trusted persons. His NSA colleagues bypassed this "firewall." They handed him the keys to the kingdom. They were passive insiders.

Let us explore another domain. For many years, the media industry had suffered from digital piracy. Pirated versions of every movie or TV show are available on P2P networks, cyberlockers, or streaming sites. The pirated pieces of content are copies of genuine versions. There are many potential sources of leakage. The most worrying one is camcording in theaters. The quality varies from poor to excellent. Ripping the original DVD is another source of pirated content. Ripping occurs either before the release date or once the content is available in the public market. In this case, the quality is excellent. For both cases, usually, the attackers are outsiders. Unfortunately, other sources also steal the movie or TV show before its public release. The theft may occur during postproduction, dubbing, or subtitling or during the promotional phase by screeners. According to a study, more than 60 % of the early content, i.e., before the theatrical release, found on the Darknet was due to insiders [230].

The common factor in all these examples is that the attackers are malicious insiders. According to CERT [231], a *malicious insider threat* to an organization is usually due to a current or former employee, contractor, or business partner who has or had authorized access to the organization's network, system, or data and intentionally exceeded or misused that access in a manner that negatively affected the confidentiality, integrity, or availability of the organization's information or information systems.

Usually, people trust what or whom they know, and what or who is similar to them.[20] People within the organization are generally naturally trusted. How can you not trust somebody with whom you collaborate every day? Edward Snowden exploited this human weakness. The trust model of our traditional perimetric defense trusts every principal within the perimeter. If you are within the perimeter, then you are trusted. In other words, insiders are trusted. According to the FBI, most insiders are authorized users doing authorized things. While insiders rarely perform any hacking activity, many things may motivate them such as greed, financial need, vindictiveness, problems at work, ideology, divided loyalties, vulnerability to blackmail, ingratiation, destructive behavior, family problems, or even a thirst for thrill [232].

---

[19]C&C channels are systems used to control a set of remote botnets, zombies, or Trojan-infected computers. For instance, the instruction set may request to exfiltrate a given type of files to a given IP address, or erase them.

[20]This is called the similarity principle. It is one of the favorite tools of social engineers. See Sect. 7.2.3.

Table 4.2 CERT classification of insiders

Insider	Target
Creator, whistle-blower	Intellectual property
Non-IT employee	Fraud
IT employee with administrator rights	Damage to resources, intellectual property

CERT defines three categories of malicious insiders depending on the target of the attack. Table 4.2 presents a slightly modified version of this classification, taking into account some recent evolutions in the security landscape.

The first category of insiders attempts to steal secret information. There are three motivations for such thefts. Often, the people who generated or produced such information have the wrong belief that this information belongs to them. If you developed an idea that was filed for a patent, if you designed an incredible piece of software, or if you built a valuable list of customers and prospects, then you may believe that it is yours, regardless of legal contracts governing their ownership. This feeling may be especially strong if you believe that you did not receive enough recognition for your work. When leaving the company, for instance, to create a start-up or join a competitor, you may take with you the confidential information that you generated, or collected.

The second motivation is business intelligence. What more knowledgeable and more reliable source of information than an insider? The insider may be an employee who the attacker pays or blackmails to get information. The insider may not belong to the company. She may be an intern [233, 234] who develops some prototypes, a subcontractor who gets access to sensitive information, or a business partner who has access to the corporate network. Business intelligence has existed for decades. Nevertheless, business intelligence has become an increasing risk in the digital industrial world. Furthermore, it does not anymore require physical on-site presence to steal digital data.

For a few years, there has been a third motivation: whistle-blowing. For moral or political reasons, people disagreeing with the decisions, behavior, or activities of their employer decide to publish secrets or to disclose misdoings. Sites such as WikiLeaks, Cryptome, and public intelligence are perfect repositories for these denunciations. Snowden and Manning are examples of persons motivated by whistle-blowing. Vindictiveness may also be high motivation for whistle-blowers. The risk associated with whistle-blowing will continue to increase in the coming years.

The second category of malicious insiders attempts to commit fraud to steal financial assets. To commit fraud, the employee may sell stolen information (as in the previous case), modify information to realize financial gains for himself or others (such as modifying the financial accounts of a company to influence the stock market), or accept payment for adding, changing, or deleting information (for instance, from outsiders, or even insiders). The motivation is purely financial.

The last category of malicious insiders attempts to break down some parts of the corporate infrastructure. The goal is sabotage. It often occurs with disgruntled employees. They may have placed logical bombs that will trigger automatically once they leave the company. They may have set hidden trapdoors to access the system remotely, even after their corporate accounts have been closed. The goal may be revenge or even blackmail. A competitor or an enemy governmental agency may also finance the sabotage. The motivation is hurting the employer.

Rule 4.5: Beware of Insiders

Unfortunately, the insider threat is not limited to malicious insiders. It also extends to non-malicious insiders. Human error is a frequent reason for leaked confidential data. For example, it can be the result of an email sent to the wrong recipient [235] or a confidential file inadvertently put in a publicly accessible directory [236].

Detecting a potential insider is difficult. How would a person who has the following characteristics be categorized? The person volunteers for extra assignments, works late, rarely takes vacations, and is interested in what his colleagues do. Human resource managers would classify this person as a dedicated person to promote. The security team would classify this individual as a potential spy. The difference in the displayed behavior may be slight, but the motivations are drastically opposed.

Detection of insider activities becomes increasingly difficult. An increasing number of people have access to corporate networks [237]. For instance, outsourcing services has become a standard business practice. These outsourced services may need access to the corporate network. Enterprises are linked with their business partners (ERPMs) to reduce costs, accelerate response times, and limit errors. Once more, these partners have legitimate access to the corporate network. The advent of cloud computing and cloud storage extends the traditional frontier of the enterprise. Instead of being centralized on a handful of managed servers, information is spread all over the world on servers that the company does not any longer control, nor own. The volume of network activity increases, making its surveillance more difficult and expensive. Furthermore, encrypting data exchange becomes standard practice, making surveillance even harder, if not impossible.

How is it possible to mitigate the risks due to insiders? Several methods may reduce the risks.

- Be sure that your employees are trustworthy and well-trained. This training is especially important for system administrators and workers who have access to critical information. Often, a background check of new employees is applied so as to hire only trusted individuals. Microsoft claims "a computer is only as secure as the administrator is trustworthy" [219]. Training and security awareness is of utmost importance. If an enterprise cannot trust its administrators, then this enterprise is doomed.
- Strictly apply the principle of least privilege (Sect. 4.3.3). It will be harder for an insider to get access to information or resources that he does not need. This

principle is also to be applied to IT administrators.[21] All IT administrators do not necessarily need access to every resource and datum of the IT system. Segregating them is a good security practice.

- Where possible, encrypt sensitive data at rest. Encrypted data are useless to an insider if she does not have access to the decryption key. Under standard assumptions, this key should be highly protected. Furthermore, she cannot modify encrypted information without being spotted. Forgery or alteration of encrypted material is impossible without having access to the corresponding decryption and encryption keys. Nevertheless, destruction of the encrypted material is still possible (Rule 1.5: Back up Your Files).
  Sensitive information has to be classified as such and also be labeled accordingly. Any employee should be unambiguously aware that he is accessing sensitive data and should act accordingly.
- Monitor activities (Chap. 9). Some tools such as DLP may potentially prevent some leaks. Nevertheless, do not put too much faith in them. Experienced skilled insiders will be able to defeat them. Logging everything is a mandatory precaution. Logs may be crucial in the case of infringement for evaluating the loss and providing proof in litigation. Sometimes, companies cannot list the stolen assets.
- Involve human resources and legal department. Sanctions should follow any violation of security policy as a strong deterrent. This policy requires having the full support of the legal department and top management [238]. The IT team and human resources have to work together to act immediately after any job termination. Security policies have to be communicated to all employees in a comprehensive way.

Bruce Schneier proposes five rules for managing trusted people [239].

- Limit the number of trusted people; the more the number of people who have access to a secret, the higher the risk of voluntary or involuntary disclosure. Reducing the number of trusted people increases the strength of the assumption that trusted people act trustworthily.
- Ensure that trusted people are trustworthy. A trusted person is a person that an entity trusts to act faithfully. A trustworthy person is a person who can be trusted to act faithfully. A trusted person may not be trustworthy. Background checking may evaluate the trustworthiness of a candidate.
- Limit the amount of trust a person has; compartmentalization is a means to reduce the impact in case a trusted person ends up not being trustworthy anymore. See Sect. 9.3.1.
- Overlap the spheres of trust of individuals; this rule is a direct application of Rule 1.4: Design In-depth Defense. If one individual fails, then another person in the overlapping sphere may detect it. This strategy is the opposite of compartmentalization. When compartmentalizing, individuals are totally isolated

[21]This limits also the liability of the IT administrators in case of leakage.

from each other. Compartmentalization is often used in secret agencies to reduce the risk of leakage. Of course, monitoring is not a key feature in the case of compartmentalization.

- Detect breaches of trust after the fact and prosecute the guilty. Auditing the system to find a breach is a mandatory part of the security posture. See Chap. 9. When a violation due to an insider is detected, it is paramount to prosecute the guilty if it was a voluntary breach. The judgment should be public to be a deterrent.

4.3.6 Isolate Your Trust Space

The trust models for digital signature have a primary assumption: The private root key remains secret. The second assumption is that unauthorized principals cannot misuse the private root key to generate a signature. A related trust assumption should be added: "the process of the signature is done correctly."

Downloading a piece of software and executing it requires either strongly trusting its issuer or being oblivious to security risks. Before a user feels comfortable downloading a piece of software, he should be able to verify [240]:

- That the issuer of the piece of software is actually who it purports to being; this proof is done by using the digital certificate of the issuer. It must be a trusted CA.
- That the piece of software has not been tampered with; this proof is done by signing the piece of software with the previous digital certificate.

Modern operating systems display a warning before executing for the first time a piece of software from an unknown publisher or an unsigned piece of software. The OS takes care of these verifications on behalf of the user.

Unfortunately, writing software is a complex task, and thus, signing software requires a more complex process. Particular care has to be taken when developing and signing software.

The first precaution is to map the signature space to the different steps of the software life cycle. Usually, the development of a piece of software goes through at least three steps.[22] The first step corresponds to the development phase. The code is written, unitary tested, and integrated. Only employees and possible subcontractors have access to it. Once integrated, the piece of software enters the beta testing phase. A limited set of identified users has access to this version. Beta test versions are not supposed to be widely used. They should never be mainstream. The final phase corresponds to the qualified product version that is rolled out in the field.

[22]Of course, the software development cycle includes many more phases such as specifications, analysis, or design.

Rather than using one unique root private key, using at least two different root private keys is recommended. The first private root key, hereafter called development-signing root key, signs the pieces of software for development teams or signs the hierarchical keys that will sign the pieces of software for development teams. A build server can automatically generate the signature when a new build of software is submitted. This is especially true when using continuous integration methodology. It is possible to use a second set of signing keys signed with the development-signing, private root keys. Beta testing will use this second set of keys when signing the pieces of software. The second, private, root key, hereafter called product-signing root key, signs the production software or signs the hierarchical keys that will be used for signing the production software. Figure 4.7 represents this recommended signature scheme.

This scheme has the benefits of isolating the different versions. Developers will receive the development-signing, public root key. Beta testers will receive the first

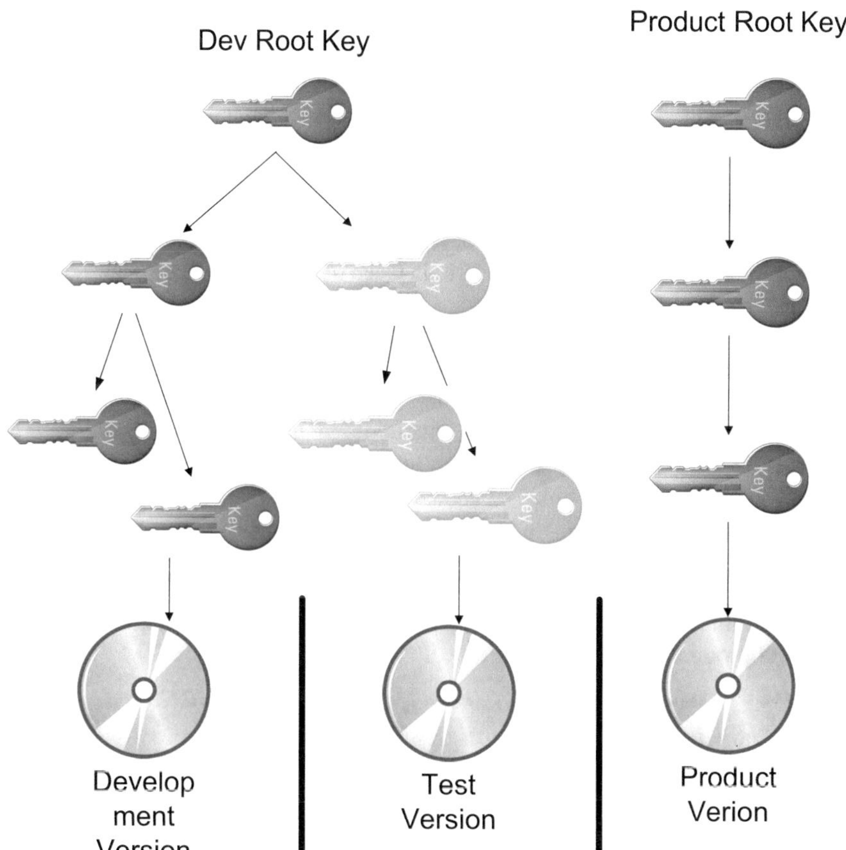

Fig. 4.7 Signing scheme for software development

public key of the hierarchy used for signing the beta test version. Users will receive the product-signing, public root key. Even if development versions or beta versions leak out, mainstream users cannot use it. As users only have the product-signing, public root key, they will not be able to validate the leaked software. Separating the signature scheme of the development environment from the signature scheme of the product environment has another advantage. The following example will introduce the problem.

In September 2012, Adobe detected two malicious utilities (`pwdump7` and `myGeeksmail.dll`) that seemingly were legitimately signed by Adobe [80]. Adobe performed an extensive forensics analysis. The investigators had found out that attackers compromised one of their build servers. This build server could submit pieces of software for automatic signature. According to Adobe, the configuration of the server was not aligned with the proper Adobe standard of security. As it was a server that was compromised, this meant that the private key stored on an HSM was not compromised. The server signed any submitted software, trusting it to be a legitimate request. Adobe also had the proof that this server required the signature of the malicious utilities. Adobe believes that the attackers accessed another server and then moved laterally to reach the build server. Once the server was reached and controlled, the attackers requested the signature of their malware. Adobe revoked the signing key on October 4, 2012. In September 2008, similarly, the servers of Fedora/Red Hat were compromised. The attacker succeeded in signing an infected package of Open SSH without compromising the signing private key [241].

These examples highlight a problem with automated code signatures. An attacker may be able to request a malicious piece of code that is legitimately signed without any control. Thus, to mitigate this risk, a few precautions are recommended. The development version and beta test version may use an automated signing process. As there are numerous builds of software to sign, an automated signature is more efficient than a manual signing process. If an attacker were to compromise the automated signature process, its impact would be negligible. The attack would affect only the developers and beta testers. Nevertheless, it is recommended that product versions be signed after a manual validation of the request. A careful manual validation should detect attempts to sign malware or an altered version rather than the legitimate version. The number of product versions is limited. Thus, the manual process may not impair the efficiency and reaction time of the development cycle. The compromising of the automated signature could not affect the product versions.

The Devil Is in the Detail In 2013, security expert Jeff Forristal disclosed vulnerability in Android OS that bypassed the signature mechanism of Android [242, 243]. This vulnerability was present in Android since version 1.6 (2008).

Android applications are delivered through Android package files `.APK` which are signed jar files. A jar file is a set of zip archive files with a file

META-INF/MANIFEST.MF. In addition, a signed jar file contains two other files: META-INF/CERT.SF and META-INF/CERT.RSA. The file MANIFEST.MF, called the manifest, contains an SHA-1 digest, using base 64, of each file in the archive file except those in the directory META-INF/. The file CERT.SF, called the signature file,[23] contains an SHA-1 digest of MANIFEST.MF and an SHA-1 digest of each file in the archive file except those in the directory META-INF/. The file CERT.RSA, called the signed signature file, is the actual signature of the signature file, i.e., CERT.SF, using PKCS#7.

When receiving an APK file, Android first checks the integrity of the package. The first step of the verification verifies the actual signature of the package. Thus, Android calculates the signature of CERT.SF and checks whether it matches the value stored in CERT.RSA. If it does not match, then the APK file should be rejected. The second step verifies whether the SHA-1 digests in CERT.SF correspond to the actual files in the jar file and whether the lists are complete, i.e., there is no missing digest and no missing file. The last step performs the same verification with the manifest file. The application is accepted only if the three controls are successfully verified.

This verification is done in Java using Harmony's ZipFile implementation. To extract an entry in a zip file, the software uses the class ZipFile.Java. When asked to extract of a file, this procedure does not check whether the zip file holds several files with the same name. Instead, it explores the archive from the end until it finds a file whose name matches. Thus, it returns only the last file with the corresponding name. In other words, the signature verification checks only this last file and no other files with the same name.

When loading a package, Android uses another library to extract an entry from the APK file. This library explores the archive file from the beginning until it finds a file whose name matches. Thus, it returns only the first file with the corresponding name. In the case of multiple files sharing the same name, the procedure returns the first one. This difference is the actual vulnerability.

The attacker starts with an existing signed package. She forges the piece of malware and gives it the name of an existing file in the archive. Using an off-the-shelf utility tool, the attacker inserts the malware in the jar file at a position before the genuine file. When Android checks the signature of the package, it retrieves the last file, i.e., the genuine file, which is correctly signed. Thus, Android believes the package has not been modified. Once verified, Android loads the package into memory. It then retrieves the first file, i.e., the forged file, thus loading the malware rather than the genuine file.

BlueBox Security informed Google of the vulnerability in February 2013. Google released a patch in March 2013. The vulnerability was publicly disclosed at Black Hat in July 2013.

---

[23]Despite its name, the signature file does not contain any signature.

Lesson: Signature is not only about cryptography and its implementation but also about the process. As more and more complex data structures use a cryptographic signature to confirm their integrity, their management becomes crucial. It is mandatory that all actors in the workflow handle the data structures, in the same way, ideally using the same software procedure.

Signature is an important element of the security toolkit. Unfortunately, it is also a complex element requiring careful design of the signing software and the verification signature. The process that governs how to generate and initiate the signature is of equal importance as attackers may fool it. The process should ensure that only authorized principals can sign. These authorized principals should verify that the request for the signature is legitimate. They should also verify that what is signed is actually what should be signed. This last verification, of course, cannot rely on cryptographic enforcement.

4.4 Summary

Law 4: Trust No One
Trust is the cornerstone of security. Without trust, there is no secure system. It is the foundation of all secure systems.

Rule 4.1: Know Your Security Hypotheses It is not possible to build a secure system without knowing the security assumptions. They define the minimal set of hypotheses that are supposed to be always true. It is mandatory to identify and document them thoroughly. Whenever one of these hypotheses is not anymore true, the system may not be secure anymore. Any change in the environment or context may invalidate the assumptions. Therefore, hypotheses must be continuously monitored as the system evolves to check whether they are still valid. If they have changed, then the design should accommodate the new security hypotheses.

Rule 4.2: Minimize the Attack Surface Attackers will probe all the possible venues for breaking a system. The more the number of possibilities available for the attacker to try, the higher the likelihood that she will succeed in finding an existing vulnerability. It is thus paramount to reduce the attack surface, i.e., the space of possible attacks available to an attacker.

Rule 4.3: Provide Minimal Access A secure system should grant a principal access only to the resources and data that the principal needs in order to perform his function. Access to any additional unnecessary resources or data is useless and creates an unnecessary potential risk. The consequence is that the role and function of each principal have to be clearly defined and thoroughly documented.

Rule 4.4: Keep It Simple Complexity is the enemy of security. The more complex a system is, the higher the likelihood is that there is an error either in its design or in its implementation. The error may turn into a security vulnerability that an attacker may exploit. Many vulnerabilities are due to software bugs.

Rule 4.5: Be Aware of Insiders While the attack usually comes from outside the trusted space, unfortunately, too often, the attacker may be an insider. She will either accomplish the attack herself or knowingly be an accomplice in it. Sometimes, she may even be tricked into facilitating the attack involuntarily. Therefore, the trust within the trusted space should not be blind.

Chapter 5
Law 5: Si Vis Pacem, Para Bellum

War is the father and king of all

(Heraclitus).

5.1 Example

"*Si vis pacem, para bellum*" is a Latin adage adapted[1] from a statement found in Book 3 of the Roman author Publius Flavius Vegetius Renatus's "*tract De Re Militari*" (fourth or fifth century). Many centuries before, General Sun Tsu has already claimed in his famous treaty "The Art of War" [123]:

> He will win who, prepared himself, waits to take the enemy unprepared.

The following examples demonstrate the truth of these citations.

After the First World War (WWI), France decided to prepare a better defense against Germany. In the 1920s, two military strategies were competing for defining the future optimal French defense. On one side, Marechal Joffre promoted a classic, defensive, static warfare. He suggested investing heavily in a trench-based defense of robust fortifications. Fortifications and trenches are essential elements of WWI, especially during the Battle of Verdun. During most of the WWI, the front line oscillated between trenches and dugouts separated by a few kilometers. On the other side, Paul Reynaud and Colonel De Gaulle advocated a more modern, mobile warfare. They suggested investing in aircrafts and mobile armors such as armored tanks rather than in static defense. Military planes and tanks were introduced during WWI. However, the full power of these new weapons had not yet been demonstrated. The French minister of war, André Maginot, convinced the government to select the first option. The outcome was the Maginot Line.

---

[1]The actual Latin citation in the book is "*Igitur qui desiderat pacem, praeparet bellum.*" Nevertheless, popular culture favored the adapted version, which is easier to memorize. The meaning stays the same.

E. Diehl, *Ten Laws for Security*, DOI 10.1007/978-3-319-42641-9_5

The Maginot Line was built from 1930 to 1939. It was a 940-mile-long line of concrete fortifications, artillery and machine gun casemates, and anti-tank defenses. The Maginot Line stretched from the Swiss border to the Luxembourg's border, thus fencing Germany from France. France extended a less fortified version of the Maginot Line along the Italian and Belgian borders. The aim of the Maginot Line was:

- to avoid a surprise attack by German troops,
- to give an alert with enough time for the 2 or 3 weeks needed for mobilization that would be necessary to have enough soldiers to defend the frontier,
- to slow down a potential invasion by incentivizing the German army to attack first Belgium or Switzerland rather than France. The expectation was that France, with its fortified frontier, would be a tougher target than countries with "open" borders.

The two first goals were satisfied. The German forces invaded Poland on September 1, 1939. In retaliation, the Allies declared war on Germany 2 days later. Until May 1940, the Western Front did not move, as the German forces were focusing on pacifying Poland. This period was called "*la drôle de guerre*."[2] The third goal was only partly reached. On May 10, 1940, Germany launched a massive attack against France. It simultaneously declared war on neutral Belgium, the Netherlands, and Luxembourg. Using innovative *Blitzkrieg* tactics, relying on fast armored tanks called *panzers*, military aircrafts, and paratroopers, Germany invaded in a few weeks these three countries. In June 1940, starting from these newly conquered countries, ten German *panzer divisions* invaded France through the wooded French Ardennes. The robust, secure Maginot Line ended before the French Ardennes. The Allies had wrongly expected that these dense Ardennes woods would be impenetrable to armored tanks. This attack circumvented the Allies forces that were massed in front of Belgium and behind the Maginot Line. As a result of this breakthrough, the mobile armored German divisions penetrated the Allied lines by getting around the Maginot Line. The Maginot Line could defend against attacks coming from the side-facing Germany. Of course, the Maginot Line was defenseless against attacks coming from the side-facing France. On June 22, 1940, 6 weeks after the beginning of the initial Western offensive, France surrendered to Germany. The French army had prepared for war but without success. France chose an obsolete defense strategy not adapted to modern mobile warfare.

Sometimes, this law is misdirected in wrong innuendos. For instance, "Who wants peace, prepares for war," sometimes turns into "Who wants peace, launches the war." The history of Pay TV has an iconic example of how this misinterpreted law became a plague for the defenders. In 1988, the Israeli News Datacom (NDC) and the French Thomson created for the British broadcaster SKY a new Pay TV system: VideoCrypt. The foundations of VideoCrypt's security were

[2]The phoney war (in French).

Fig. 5.1 Line cut and rotate scrambling

NDC's smart card-based CAS and Thomson's analog video scrambling system. The analog scrambling used a scheme called line cut and rotate [85]. It defines 256 points in the video line. For each successive video line, a PRNG, seeded by the CAS, selects a random cutting point from among these 256 points. The cutting point splits the video line into two segments. Scrambling inverts the order of these two segments. Descrambling reverses the order of the two parts to their initial position. The cutting point changes every successive video line. Figure 5.1 shows the corresponding visual effect.

The CAS used the security of the smart card and the cryptography expertise of Professor Adi Shamir, one of the designers of RSA. A battle started with the hackers. As forecast by NDC, rather than attacking the scrambling algorithm, the hackers went against the CAS. This attack was foreseen and was the rationale for using expensive detachable smart cards rather than traditional embedded secure processors. The replacement of security modules or cryptographic elements becomes easier. This design choice proved to be wise. SKY could send revocation commands to a smart card. The revoked smart card could then update its internal Electronically erasable programmable read-only memory (EEPROM) to memorize the cancelation of the subscription. Unfortunately, in the early 1990s, programming EEPROM required a programming voltage of 21 V. Hackers found out that if the set-top box did not provide this programming voltage, then the smart card could not memorize the cancelation. The attack just removed the power of pin 6 of the smart card, which delivered the programming voltage. This attack was the first hack into the system. Soon, News Data Systems (NDS), the new name of NDC, set up a team of experts. This team monitored the hackers' forums, such as TV List or DIR7, and designed new electronic countermeasures to defeat the latest successful hacks of Pay TV pirates.

This effort did not focus uniquely on understanding the activities of pirates trying to hack their smart cards and fighting them [146]. Soon, NDS hired two well-known hackers as consultants: Oliver Kommerling and Chris Tarnovsky. In 1997, NDS put together a black hat team. In 1998, this team cracked the smart cards of Pay TV competitors: Canal+ (SECA) and Nagra (for EchoStar). The team used expensive, sophisticated tools such as focused ion beam devices. In March 1999, the ROM code of the SECA card was posted on the DR7 site. Some indices seem to link the ROM code to NDS Black Hat team. Unknown sources regularly published on specialized pirate sites; the other exploits against IRDETO and EchoStar cards. On March 11, 2002, Canal+ filed a suit against NDS for the hack and leak of SECA cards. In April 2002, Kommerling signed an affidavit that NDS was responsible for the SECA card hack. In June 2002, Canal+ dropped the case by after an undisclosed settlement with NDS.[3] Many pieces of evidence seem to link the NDS Black Hat team to exploit on competitors' smart cards. The war extended from pirates to competitors.

The Devil Is in the Detail In 1995, the US broadcaster DirecTV deployed a smart card, the H card, designed by Nagra, to protect its Pay TV channels. Unfortunately, the designers of this smart card reused a design already in use in Europe. European pirates had already hacked this design. Thus, US crackers had access to the precious knowledge collected previously by European hackers. Deep knowledge of the protection and behavior of the smart card and part of its software bootstrapped the hacking community. Soon, they designed a card writer that could modify the subscription rights managed by the smart card.

The arms race with the hackers had started. DirecTV developed a mechanism that allowed the STB to detect hacked cards and to block them by forcing an update. The hacking community soon delivered another piece of hardware that healed the updated blocked cards. Later, the hacking community wrote a piece of software that definitively disabled the ability of DirecTV to update the pirated smart cards. Then, DirecTV launched some software patches that required the update to be present on the smart card for the decoder to work. Each time, within a few hours, the hacking community released pieces of software that overcame the new countermeasures.

After a while, DirecTV launched one new set of successive update. Surprisingly, the source code of these update was meaningless and useless. The bytes had just to be present in the smart card. They had no visible effect. It was like meaningless garbage. The hacking community adapted the software of the hacked cards to accommodate this seemingly meaningless evolution. After about thirty update, the last uploaded bytes turned the bytes already present into a dynamic programming capability.

[3]It is rumored that the settlement was around one billion of Euros (1.1 billion US dollars).

On Sunday, January 21, 2001, DirecTV fired back. The last update checked the presence of some bytes that would only be present in hacked smart cards (we call that the signature or fingerprint of hacked cards). If the newly built software detected a pirate smart card, the software modified the boot section of the smart card and wrote four bytes to the programmable read-only memory (PROM) area. Among them, the value of memory 0x8000 was set to 0x00. The smart card's boot checked whether the value of memory 0x8000 was 0x33 (as in genuine cards). If this was the case, the program continued and sent the regular Answer-To-Reset. Else, the program stayed blocked in an infinite loop, never starting. The hacked smart cards no more answered the STB or any card reader. The checked byte was in the PROM area, i.e., non-modifiable once written. Thus, it was not possible to reverse it to the expected value. The game was over. The pirate cards were definitely bricked.

This was on Black Sunday. Thousands of pirate cards were definitively dead. One interesting fact is the date that DirecTV chose to strike back: 1 week before the Super Bowl (one of the biggest annual US sport TV events) [244]. Football fans had no alternative. They had to subscribe to DirecTV to watch their annual, favorite game. The smart choice of the date was part of the success of the counterstrike.

Lesson: It is paramount to prepare the war. It is also highly important to decide when to counterstrike. If possible, the timing of the counter attack should be when the counterstrike hurts most the attackers.

5.2 Analysis

5.2.1 Security Is Aging

All through history, warfare has demonstrated that security is continuously aging. A cycle exists of defense and offense. A new method of defense systematically follows a new weapon. Then, a new weapon circumvents the new defense. Following is a non-exhaustive list of such evolutions in warfare. Cave dwellers used bone or flint-headed axes or spears. Then, the blades or heads of these weapons were made of bronze and later from iron. With the progress of metallurgy, edges became sharper, and the blades could pierce simple armors. Leather and iron-padded armors offered enhanced protection against these weapons. Breeding of horses enabled horsemen and cavalry. Cavalry easily slaughtered footmen. Greeks invented hoplites. These disciplined warriors used long specialized spears to defend against cavalry charges. Horses were impaled by the spears before reaching the soldiers. Archery became more sophisticated and could kill at a farther distance than blades. Plated armors were an answer to this increasing sophistication of archery. Knights, exclusively members of the nobility, were mostly immune to arrows. Arrows only killed footmen, who could be sacrificed as they were not members of the nobility. Unfortunately for the nobility, bolts of crossbows pierced their heavy-plated armors. Crossbows announced the doom of mounted knights, and of medieval chivalry. Gunpowder accelerated the obsolescence of castles. They could destroy the walls of medieval castles. In the seventeenth century, French engineer Sébastien Le Prestre de Vauban designed new forts that defended against cannons. For many centuries, forts and trenches had secured defenders and besieged cities against firing weapons. WWI changed warfare with aircrafts and armored vehicles. Open space warfare replaced trench warfare. With WWII, mobility became a critical strategic factor. The nuclear bomb annihilated all previous defenses by enabling the complete destruction of the planet for the first time in the human history. For several decades, dissuasion was the only suitable defense against nuclear weapons. The strategy was, "If you nuke my country, I will eradicate yours also." In 1983, war moved to space with the American Strategic Defense Initiative project, coined Star Wars. A network of satellites with high-power lasers would destroy any Russian nuclear ballistic missile attacking the USA. At the beginning of the twenty-first century, the context of war once more changed. It was not anymore about countries fighting each other with conventional weapons. It was, rather, small localized conflicts. Guerilla warfare started to replace conventional warfare. Drones seem to have become the new preferred type of weapon for this new kind of warfare. What will be the corresponding defense? The next revolution may be autonomous weaponry driven by Artificial Intelligence.

The same race occurs in cyberspace. Each new type of malware or cyberattack generates a new kind of defense and vice versa. In 1986, the first virus, "Brain," appeared. In 1988, Robert Tapan Morris designed the first auto-replicating and auto-transferring piece of software, coined the Morris Worm. The first anti-virus software appeared a few years later, giving birth to a new industry. Then, the first DoS appeared. In 1996, the first DoS attack occurred against the ISP Panix. In 1998, CERT reported the first DDoS tool: FAPI. Many sophisticated DDoS tools followed, such as Stracheldracht, Trinity, and Tribe Flood Networks. The security industry created new instruments to detect ongoing DDoS attacks (signature detection, anomaly detection) and to mitigate them (e.g., automatic load balancing). In the mid-1990s, the first phishing attacks appeared against AOL. Attackers, posing as AOL representatives, requested users to verify their account information. The countermeasure resulted in better security practices. The defense helped to generalize the use of HTTPS, which makes Web site phishing harder. An authenticated Web site is harder to impersonate than a simple HyperText Transfer Protocol (HTTP) Web site. The deployment of smartphones enabled a new generation of friendly two-factor authentication methods. There was not anymore the need to deploy physical tokens: The smartphone acted as one.[4] In 2000, the first

[4]It is interesting to see that, with the FIDO alliance, there is a return to physical tokens. Smartphones may not be as secure as physical tokens embedding a secure processor.

spyware appeared. "Reader Rabbit," an educational software marketed to children by the company Mattel was surreptitiously sending data back to Mattel. Anti-virus vendors added anti-spyware modules to their offering. Interestingly, none of these new defenses have rendered obsolete any category of attack. Viruses, DDoS, phishing, and spyware are still around, operating with increasing sophistication. It is a constant cat-and-mouse game between attackers and defenders.

The lesson is that a system never remains secure. Its security decays with new attacks. Then, new defenses to prevent or mitigate these attacks have to be deployed to reinforce the declining security.

This race has an impact on the software development life cycle. Usually, the development of one piece of software follows different steps.[5] The first step encompasses the specification phase and the design phase. It defines what the piece of software should do and how it should do it. The second step, the coding phase, and the testing phase, encompasses the implementation of the actual software. After this step, the piece of software is released and deployed. Software maintenance is the final step. Usually, for business reasons, the maintenance takes limited effort and is reduced to a minimum.

Indeed, it is assumed that a proper initial design and serious testing reduce the likelihood of critical bugs. Once a piece of software is bug-free, it remains bug-free for the rest of its life. Unfortunately, this is not true for security vulnerabilities. For the sake of demonstration, let us assume that a piece of software is free of any security vulnerability at its launching.[6] This status may not remain valid until this piece of software becomes obsolete. Indeed, a new type of attack affecting this piece of software may be discovered during its lifetime. In that case, the piece of software turns from being secure to being vulnerable. Therefore, it should be redesigned to be protected against this new attack. This is the role of the security patch.

As the complexity of software increases, software designers use more and more libraries written by other teams. A large number of pieces of software use open-source libraries such as OpenSSL, `libpng`, `Bash`, `FFMPEG`, and FreeType.[7] Whenever vulnerability appears in one of these libraries, every piece of software that employs the now vulnerable library also becomes vulnerable. In September 2014, Stéphane Chazelas discovered security vulnerability in the shell command line interpreter Bash [245]. Bash is part of Linux distributions and Mac OS. The vulnerability, coined Shellshock, similar to a code injection fault, allows the execution of arbitrary code in the shell mode on vulnerable hosts. Bash did not handle environment variables correctly. By inserting the code of a function as part of a variable, it was possible to execute commands when Bash evaluated this variable. The threat was extremely severe, as Bash accepted any arbitrary command. For

[5]These steps are typical for waterfall methodologies. For agile methodologies, the two first steps are integrated into each sprint period. Maintenance extends over all sprint periods and continues after the final release.

[6]This is probably never true with current complex systems.

[7]FreeType is a portable library that displays vector and bitmap fonts.

instance, a CGI script could dump files that would normally not be publicly accessible, such as passwords or personal data. Many systems use Bash. An example is Apache, which accounts for more than half the Web servers in the world. The vulnerability was present in the source code since 1992. A few days after its public disclosure, the first malicious botnets were scouring the Web to locate vulnerable servers [246]. Four months after the release of the patch, attackers still actively exploited Shellshock. Three vulnerabilities using Shellshock accounted for about 87 % of the incidents due to the ten top vulnerabilities of the month [247].

A piece of software may use 50–150 third-party libraries [248]. In 2013, the Open Web Application Security Project (OWASP) ranked the use of components with known vulnerabilities at the ninth position in its top-ten list of issues [249]. In the optimal case, the time to cure the vulnerability will take at least twice the usual duration of patch development. Indeed, first the developer or vendor of the vulnerable component has to issue the patch for its library. Then, the developer of the piece of software using the vulnerable component has to apply the patch to its product and generate the corresponding security patch for its product and distribute it.

The patching process is well established in the IT world (Sect. 6.3.2). Unfortunately, this is not the case for the IoT. The environment of the IoT is not favorable for security patching. There are several reasons.

- Unfortunately, not all devices were designed to be renewable. Many devices in the field have no method of updating their firmware. This means that once vulnerable, they will remain vulnerable indefinitely.
- Patching may also affect and draw on equipment resources. Some IoT devices will have scarce resources, stripped to the minimum when deployed. Manufacturers attempt to reduce costs as much as possible. Such devices may not be able to support the application of several successive patches. The patch may increase the footprint of the code and require more calculations.
- The full effect of a patch is often efficient if and only if all the devices of the ecosystem apply it. Sometimes, if one device is not patched, it becomes the entry point for attacks, compromising a complete network (Chap. 6).
- Even if the device is ready for patching, consumers will not necessarily apply these patches. Many reports emphasize that the software running on home computers of users are most of the time not patched. We may infer that the situation will be even worse with CE devices. For decades, customers were used to purchasing a CE device, installing it once, and then not worrying anymore about it. Applying patches will not become standard practice for such devices quickly. It will require educating users.
- The last reason is an economic one. The current business model for CE devices is that the consumer pays only once when he purchases the equipment. Designing patches and testing them is a costly process for the manufacturer. With the current business model, the manufacturer would bear the entire cost of patching without any expected ROI. Hence, there is no incentive for the

manufacturer to bear this additional cost. Furthermore, as most CE devices have a lifetime of several years, this security maintenance should extend to the full life of the apparatus. Proposing a paid maintenance service is not realistic. The likelihood that consumers would purchase a security upgrade is small.

This analysis of IoT patching is worrisome for the future of IoT. There seems to be no simple, trivial solution to this problem. Unfortunately, without a solution, the future of IoT may be impaired. IoT devices could become the preferred targets for some types of attack. For instance, due to the large-scale deployment, turning IoT devices into zombies for botnets could be a profitable operation for organized crime.

5.3 Takeaway

5.3.1 Active Defense

As shown by previous examples, static security is doomed. Security requires evolving continuously to adapt to new threats and become a moving target. A moving target is harder to hit than a static target.

Rule 5.1: Be Proactive

Preventing a problem has always been a superior strategy to fixing the same problem after it has occurred.[8] Therefore, the strategy defining the security policy should be proactive rather than passive or reactive. A static security stance means sure death in the digital world for several reasons.

- Security is aging (Sect. 1.2.1). Both white hat hackers and Black Hat hackers routinely discover new exploits. Skilled hackers design new tools routinely and distribute them on specialized forums. The distribution of these tools embraces even the latest technological trends, such as the cloud. The concept of Malware as a System (MaaS) has been spreading for many years [250]. Like the lawful cloud-based solutions, MaaS enables fast, easy, and inexpensive deployment of large-scale attacks without the need of hacking skills and without the need to invest in a computing infrastructure.
- Attackers are relentless. They use automatic tools to search for vulnerable targets on the Internet. A recent study from the University of Michigan demonstrated that large horizontal scans, exploring a large address space for a given set of ports, occur routinely. The researchers analyzed the scanning landscape after the public disclosure of an undocumented backdoor in Linksys routers. Within

[8]The same is true for software development at large. The earlier a bug is discovered, the easier and the cheaper it is to fix it. A thorough early design phase reduces many issues later in integration phase.

48 h following the disclosure, 22 sources started 43 scans of the full IPv4 address space looking for the hidden backdoor port (32,764). The scan traffic increased for 1 month. Security firms and scholars were the sources of some of these scans. Unfortunately, many other scans were from unidentified sources, most probably malevolent [251].
Once attackers have located a potentially vulnerable target, they will spend hours breaking into the system if the target can be valuable.
- New hacking tools appear routinely. They are more sophisticated than their predecessors. For instance, new scanning tools, such as `ZMap` and `masscan`, can scan the entire IPv4 address space in less than 10 min, provided they have access to sufficient bandwidth and enough computers. They replace the traditional `nmap` that is far slower. Interestingly, these tools were developed by security researchers and not by hackers.[9] Nevertheless, Black Hats will use them for nefarious purposes. IDA Pro is the preferred disassembler for reverse engineering. Continuous contributions are enhancing and providing new plug-ins for this tool. This constant effort strengthens this powerful tool and thus weakens software tamper resistance.

The security of a system must continuously be updated to address the progress of the hacking scene. For that purpose, the security expert must be aware of the new advances. The security expert must continually train himself in the new techniques and attacks. The security of a system must be adapted to answer to new relevant attacks. As described in Sect. 2.2.3, whether or not a new threat must be addressed depends on the threat model, the likelihood of its occurrence and the cost of implementing a countermeasure. On a regular basis, organizations must review their security policies to verify whether they are up-to-date with the current landscape of attacks and address new rising threats correctly. The revised version of policies must be communicated to the users. Updating the security policies without disseminating the update would be useless. The focus of the communication should be on the modifications. Explaining the rationale behind these changes increases the users' awareness and their eagerness to comply with the new policies (Sect. 10.3.3).

Designers of secure systems also have to be proactive. Once his system is deployed in the field, the designer's work does not end. The next generation of the system must be prepared. Obviously, the designer does not yet know against which new attack he will have to defend the system. Nevertheless, the designer may start by creating diversity for the new generation of product. Diversity is an important component of warfare, as stated by General Sun Tzu [123]:

> Do not repeat the tactics which have gained you one victory, but let your methods be regulated by the infinite variety of circumstances.

Diversity makes the work of an attacker harder. If the same functionality uses different implementations for each successive version of a product, the attacker will

[9] `ZMap` is an open-source project supported by the University of Michigan. `Masscan` is an open-source project supported by the Errata Security team.

have to start from scratch with each new release. The impact of the knowledge that the attacker acquired from the previous versions will be minimized. This diversification is paramount if the trust model relies on some resistance of the software to reverse engineering. If a software has to resist to reverse engineering, then each implementation of tamper-resistant software has to be different from previous versions. If this were not the case, the attacker would reuse her knowledge acquired from the hack of the previous version. Diversity slows down the attacker.

An interesting question is that of deciding when to publish the new version of the protection. Should this new version be released only once the current version has been hacked successfully? Alternatively, should the new version be released before a hack occurs, for instance, as soon as the new version is available? This decision is mainly driven by the response to the following question: "Can the system or the vendor afford to be hacked?" For instance, if the system protects national or military secrets, then the hack must be avoided at any cost. Deploying new, well-tested versions of the same product before any hack occurrence may be a wise strategy. The attacker will have to start from scratch on the new release, losing all the work done on the previous version. The attacker will have lost her current investment without significant gain. This strategy is valid only if there is a way to enforce the obsolescence of the old version once the new version is deployed. Else, the attacker will continue to concentrate her attacks on the previous version. This strategy is an application of Law 6: Security Is No Stronger Than its Weakest Link that the next chapter will explore. If the ROI triggers the date of responding to hack, then the new version will have to be deployed after the occurrence of a hack, and the new version needs to cope with this hack. Furthermore, the launch of this deployment will occur only once the hack has generated a substantial monetary loss. For an optimal ROI, some monetary loss may be acceptable. Nevertheless, in both cases, the design of a new version should start before the occurrence of a hack.

The focus should be on detection and reaction rather than on protection [82]. The current trend is to implement many protection tools. Usually, these defensive tools protect a given perimeter. Unfortunately, because of Law 1, these tools will be circumvented ultimately. If the defenders focus uniquely on protection, either they will never detect a successful attack or they will discover the successful attack too late. Proactively trying to identify compromises is mandatory. Similarly, reacting quickly once the attack is detected is also a strategy for mitigating the consequences of a successful attack. Ideally, the reactions should be prepared and documented as much as possible.

5.3.2 *Renewability*

The previous examples had a strong assumption: the security of the system could be securely renewed. Unfortunately, this assumption is not always accurate. Many examples illustrate this mistake. CSS, which protects DVDs, could never recover from the publication of the decrypting software DeCSS. Its security could not be updated. The specifications made provisions only for the revocation of keys. Unfortunately, the complete system, with all its static keys, had been compromised. In June 2015, Neil Smith of Riskive Security identified hard-coded SSH and HTTPS encryption keys in N-Tron's 702-W Industrial Wireless Access Point device [252]. The SSH and HTTPS private keys were hard-coded on the access point and were common to all devices. Hence, an attacker could use these keys from one device to decrypt traffic from any other device. Users could not modify these keys. Thus, if these keys were to be published, then the corresponding traffic would not be anymore protected. The proposed countermeasure was to isolate the wireless access points from the global Internet or to use a VPN to superencrypt the communication. The manufacturer seemed unable to offer a solution to cope with this vulnerability.

These examples illustrate that all security systems should be designed to be renewable to survive severe attacks.

Rule 5.2: Design for Renewability

What is necessary for a security system to be renewable? A successful hack may compromise any of the following three components:

- The implementation: The hack may exploit a bug, for instance, a buffer overflow, a cross-site scripting error or a race condition. The hack may use a weak implementation, for example, keys in the clear in the memory of contiguous bytes. The hack may exploit a mistake in the utilization of a cryptographic algorithm.
- The algorithm or the protocol: The attacker may have discovered a conceptual vulnerability in the algorithms used. This type of hack often occurs when using proprietary algorithms that the security community did not examine. This kind of vulnerability may also appear after a new kind of attack has been designed, or significant progress in mathematics occurred. For instance, this was the case when differential linear attacks were published for the first time. Many encryption algorithms were not immune to this new class of attack. The vulnerability may also come from an incorrect or unexpected use of the algorithm or protocol.
- Cryptographic material: The attacker discovered some long-term secret keys.

Therefore, renewability requires coping with these three types of attacks. The response to the two first types of attack is similar. A new version must replace the software implementing the compromised security. The first requirement is that the security algorithms and protocols be not hardwired. It is not possible to change hardware once it is deployed in the field. Unfortunately, hardwired implementations of cryptographic algorithms are much faster than the corresponding implementations in software. The second requirement is that the compromised principal can receive an update. Updating a target is easier today with the wide deployment of the Internet and connected devices. Nevertheless, not all principals are connected to the Internet. In some cases, principals have no possibility of connection to the Internet.

In such cases, there is a need to orchestrate the update via a physical means such as a recordable medium or a USB drive. The third requirement is that the upgrade be possible only when it originates from a trusted authority, and nobody has tampered with it. Usually, this last requirement is fulfilled by using a secure loader. As the secure loader is the only piece of software allowing us to update the device's software, it cannot be replaced with confidence once deployed in the field. Thus, the design and the implementation of the secure loader have to be extremely carefully done. No error is accepted; else renewability would be impossible. Fortunately, a secure loader is a rather small piece of software and thus can be carefully designed and thoroughly tested to detect any existing bug.

Having the possibility of replacing the security of a compromised piece of software by a new version is not sufficient. Enforcing the update of the new software is mandatory. The compromised piece of software has to be phased out. The usual solution introduces some incompatibility between the two versions so that the older compromised version cannot operate anymore. For instance, the initial version may use a given implementation D of decryption [253]. The decryption of message m' becomes $m = D(m', k)$ where k is the decryption key, which is sent over a secure channel. The new version uses the same encryption function, but the new decryption implementation uses the derived key $k' = f(k)$. The new implementation D_{new} produces the same final result as D, but it takes into account the following modification: $m = D_{\text{new}}(m', k') = D(m', f^{-1}(k'))$. The first version D cannot decrypt the message. Thus, the user has to upgrade to the newer version. If the compromised version continues to work correctly, then the attacker will continue to exploit the compromised version rather than upgrading to the new protected version. Having an enforcement method for upgrading is a good security feature. In some instances, some level of compatibility is mandatory. This constraint may make the enforcement of renewability more difficult.

Renewing cryptographic material is a more complex task. The primary requirement is that the system should not use any global secret that cannot be replaced. Unfortunately, to deliver new cryptographic material online to a deployed principal, it is mandatory to create a secure authenticated channel. A secure authenticated channel usually employs an asymmetric cryptosystem to authenticate and to create a shared symmetric session key. The authentication uses a signed certificate to prevent man-in-the-middle attacks. If the private root key used to sign certificates leaks, then the system cannot be renewed anymore. The new signed versions cannot then be trusted as an attacker can sign her own piece of software.

5.3.3 *Be Vigilant*

Rule 5.3: Do Not Rest on Your Laurels

How can a security practitioner know whether his system is still secure? The system may seem safe because it has not yet been compromised. Unfortunately, this is a wrong belief, and it generates a false feeling of security. Nothing is worse than the wrong belief that a system is secure while it is not secure. It is impossible to demonstrate that a system is secure. There is no such available tool or proof for it. It is only possible to prove that a system has a set of vulnerabilities or that it has resisted a given amount of effort. If a system has not been compromised, this does not mean that the system is secure, for several reasons.

- The system may already be compromised, but its administrators and users are not aware of it. Many attacks rely on stealthy operations and attempt to avoid detection by the administrators or users. Only in some rare cases, such as ransomware, is the goal of the attack to be noticed. In most cases, the attack will try to hide, even after its successful completion. Only careful monitoring of the behavior of the system and exhaustive analysis of logs can help to detect an ongoing attack. Alternatively, the impact of the attack may give a hint that it occurred. For instance, if the purpose of the attack was the theft of confidential information, then the publication of this information or the unexpected use of this information may indicate that an attack occurred in the past. In 2010, the US Department of Defense discovered that one of its soldiers, Bradley Mannings, exfiltrated classified data only once the Web site WikiLeaks started to publish them. The US Army did not identify the leak preventively.
- The system may be under attack, and the defense has successfully thwarted all previous attempted attacks. Nevertheless, the previous attacks may not have been the right ones. Existing weaknesses may not yet have been discovered, revealed, or exploited. A new type of attack or a more skilled attacker may defeat the current system.
- The system may not have been yet under attack. All deployed systems are not under fire. The likelihood of being attacked depends on many factors. The first factor is the interest in this system for attackers. If the system holds assets that have high interest for a potential enemy, then the likelihood rises significantly. The enemy may want to steal this valuable asset. The emergence of APTs is an illustration of this point. The interest may also just be general curiosity. Curiosity and the appeal of the challenge are usually the primary motivations for hard-core hackers. What other motivation than pure curiosity can lead security teams to analyze, on their own initiative, a hard-coded Bluetooth PIN vulnerability in a connected toilet [254] or a Wi-Fi-enabled kettle [255]? Or an academic researcher to hack a wireless light bulb system [256]?

  The second factor is the scale of deployment of the system. The more widely deployed the system, the higher the likelihood that some attacks will target it. For instance, some analysts claimed that there were more viruses for Microsoft Windows than for any other operating system mainly because its deployment was the largest, rather than due to a weaker implementation. One virus designed for Windows could target more vulnerable victims than one virus developed for another platform that was not as widely deployed as Windows. This explanation is highly controversial. Nevertheless, the economic reasoning behind it is sound.

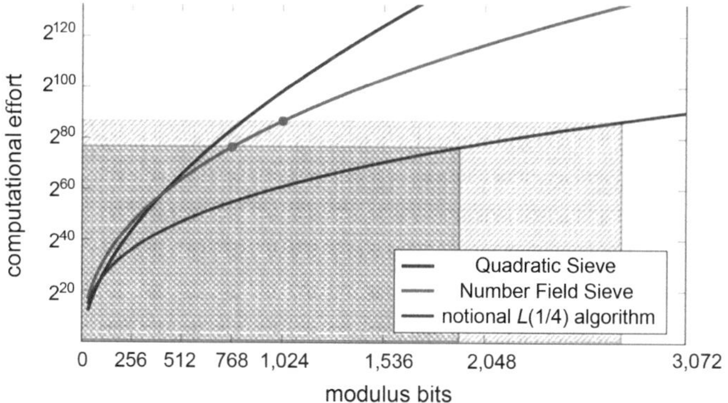

Fig. 5.2 Evolution of the factorization problem

A similar rise in the number of deployed malwares dedicated to Android platforms also accompanied the market success of Android.

Another explanation of the link between deployment and likelihood of an attack is the accessibility of the platform for experimentation. The attacker will need to get access to a host operating the system. This physical access may not be a trivial task for a proprietary system. Furthermore, the documentation is scarcer than the documentation of popular platforms. Collecting knowledge about a proprietary system is tougher than it is for a widely deployed system. Without a target on which to experiment the attacks, and without corresponding documentation, the attack is out of the reach of most hackers, except the most determined ones.

Since 2014, an annual security conference, cataCRYPT, has been dedicated to studying what might happen if a breakthrough in security occurs and how we might cope with it. The aim of this workshop is to discuss not how this breakthrough may occur, and the likelihood of such breakthrough, but rather the consequences of such a breakthrough.

The factorization problem is an example of such a mind-set (Fig. 5.2). The security of RSA relies on the difficulty of factoring the product of two large prime numbers. The security of Elliptic Curve Cryptography (ECC) relies on the difficulty of factoring the product of two numbers within a group of a finite field. Each progress in the area of factorization potentially weakens the strength of these cryptographic algorithms.

This risk of a breakthrough is a valid risk to any secure system. In 1993, the NIST published the specifications of a cryptographic hash function, SHA-0. In 1995, the NIST replaced it by an enhanced algorithm called SHA-1. SHA-0 was supposed to have weaknesses that would allow the generation of collisions. A collision for a cryptographic hash function means that it is possible to generate two documents that will present the same hash value without using brute force.[10] The world seemed safe with SHA-1. Until February 2005, when Xiaoyun Wang, an unknown Chinese cryptographer, and her team announced an attack on SHA-1 [158]. Instead of the typical attack based on the birthday paradox that requires an average of 2^{80} trials, the new attack required only 2^{69} trials. In August 2005, at the Eurocrypt conference, the same team announced a new attack that further reduced the number of trials to 2^{63}. SHA-1 was doomed. In 2008, the NIST initiated a process to select SHA-3, the successor of SHA-1 (Sect. 3.2.1).

Security experts should never be complacent. They cannot afford complacency, as their past successes do not imply their future victory. Even if highly unlikely, the unthinkable may always happen in the cyberworld.

5.4 Summary

Law 5: Si Vis Pacem, Para Bellum
Security is a war between two opponents. On one side, the security designers and practitioners defend assets. On the other, cyberhackers attempt to steal, impair, or destroy these assets. Most of the traditional rules of warfare apply to cybersecurity.

Rule 5.1: Be Proactive A static target is easier to defeat than a dynamic one. Security defense should be active rather than reactive where possible. The defenders must prepare new defenses and attempt to predict the next attacks. The new defense mechanisms do not need to be deployed immediately. In most cases, their deployment may be delayed until their impact will be optimal.

Rule 5.2: Design for Renewability According to Law 1, any secure system may be compromised 1 day. The only acceptable method to address this risk is renewable security. Every secure system must be renewable in the case of a successful hack. Without renewable security in its design, a system is doomed. Nevertheless, to ensure secure renewability, the kernel that manages renewability cannot be updated in the field. This kernel must ensure that attackers cannot misuse this renewability mechanism for their own purpose and that attackers cannot prevent the renewal. This kernel must also ensure that renewability cannot be reversed.

Rule 5.3: Do Not Rest on Your Laurels Complacency is not an acceptable mind-set for security practitioners. They must constantly be vigilant. The attackers are adapting quickly to new defenses and are creative.

[10]In 2004, French cryptanalyst Antoine Joux generated such a collision for SHA-0 [257].

Chapter 6
Law 6: Security Is no Stronger Than Its Weakest Link

So in war, the way is to avoid what is strong and to strike at what is weak

Sun Tzu, The art of war [123]

6.1 Examples

In 539 BC, the Persian king Cyrus conquered the town of Babylon. As the Babylonians had feared Cyrus's appetite for conquest, they had readily prepared the city for a long siege. Strong fortifications protected the city. For many years, the city had been storing plenty of provisions in expectation of a siege. According to the Greek historian Herodotus [258], the way that Cyrus defeated the city was original. The Babylonians were confident of their fortifications. They even held festivals during the siege. The deep, uncrossable Euphrates River traversed Babylon. Cyrus's troops dug a channel from the Euphrates to a nearby marsh, redirecting the water flow away from its natural riverbed. Consequently, the level of the river sunk low enough for the Persian army to invade the city through the riverbed under the Babylonian walls. Of course, the internal banks of the Euphrates were not defended, and the city fell quickly.[1] The Persian attackers used the weakest point of the defense: the river that also served as water supply.

As in every year, confetti enlivened the 2012 Macy's Parade of Thanksgiving in New York. Surprisingly, the confetti were made from shredded confidential documents of the Nassau County Police Department. Furthermore, on the confetti, it was possible to read complete names, dates of birth, SSNs, banking data, and other personal information about Nassau County police officers and detectives. Since some of these officers and detectives were believed to be undercover agents, this put their lives at risk [259]. Though the confidential documents were shredded, the confetti were not small enough to conceal the printed data. Although the Nassau

[1]Unfortunately, modern historians do not believe Herodotus's story is reliable.

E. Diehl, *Ten Laws for Security*, DOI 10.1007/978-3-319-42641-9_6

County police had a proper process for destroying sensitive documents, the last step of this process was inadequate. It did not employ the appropriate type of shredder. Indeed, there are three types of commercial shredders.

- Strip-cut paper shredders cut paper into thin strips. A page ends up shredded into about 40 pieces. Nevertheless, the stripped pieces of paper remain readable. Overall, readability is even greater if the printing orientation is parallel to the shredding (usually when using landscape orientation). In this case, a strip may contain a meaningful sentence or complete data. Reconstruction of the original document becomes rather easy [260].
- Cross-cut paper shredders cut paper in both horizontal and vertical directions. A page ends up in about 400 pieces. It becomes harder to reconstruct the original document. Nevertheless, some advanced software can reconstruct shredded or torn documents [261].
- Micro-cut paper shredders cut paper into extremely tiny pieces. A page ends up in about 3000 pieces. Currently, it is impossible to reconstruct the original document from the shredded version even with the assistance of software.

The previous example was about destroying physical information. The same problem exists in the digital world. Digital data wiping is often a weak link when handling sensitive information. For instance, in April 2013, a decommissioned Japan Coast Guard patrol vessel was sold to a Japanese demolition company. The Japan Coast Guard regional headquarters could not confirm whether the boat's navigation data were deleted prior to the delivery of the vessel [262]. These navigation data are sensitive as they describe the patrolling paths. This information is valuable to enemies and smugglers. Unfortunately, the demolition company was run by a senior regional member of the General Association of Korean Residents in Japan, a pro-North Korean organization. If ever someone could recover this information, the data would be invaluable to North Korea.

Captcha stands for "Completely Automated Public Turing test to tell Computers and Humans Apart." The objective of Captcha is to perform a test that should differentiate a human operator from a computer. The test proposes a problem that should be hard for an artificial intelligence but easy for a human being to solve [263]. Usually, the test displays warped or cluttered letters that the end user has to identify to prove that he is not a machine. The warping or cluttering must be complicated enough for an automatic character recognition program to fail and be still readable enough for a human being to decipher. This equilibrium is difficult to achieve. There are other forms of Captcha than character recognition. For instance, the Croatian Ruder Boskovic Institute proposes the services of a quantum random bit generator. Its registration page introduces an innovative version of Captcha [264]. The Institute's test distinguishes not only human beings from machines; it also discriminates mathematicians from non-mathematicians. Its Captcha requires solving a non-trivial mathematical challenge, as illustrated in Fig. 6.1.

reCaptcha is the çaptcha version of Google (Fig. 6.2). In May 2012, the hacking team DefCon 949 (DC949) disclosed a method that defeated the current version of

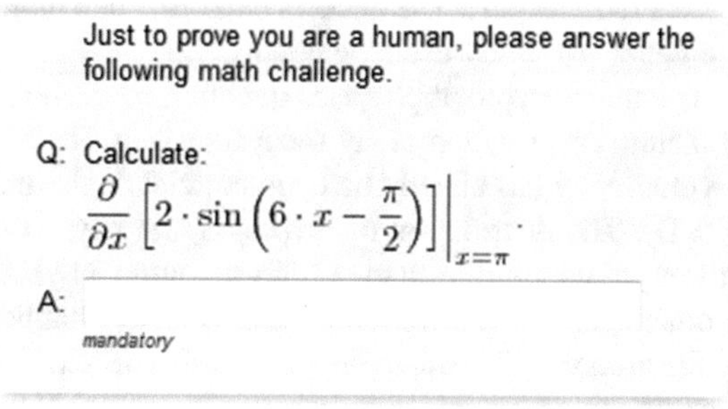

Just to prove you are a human, please answer the following math challenge.
Q: Calculate:
$\frac{\partial}{\partial x}\left[2 \cdot \sin\left(6 \cdot x - \frac{\pi}{2}\right)\right]\Big|_{x=\pi}$.
A:
mandatory

Fig. 6.1 A mathematical form of Captcha

Fig. 6.2 Google's reCaptcha

reCaptcha [265]. The accuracy announced was an astonishing 99 %. Rather than targeting the visual challenge, the process to break reCaptcha attacked its audio part. reCaptcha proposes challenges coming from warped words scanned from books. Thus, the system has at its disposal an enormous sample of potential challenges. Unfortunately, there was a weak link. reCaptcha has a mode dedicated to visually impaired people (triggered by the loudspeaker icon in Fig. 6.2). Audio challenges play recorded words with background noise to replace the visual problem. Unfortunately, the number of audio challenges was limited to 58 words, and the background noise was generated by mixing a limited number of audio sequences. Thus, there were fewer audio challenges than visual challenges. Hence, DC949 targeted the easier problem. It designed a system optimized to extracting the 58 known uttered words from background noise. Before the 2012 LayerOne conference, Google updated its algorithms, defeating the hack.

On February 22, 2014, hackers succeeded in taking control of the Domain Name Service (DNS) of the European Commission (EC) Council [266]. Then, they poisoned the EC Council's DNS. Once the hackers controlled the domain, they used this privilege to further exploit their attack. The EC enterprise email service uses a cloud provider. With their stolen privilege, the hackers issued a password reset command to this email provider. The email provider accepted the request. The EC Council's security policy mandates the use of two-factor authentication when end users access their emails. However, the cloud provider does not require such a level of security for the management of its own service. Therefore, the attackers were able to use a single-factor authentication for the management of the email system, despite the fact that this email system requested two-factor authentication from its end user. The administrative part of the system used weaker security practices than the operational part. The hackers exploited this more fragile security to get access to the better protected users' accounts.

Quantum cryptography uses quantum mechanics to perform cryptographic tasks. Quantum cryptography may seem futuristic. Nevertheless, one domain of quantum cryptography has already had commercial deployments: Quantum Key Distribution (QKD). The distribution of cryptographic keys is one of the most critical tasks of any cryptographic system. Using quantum physics to distribute keys securely is promising. QKD allows Alice and Bob to establish a shared key securely with a bidirectional link exchanging quantum particles. The shared key is then used to encrypt the communication between Alice and Bob with traditional cryptosystems. QKD has the benefit that if ever Eve attempts to eavesdrop on the communication channel used to define the shared key, then Alice and Bob will be aware that an entity in spying on their shared key. If they are spied upon, Alice and Bob drop the shared key and attempt to establish another shared key. This property is due to an intrinsic quantum characteristic. According to quantum indeterminacy, observing a quantum particle affects its quantum state. Thus, whenever Mallory watches the exchanged particles (photons), she modifies their state (qubit). QKD protocols, such as BB84 or E91, detect this modification of the qubit. Therefore, in theory, secure communication using QKD for key management cannot be successfully eavesdropped on.

In 2010, Las Lyderen et al. successfully eavesdropped on two commercial QKD systems [267]. The system uses gated photodetectors to retrieve the qubit. Usually, these detectors are avalanche photodiodes. Mallory can blind the gated detectors in the QKD system by brightly illuminating them. In the blind state, the detectors act as classical, linear detectors. The detectors are then entirely controlled by conventional laser pulses superimposed on the bright continuous-wave illumination. The controlled detectors measure what is dictated by Mallory; with matching measurement bases, Bob detects exactly the bit value sent by Mallory, while, with incompatible bases, Bob does not detect the bit. Mallory bypasses the quantum transmission. With such a system, Mallory can mount a man-in-the-middle attack. Mallory reads the qubits sent by Alice. Of course, the observation alters the qubits. Nevertheless, Mallory simulates the read qubits on the blinded, controlled gated detectors. The blind detectors return Mallory's qubit rather than Alice's modified qubit. Lyderen attacked the weakest point of the system: the quantum detectors.

6.2 Analysis

6.2.1 Design Issues

Most attackers look for the easiest target. Section 1.2.4 introduced APTs. These attacks are high profile ones. They are the less frequent type of attack. The most common attacks are the easiest ones. In 2012, about 80 % of the cyber incidents implying data breach were opportunistic. Furthermore, they did not require proficient hacking skills, as their victims were insufficiently protected. The proportion of small-to-medium enterprises targeted by these opportunistic attacks is far higher than the percentage of large enterprises. Larger enterprises are expected to invest more in their security than smaller enterprises. They have better and more tools. Their security teams are larger. Thus, they are less attractive targets for "lazy" attackers. Of course, if security increases, the chance of attracting an opportunistic attacker decreases.

The following equation summarizes the second outcome:

$$(\text{Secure system A}) \cup (\text{Secure system B}) \neq (\text{Secure system} (\text{A} \cup \text{B}))$$

The concatenation of two secure systems usually does not result in a secure system. There are several reasons for this. Secure system A and secure system B do not necessarily share the same trust model. Nevertheless, the trust model of the final system will, at least, encompass all the assumptions of both trust models. An assumption that was true in system A may not be valid in system B or may be weaker in system B. Thus, the concatenation might reduce the robustness of the trust model of the global system, as one assumption may be invalid or weaker. This reduction is an illustration of Rule 4.4: Keep It Simple. Furthermore, the interface between two subsystems is a potential weak point. The wise attacker will place her attack at this interface. Of course, the astute designer will secure this interface. Nevertheless, securing this interface adds a new layer of complexity, increasing the possibility of weaknesses, for instance, through implementation flaws.

To optimize a system, the designer may have to suboptimize its subsystems.

Rule 6.1: Know the Hardware Limitations

The designer should be aware of the physical limitations of the system he is building. Any system has intrinsic, non-modifiable constraints. If the designer does not consider these limitations, then there will often be some security issues. The most common problem is the wrong belief that a piece of information has been erased while it has in fact not been properly cleared. An attacker may retrieve the non-erased information, whereas the designer assumes that the information is not anymore available.

The Devil Is in the Detail In 2008, most popular computer disk encryption systems could be defeated if an attacker had physical access to the computer while in suspended or in password-protected standby mode. Alex Halderman and his eight colleagues compromised full hard drive encryption systems such as BitLocker on Windows Vista, FileVault on Mac and TrueCrypt on Linux [268].

The attack was simple and used two steps:

- Dump the main memory into a file,
- Search this memory dump for encryption keys.

It is mandatory to understand how memory works in modern computers to understand the attack. Computers use dynamic random-access memory (DRAM) to store non-persistent information. Each bit is stored as the value of the charge of a capacitor forming a cell (Fig. 6.3). Since all capacitors leak charges, the charge has to be regularly refreshed, i.e., the current value of the cell has to be rewritten in the cell. A typical refresh rate is 64 mS. Without refreshing, the voltage of the cell would decay slowly to null, which would result in the logical value of the cell decaying to 0 or 1, depending on its wiring. The time during which the bit value remains unchanged is called the retention time. The typical retention time exceeds the average refresh time.

To visualize this decaying phenomenon, Halderman stored a bitmap image on DRAM, switched the computer off, and after a while powered it back on. Then, he watched the image remaining in memory. The image vanished entirely only after 300 s of idle time. Figure 6.4 illustrates the retention capabilities. This example illustrates that most of the memory remains unaltered after a few seconds.

The attack exploits this physical phenomenon. If the attacker acts fast enough, she has access to a quasi-unaltered memory. The first step of the attack requires dumping the memory. The attacker reboots the target computer using an instrumented boot image that reads the entire memory and stores it in an accessible file, for instance, on a USB drive. Of course, this requires that the computer be bootable from an external (or remote) drive. The boot software does not need to be a full-fledged OS. It needs just to execute a memory dumping program. The BIOS uses only a small fraction of the memory, and the dumping program can have a tiny footprint. Therefore, almost all of the entire original memory can be dumped.

Another method for dumping is to remove the physical DRAM from the target computer and install it on a computer operated by the attacker. The ruse is to cool down the physical component. The retention time increases when the temperature lowers. Refrigeration dust-air cans are available in any electronics shop. Spraying the air on the memory can reduce the temperature down to −50 °C, which increases the retention time by several minutes, providing enough time for the attacker to swap the memory chips.

The second step of the attack retrieves the decryption key used by the encryption program in the dumped memory. Recovering the key from the memory requires the attacker to find its location in the DRAM. A simple method would be to parse all the memory and to try to decrypt data with each four-byte aligned word as a key. For 1 GB of DRAM, 2^{28} keys might be tested.[2] More sophisticated strategies are possible. For instance, Shamir and

[2]This was the method used by hacker Muslix to defeat AACS, the content protection of Blu-ray discs [269].

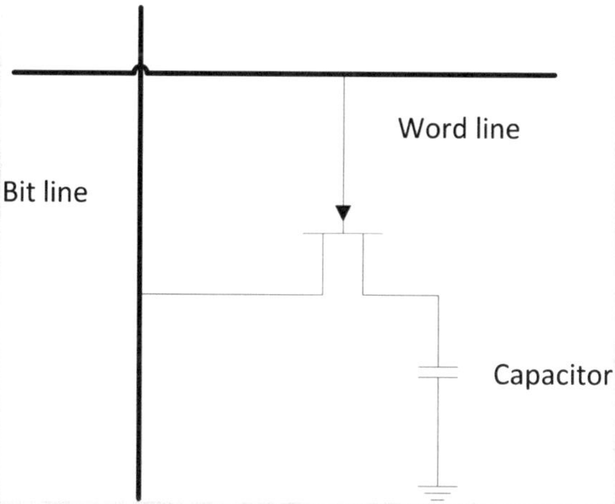

Fig. 6.3 Typical DRAM cell

Fig. 6.4 Degradation of a bitmap image in memory after five, 30, 60, and 300 s without power supply [268] at ambient temperature

van Someren [163] proposed visual and statistic tests of randomness to identify regions in memory that could contain some critical material. Unfortunately, in this context, the memory may be corrupted due to the decaying. Therefore, the DRAM may contain large areas of random data, which would lead to many false positives.

As memory dumping cannot be lossless, we assume that the stolen key may contain a few wrong bits. Halderman and his colleagues proposed algorithms to recover the correct key from the dumped memory. They can fix error probabilities in the range of 5–50 %, depending on the type of key. These algorithms recover symmetric and asymmetric keys.

Error correction cannot be conducted with basic brute force search. For example, if 10 % of the 1 s have been decayed to 0 on a 256-bit key, an amount of 2^{56} keys have to be tested. The proposed algorithm significantly decreases the computation needed to retrieve the key.

Most encryption software precomputes the intermediate keys. This precomputation increases the speed of encryption and decryption. Typically, in RSA algorithms, some data are derived from the secret key. This extra information can be easily detected, and error correction algorithms can be easily applied. To improve the efficiency of the recovery algorithm, Halderman and his colleagues modeled the decay and used it to recover errors.

For 56-bit DES algorithms, the DES key is decomposed into 16 subkeys. Each of these subkeys contains 48 bits of the original 56-bit key. In coding theory terms, subkeys are repetition codes. A repetition code allows the efficient correction of errors. With a 50 % error rate, the probability of correcting subkeys is more than 98 %. For triple DES and 112-bit keys, the probability is at least 96 %.

The derivation of the AES key is more complicated than that for DES. Nevertheless, some techniques have been used to reconstruct the key. Results of Halderman show that with a 15 % error rate, a 128-bit key can be restored in a fraction of a second, and in about 30 s with a 30 % error rate.

Countermeasures

The first countermeasure is to erase the key when it is not anymore in use. However, this is not applicable to drivers that encrypt the full hard disk as they are continuously in use. The second countermeasure is to obfuscate keys in memory smartly. At least in the case of DES algorithm, the 16 subkeys should not be adjacent in the memory. They should be spread randomly in a large space. In other words, the design must assume that the attacker will have access to the entire RAM. This assumption would prevent this attack and many others.

The BIOS should erase the entire memory while booting and prevent booting from removable media, or from network drives. As the BIOS can be password-protected, it would not be possible for an attacker to bypass this security level. If the BIOS is not password-protected, an attacker can modify the order of booting drives, prioritizing the USB drive as the first potential one.

Nevertheless, the most efficient countermeasure is under user's responsibility. Before leaving the computer, he should switch off the computer and keep an eye on the machine for a minute. After this, we can assume that the memory content will be too much degraded. Of course, this is not convenient.

In the future, new hardware architectures could prevent this kind of attack, such as DRAM that erases data quickly after the power has been turned off. Specific, secure memories that could store keys would also be an attractive solution.

Lesson: The possibility of recovering encryption keys from PC memory, even if this PC is in a seemingly secure state, as in hibernation or locked screen state, highlights several security principles. The first principle is that a system is only as secure as its weakest module. The second principle is that securing a system is much harder if an attacker has physical access to it. The physical access enables many additional attacks.

Even if a solution uses proven cryptographic algorithms or protocols, it is useless if its implementation is not secure. Often, breaches are the result of a weak implementation. The hack of AACS is a good example [269]. Developers must be aware that it is tough to hide sensitive data on a computer. All confidential data should be protected, hidden, or obfuscated in memory. Software-based data secrecy is a complex challenge. The task is far easier with tamper-resistant hardware.

DRAM is becoming a vulnerable component of modern devices. The previous attack mandated physical access to the memory. This is not always mandatory. In 2014, a new type of attack based on other physical limitations of DRAM appeared. A group of researchers from Carnegie Mellon University and Intel published a new kind of disturbance attack on DRAM: `rowHammer` [270]. Rows of cells organize DRAM. Indeed, when reading from or writing to an address, the circuit accesses the entire row containing the cells of the address rather than only the particular cell. When reading a cell, the chip activates the corresponding word line and reads the value of the capacitor on the corresponding bit line (Fig. 6.3). Memory cells are susceptible to intercell cross talk. The manufacturer specifies the maximal reading speed. The researchers discovered that the fast, repetitive reading of two consecutive rows could generate a high rate of disturbances that produce errors in the memory. In other words, the value of a cell may degrade due to cross talk. The actual software to generate such errors is a simple and short set of assembly instructions. It is a loop that reads two addresses from DRAM into registers, flushes these registers and the instruction cache, and then loops back to the beginning. Thus, two consecutive rows are accessed repetitively at a speed faster than the manufacturer's specifications. The researchers applied loops of about one million iterations. The program does not need to execute as root. It works in user mode. They tested 129 different DDR3 DRAM commercial modules. They induced errors in 110 modules. Thus, the researchers demonstrated that with simple software, it was possible to wreck DRAM memory, creating, for instance, a DoS attack.

In 2015, Google researchers went one step further. They used the `rowHammer` technique to create an actual fault injection attack. On a standard x86–64 bit machine, they demonstrated two exploits [271]. Native Client (NACl) is a sandboxing system for Google Chrome that authorizes only a limited subset of instructions of the machine. The researchers succeeded in having "blacklisted" instructions execute in the NACl sandboxed environment. In the second exploit, they succeeded in escalating the privilege to Kernel privilege on a standard Linux distribution. Of course, these exploits have some limitations. The escalation was performed on a Linux machine with not all sandboxing mechanisms enabled. Nevertheless, they highlight that `rowHammer` may become a powerful fault injection tool. The interesting characteristic of `rowHammer` is that it is purely software-based. It does not require physical access to the target machine. This may be an interesting vector of attack in multi-tenant environments such as public clouds. The goal would be to breach the isolation provided by two Virtual Machines executing on the same machine.

6.2.2 Side-Channel Attacks

At Eurocrypt 2014, Daniel Genkin, Adi Shamir, and Eran Tromer demonstrated that they could extract a 4096-bit RSA private key by listening to the executing computer [272]. Due to their vibrations, electronic components, mostly the capacitors, emit a high-pitched noise known as coil whine. The researchers used highly sensitive professional microphones to capture the sound emitted by the computer. Using an adaptive chosen-ciphertext attack, they extracted the private key in less than 1 h for GnuPG, a popular open source implementation of PGP. The chosen ciphertexts were such that they leaked information on this low bandwidth channel. The emitted sound is of frequency less than 300 kHz, whereas the operations are at several GHz. The attack is specific to a given computer model and a given piece of software. The sound emitted by the computer leaks out information about the private key. This attack is an example of a side-channel attack.

In 2010, a team of researchers at the University of Pennsylvania used a surprising side channel to guess the passwords protecting smartphones [273]. When using a touch screen, the user's finger leaves smudges of human oil. These smudges may reveal the pattern employed by the user. The surface of a clean touch screen reflects light, whereas a smudge mainly diffuses light. This difference makes the smudges visible under certain lighting conditions. The researchers were able to guess either the full password or, at least, a good chunk of it. The side channel was the human oil left on the screen. More recently, researchers demonstrated a similar attack using the thermal trace left by the user's finger on the screen.

Side-channel attacks are devastating, non-intrusive attacks that reveal secret information. The information leaks through an unintentional channel in a given physical implementation of an algorithm. These channels are the result of physical effects of the actual implementation. They may, for instance, be timing characteristics, power consumption, generated audio noise, or electromagnetic radiation. The importance of these types of attack is due to three key features:

- They are not intrusive; timing attacks require measuring times. Power attacks need to measure power consumption. Electromagnetic attacks require measuring emitted radiations with antennas. None of these measurements requires destroying the analyzed device and entering the device. They are external and nondestructive.

- They are passive attacks. They exploit physical phenomena. The attacks do not perturb the attacked device. In many cases, the attacked principal is not aware of being under attack.
- They do not require sophisticated, expensive tools. Timing attacks need high-speed digital oscilloscopes. In their simplest forms, power attacks require a high-precision voltmeter in derivation of a resistor. Electromagnetic attacks require an electromagnetic probe in a Faraday cage. All these attacks need an acquisition device, such as a high-speed digital oscilloscope. This type of equipment is commonly available in any electronics laboratory.

The expected outcomes of a side-channel attack are not necessarily the disclosure of a secret. They may be triggering events for a fault injection attack. For instance, such an attack may reveal the precise moment when a given comparison occurs. The perturbation of the fault injection attack may derail the targeted comparison by happening at the exact moment identified beforehand by the side-channel attack.

The core of side-channel attacks is the observation of physical phenomena. Indeed, operating processors leak many physical signals. They consume power. They take some time to perform operations. They emit electromagnetic radiation. They generate audio noise. All these physical signals can be measured. These phenomena may disclose valuable information. Side-channel attacks study this leaking of analog information to guess unexpected information. The observation is done at a frequency several times higher than the clock frequency of the system to have a precise analysis of analog information.

To explain side-channel attacks, this section briefly describes a simple power analysis (SPA). Most of the modern processors employ static Complementary Metal-Oxide Semiconductor (CMOS) technology. A CMOS inverter is composed of a pair of a negative metal-oxide semiconductor (NMOS) transistor and a positive metal-oxide semiconductor (PMOS) transistor. The inverter consumes power mainly when changing its state, i.e., when it switches from logical 0 to logical 1 or vice versa. When changing from logical 0 to logical 1, the current is used to charge the inverter's capacitance. When switching from logical 1 to logical 0, the current is used to discharge the inverter's capacitance. These two power consumptions are different. Thus, this asymmetry could be used to gain information on the status of the inverter. Obviously, the analysis is not possible for one single given inverter as it is not feasible to isolate it from the other inverters. Nevertheless, statistical analysis of many samples may give a good hint of the value of a bit of a secret key. If for a particular status, more inverters should switch from logical 1 to 0, or vice versa, then the profile of power consumption should reflect some asymmetry that may leak out the value of this bit. Although the attacker does not need to access the inside of the processor, she has to know the details of the implementation of the cryptographic algorithm. With the knowledge of the implementation of the algorithm, the attacker can select some sensitive, intermediate part of the computation. In this sensitive, intermediate part, some elements of the subkeys are known if one bit of the key is 0 or 1. The attacker injects a known text and computes the outcome by measuring the power during the operation. The collected traces are partitioned into two sets where the value of the intermediate part is substantially different. Statistical analysis may extract the highest probability for the analyzed bit. With this bit disclosed, the attacker starts again to guess the next bit of the key.

Preventing side-channel attacks is tough. It is a typical arms race. Usually, the countermeasure requires the attacker to collect more data, to enhance the accuracy of her measurements and to increase the complexity of the algorithms used by the analysis. Most protections are unique to the implementation and the exploited side channel. Nevertheless, some common principles drive these countermeasures. The signature of calculations should be independent of the data used. For instance, the calculation time and the power consumed for encryption should be independent of the input data and the keys. Conditional branching should be avoided as it may generate differences in the execution time or consumption time. If conditional branching is used, their existence should not be observable from outside. It is a common misunderstanding to believe that adding random delays or random consumptions prevents side-channel attacks. This is not the case, as randomization adds white noise to the measurement. It only increases the number of samples needed. Filtering out the noise of samples defeats the randomisation.

Side-channels attacks are powerful tools for advanced attacks. This field of security receives much interest from the security community. Many publications focus on this area.

6.2.3 Rollback and Backward Compatibility

SSL V3.0 is a notoriously insecure protocol. TLS has replaced it. Unfortunately, to maintain backward compatibility with SSL, TLS clients must implement a downgrade strategy. In 2014, Google researchers disclosed POODLE, an attack exploiting this feature [274]. The usual approach for this type of backward compatibility is that, at the beginning of the session, the server first proposes the most recent protocol. If the client does not accept it, the server suggests the previous version of the protocol, and so on until the client agrees on one version. Unfortunately, TLS V1.2 implements additional possible conditions, such as some network conditions, to trigger the corresponding downgrading handshake. Thus, an active attacker, who controls the communication between the host and the client, may trigger such a handshake and downgrade to SSL V3.0. Once the SSL V3.0 session established, the researchers demonstrated that they could, using the weakness of SSL V3.0 (padding condition with CBC encryption), decrypt cookies and other valuable assets. SSL V3.0 was designed around 1999. Thus, a more than 15-year-old, revoked protocol defeated the most recent revision of this protocol, i.e., TLS V1.2.

In the 1990s, the US government had to authorize the exportation of cryptographic algorithms in products targeted to foreign countries to satisfy its software industry. Nevertheless, to keep the ability to decrypt messages protected by theses foreign versions, the US exportation rules restricted the maximal length of asymmetric keys to 512 bits.[3] Thus, early implementations of SSL for foreign countries used weaker keys that could be brute-forced. In 1992, US domestic browsers could use RSA-1024 and DES with 56 bits, whereas the international version was limited to RSA-512 and DES with 40 bits. Corresponding foreign libraries were called export cipher suites. Therefore, many Web sites and browsers are still programmed to support RSA-512. Today, cracking RSA-512 is a matter of hours. In March 2015, a French INRIA team discovered that several current implementations of SSL incorrectly allow the message sequence of export cipher suites to be used even if a non-export cipher suite has been negotiated [275]. The attack was coined Factoring RSA Export Keys (FREAK). As it is easy to factor the modulus used for 512 bits, it is possible to retrieve the RSA-512 private key of such servers. With this private key, it is easy to mount a man-in-the-middle attack when forcing the transaction to use the export cipher suite rather than the non-export cipher suite. Likewise, most browsers were vulnerable to FREAK attacks. For example, OpenSSL, Apple Secure Transport, and Microsoft SChannel were exposed to FREAK. Meanwhile, all browsers and libraries have implemented patches. Unfortunately, many Web sites have not been patched. A dedicated Web site lists the most visited sites that are still vulnerable [276]. Two months after the disclosure of FREAK, the same team reported a similar vulnerability in the SSL/TLS protocol, coined Logjam. The attack forces the handshake to switch to a 512-bit export key when using the Diffie–Hellman protocol. As in FREAK, it is possible to factor this key. Rumors indicate that the NSA may have exploited this vulnerability for many years to thwart TLS-based VPN connections [277].

A typical attack is the rollback attack: return to the past where a vulnerability is still present. The two previous examples illustrate such a rollback attack. Backward compatibility is one source of rollback attacks. A backward-compatible product is a product that will work with data that were produced with older versions or with services that operated with older versions. If the earlier versions had vulnerabilities that were fixed only in newer versions, then an attacker may exploit these vulnerabilities. For instance, earlier versions of a product may have accepted weak cryptographic protocols or keys of short length that are not anymore considered secure. This cannot be fixed by patching older versions; it requires changing the version. For instance, in the case of FREAK, a solution was to ban the weak export

---

[3] In 2000, the US exportation rules relaxed this limit. Currently, the restriction is mainly for the countries that are declared enemies of the US or considered as supporting terrorism. The Waasenar Arrangement regulates the international exchange of conventional arms and dual-use goods and technologies. This arrangement encompasses cryptography as a weapon. Since 2014, it also forbids the export of:

Software "specially designed" or modified to avoid detection by "monitoring tools," or to defeat "protective countermeasures," of a computer or network-capable device, and performing any of the following:

(a) The extraction of data or information, from a computer or network-capable device, or the modification of system or user data; or (b) The modification of the standard execution path of a program or process in order to allow the execution of externally provided instructions.

cipher suites. Unfortunately, such a decision breaks the full backward compatibility. Thus, it may be difficult to sell this evolution of the product to product management and marketing departments. In June 2015, the publication of RFC 7568 officially deprecated SSL V3.0. TLS clients and servers must not send a request for an SSL V3.0 session. Similarly, TLS clients and servers must close any session requesting SSL V3.0. This decision suppressed dangerous backward compatibility.

There are other potential sources of rollback attacks. The attack on Microsoft Xbox described in Sect. 1.2.3 offers an example. The attacker attempts to retrofit the device to an earlier version. A good attack is to spoil the upgrade to a newer version or to spoil the patching. Therefore, proper protection requires that a piece of software verify whether it is the latest version. For instance, a piece of data to be processed by a software may indicate the minimal version of the software needed to process the piece of data. Whenever the requested version exceeds the current version of the processing system, this system should not handle the incoming piece of data. This requirement may be enforced by introducing some incompatibilities of data with previous versions. Of course, the requested version number should be protected in integrity.

Thus, sometimes, the weakest points of a system haunt us from the past. They are versions that were vulnerable or became vulnerable and that are still in use despite the existence of a more secure recent version.

6.3 Takeaway

6.3.1 Test

To identify the weakest point, the designer should know all the weaknesses of the system. He acquires this knowledge through a thorough process of testing and security assessment. Security should be integrated with the complete development process and should be present in every step of this process. It is common knowledge in software development that the earlier a bug or a mistake is identified during a waterfall process, the less expensive its correction will be. Likewise, the earlier a security vulnerability is identified, the cheaper and easier its correction will be.

The traditional waterfall development cycle has five steps: requirement, design, coding, integration and validation, and maintenance.

The requirement phase must take security into account. Security requirements should be a part of the overall requirements of a product or system. The security requirements are outcomes from the threat analysis (Sect. 2.2). They define the assets that have to be protected and the security characteristics, i.e., confidentiality, integrity, identity, authentication, or availability. They may also define the expected level of attackers.

The design phase uses these security requirements to develop the security solutions. A group of security experts should review the design to check whether it fulfills the security requirements. The analysis also assesses the sanity of the proposed security solutions. This analysis is theoretical and checks the algorithmic security. Designers have to fix all vulnerabilities that the review highlighted at this phase regardless of their criticality. As any necessary change takes place early in the cycle, dealing with the vulnerabilities has minimal impact on the cost and duration of development.

Coding is a critical phase, as many vulnerabilities are the result of bad implementations. Impacted developers should be trained on how to write secure software. They should be provided with best practices and guidelines. Also, reviewing the source code of critical, secure pieces of software should be systematic. During a peer review of secure code, several security experts study the source code together to detect vulnerabilities. Vulnerabilities may come from bugs, from errors in parameters such as buffer overflow or from bad implementation. The reviewers have to be security experts skilled in software implementation. Once more, at this phase, every discovered vulnerability should be fixed, regardless of its criticality.

During the validation phase, security testing should be performed. Obviously, the testing team must be independent of the development team. There are two types of vulnerability tests: black box and white box.

- Black-box test or penetration test means that the evaluator has only public information about the tested system. The evaluator views only the external interfaces of the system. The evaluator operates in the same conditions as an attacker who does not collude with an insider. A thorough evaluator may attempt to collect additional information either by social engineering or by mining the Deep Web. If some confidential information has been leaked, then the reviewer may exploit it whether or not it is public information. An attacker may gain a similar advantage by collecting the same information.
- White-box test means that the evaluator has access to the entire documentation of the system, including design and implementation information. The penetration tester sees the inside of the box. Usually, an attacker has no access to this level of information unless she is an insider.

Black-box testing mimics the possible behavior of external attackers. It is rather simple to deploy and does not imply serious financial investment. It may give an evaluation of the resistance of the system against attackers. As such, it is often preferred over white-box testing. White-box testing requires greater initial investment than black-box testing. For instance, the development team must prepare the documentation to be handed to the penetration team. Normally, most of the information should be already available and have been produced during the development process. It should be mostly a packaging effort. Nevertheless, it represents additional effort. In some cases, the development team may even have to instrumentalize its system to enable some specific tests. The analysis is deeper than that of the black-box testing. Which type of test should be preferred?

If the test has to be fast, then black-box testing is the best solution. If the test does not have to be fast, then white-box testing is ideal. For the same penetration effort, the evaluation team will find more vulnerabilities, and more critical ones, in white-box testing than in black-box testing. Indeed, it may find vulnerabilities that are unlikely to be found by external attackers. Furthermore, it may find weaknesses that an experienced insider may exploit. In general, the system, once fixed, should be more robust than when black-box tested. Black-box testing is not a good evaluation of the robustness against insiders.

The Common Criteria (CC) is a security product certification that is government-approved. CC is an international standard, ISO 15408. The CC process is extremely rigorous and uses exclusively white-box assessment. Only a limited number of accredited testing laboratories are approved to deliver such a certification. The CC process is strictly defined. The first step is the definition of a Protection Profile (PP). A PP states a security problem or a set of problems. It specifies security requirements to address that problem without dictating how these requirements will be implemented. The robustness of a product, called Target Of Evaluation (TOE), is assessed at a given level. The difference in the evaluated levels depends on the amount and quality of the provided documentation and the scale of effort applied to the evaluation [278]. The CC defines seven Evaluation Assurance Levels (EALs) as listed in Table 6.1.

EAL1 verifies that the TOE operates as described in the provided documentation. EAL1 is not related to security, and the associated threats are not supposed to be serious. For instance, EAL1 may be used to certify whether enough effort has been applied to protect personally identifiable information. EAL2 requires the design documentation as well as test results. The evaluation verifies whether the system generates the same results. EAL3 requires access to the design documentation, and the system is thoroughly tested for vulnerabilities. EAL4 also confirms the quality of the design and the evaluators provide feedback to enhance it. The testing targets known vulnerabilities. EAL5 considers a rigorous design (semi-formal). The robustness of the implementation is assessed. Resistance to side-channel attacks may be evaluated if the PP requests such a protection. The attacker is expected to have reasonable skills (IBM Class II). EAL6 verifies that the development follows a rigorous process that ensures that the implementation complies with the design. Resistance to known side-channel attacks is evaluated systematically. The attacker is expected to have high skills (IBM Class III). EAL7 corresponds to the most rigorous testing. The design has to be formally verified.

Table 6.1 EAL levels

Level	What is evaluated?
EAL 1	Functionally tested
EAL 2	Structurally tested
EAL 3	Methodically tested and checked
EAL 4	Methodically designed, tested and reviewed
EAL 5	Semi-formally designed and tested
EAL 6	Semi-formally verified, designed and tested
EAL 7	Formally verified, designed and tested

The cost and effort increase with the EAL value. The higher the level of an EAL, the greater the effort by the accredited testing laboratory to assess that the TOE complies with the expectations. Security evaluation starts being consequential with EAL4. Some operating systems are certified at EAL4. Many smart cards and HSMs are certified at EAL5 or higher. The higher the level is, the greater the confidence in the product will be. Nevertheless, the level is assessed against a given PP. If the PP encompasses a limited set of requirements, then the certified level applies only to this limited set of requirements. For instance, in the case of OS, the PP is reduced and does not cover all potential security requirements that we may expect from an OS. In 2005, Microsoft announced that Windows XP and Server 2003 have been awarded CC EAL4. The PP used for the evaluation made several "simplifying" assumptions [279]. For instance, it assumed that only trusted operators access the TOE and that the TOE is not connected to any external network. Furthermore, the TOE did not assess the soundness of the cryptographic library.

Security is not absolute. Therefore, the rating of the attacks in the CC depends on multiple factors, as described in Table 6.2.

CC concatenates rating for six factors.

- Elapsed time evaluates the effort needed to breach the system. It ranges from less than 1 h to more than 1 month.
- Expertise has four categories. The two first levels correspond to IBM Class I and Class II. The upper levels are expert or multiple experts, meaning that the attack has used different expertise fields.
- Knowledge of the TOE defines the level of confidentiality of information about it. It encompasses public, restricted, sensitive, critical, and very critical design. It describes the likelihood of an external attacker getting access to such information.
- Access to TOE defines how many samples of TOE are needed to breach it. It ranges from less than ten to more than 100. It qualifies how many different samples the attacker will need to access.
- Equipment characterizes the complexity of the equipment needed. It encompasses standard, specialized, custom-made, and multiple custom-made tools. A custom-made tool is a tool that the attacker specifically designs and develops for this target.
- Open samples define how much of the hardware of a TOE is locked. This factor is used when evaluating both the software and hardware of a system.

Table 6.2 Rating table for CC

Factor	Identification	Exploitation
Elapsed time		
Expertise		
Knowledge of the TOE		
Access to TOE		
Equipment		
Open samples		

Each factor is assessed with two parameters: identification and exploitation. Identification qualifies the effort needed to demonstrate the feasibility of the attack. This is a theoretical evaluation. Exploitation quantifies the effort required to perform the actual full-fledged attack. This rating shows the complexity of evaluating the security level of a target.

6.3.2 *Fix Security Issues Adequately*

Security is continuously decaying once a system is deployed. A system that was secure at its launch time most probably will become vulnerable after being some time in the field. Attackers find new classes of attack. New vulnerabilities appear every week. Usually, security patches developed by product or software manufacturers cure these new vulnerabilities. If these security patches are not applied, then the system becomes an easy target for attackers. Usually, when an attacker goes after a given target, her first step is to identify the complete system, its operating systems, and what pieces of software the system executes. OS fingerprinting tools such as `SinFP` greatly facilitate this task. The identification provides also the version of these elements. With this information, the attacker checks whether there is a set of known exploits that applies to the targeted system. Such a situation would occur if the system were not properly patched. Using the set of appropriate exploits, the attacker may take the control of the system. Unpatched systems are often the weakest point in a corporate IT system. Therefore, patching is of the utmost importance.

Rule 6.2: Patch, Patch, Patch

Figure 6.5 describes the typical life cycle of a vulnerability [280]. The different steps are as follows:

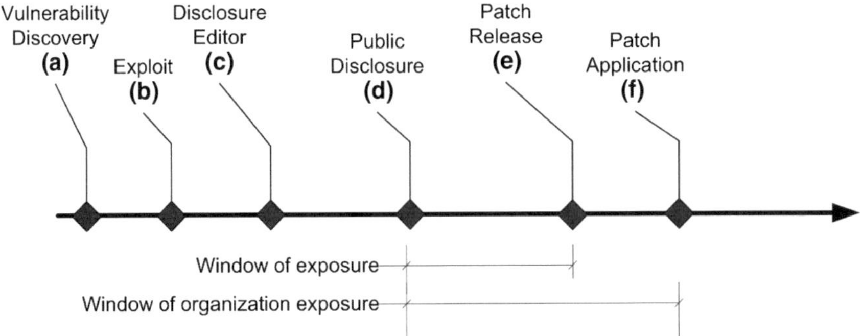

Fig. 6.5 Vulnerability life cycle

(a) An individual or a team discovers a new vulnerability. The discoverer may be a security researcher, a team member of the vendor or a white hacker. The discovered vulnerability will become a 0-day vulnerability. The following description assumes that the discoverer is not a blackhat. Usually, a blackhat keeps her discovery private. The blackhat may use this vulnerability for her own advantage or sell it on the black market.

(b) Hackers exploit the vulnerability in the wild. There are two cases.

   1. Attackers knew about this vulnerability before the discoverer did. Then this exploitation may occur before step *a*. The discoverer may have spotted an ongoing attack and, through reverse engineering, found out how the attackers proceeded. From that point, the discoverer infers the vulnerabilities.
   2. Attackers had not yet discovered the vulnerability. Then its exploitation occurs after step *d*, once it has been announced publicly. The development speed of new malware using freshly disclosed vulnerabilities is astonishing. New 0-day vulnerabilities are sometimes exploited within hours after their public disclosure.

(c) The discoverer reveals the vulnerability to the vendor. Usually, the vendor negotiates a period of silence before public disclosure (Sect. 1.2.1). During this negotiated period, the vendor should design the corresponding security patch.

(d) The vulnerability is publicly disclosed. There are many possible venues: security conferences, a CERT, disclosures of hacked servers of specialized companies [281] or security blogs. Usually, the disclosure comes with source code demonstrating the exploitation of the vulnerability. From this moment, the exploitation of this vulnerability may drastically increase. This case is especially true if the vulnerability is easy to exploit and if the attack provides excellent benefits to the attackers.

(e) The vendor releases the patch. Ideally, from the researcher's point of view, step *d* and step *e* should occur concurrently. This sequence has the optimal media impact. The worst case for security is when step *e* occurs after step *d*. In that case, all targets are vulnerable during this interim period. Usually, the vendor defines a priority rating, depending on the severity of the vulnerability, the rate of deployment of attacks in the wild, and the exposure to attacks. In some cases, the patch should be applied in less than 3 days [282].

(f) Both enterprise and end user apply the security patch to their systems.

The period between the public disclosure (step *d*) and the release of the vendor's patch (step *e*) is called the window of exposure. The period between the public disclosure (step *d*) and the release of the application of patch (step *f*) is referred to as the window of organization exposure. The goal is to minimize these windows of exposure. During these windows of exposure, the systems are in their most vulnerable phase, as all attackers are aware of the vulnerability, and the remedy has not yet been deployed.

Obviously, the actual realization of the last step is crucial for minimizing the window of organization exposure. Most attacks use vulnerabilities that are not 0-day vulnerabilities, i.e., occurring after step *d*, before the systems have been patched. This observation highlights that the delay between step *e* and step *f* is too long. A 2015 study listed the advice that security experts would give to non-technical users. The first advice was to keep systems and software up-to-date [283]. This outcome highlights the importance of patching. Many studies demonstrate that patching is not rolled out as fast as is necessary. Why are so many systems not patched in real time?

Indeed, patching in a corporate environment is more complicated than just running a patching application [284]. Many documents describe best practices of the patch management process [285]. A proper patch management process requires the following elements:

- Good asset management; it is essential to have a detailed, accurate catalog of all systems managed by the company. It must list for each asset its configuration, including its location, IP address, MAC address, hostname, OS and its current version; the already applied patches; and the applications running on it. Without such a thorough inventory, it is not possible to ensure that all systems are patched. One unpatched critical system may be sufficient for an attacker to breach the secure perimeter.
- Analysis of the vulnerability or set of vulnerabilities that the patch should cure. The review should assess the level of exposure and the potential impact of the vulnerability. This analysis defines whether the patch should be applied and the degree of urgency of implementing it. Security experts who are knowledgeable of the systems and their environment should conduct the review. Without such broad knowledge of the context, the analysis would be inefficient and useless. Furthermore, vulnerability-scanning tools, such as Nessus or Qualys, draw a good map of the potentially vulnerable targets (provided that the tool has access to all potential targets). These vulnerability-scanning tools are invaluable if experts properly review the generated reports. These tools are useless if not used regularly or if their outcomes are not analyzed and exploited later. Having a list of all the unpatched systems without curing them is of no value. It is even a waste of money (Sect. 10.3.2).
- For each deployed configuration, the patch should be tested on a test bed system. This test bed system should be dedicated for test purposes only, to avoid impacting production systems in the case of failure. Of course, the test bed system must accurately simulate the targeted configuration. The aim of this test is to verify:
  - Whether the patch cures the vulnerability,
  - Whether the patch is innocuous, i.e., it does not crash the system or impair some functionalities.

If ever the patch cannot be tested on a test bed system, it is recommended to apply it first to one system that has been carefully backed up and is not critical for production and business. If ever the application of the patch would crash this initial system, then it would be easy to restore it to its previous state. The crash would not impact the business severely.

For critical systems, a back-out plan should be developed and tested beforehand. A back-out plan defines how to reverse the production system to its original state if ever the patch were to address adversely the critical system. It is paramount that we verify the effectiveness and applicability of the back-out plan before launching the patching process.

- Rolling out the patch; there are several strategies.
  - All-node patching; all the nodes are shut down. The patch (or a set of patches) is applied simultaneously on all of them. Once patched, the nodes are brought up again. This method is fast and simple and has minimal compatibility issues with the previous version. The window of organization exposure is at its minimum with this strategy. Unfortunately, the downtime is at its maximum, and if there is a problem with the newly patched system, the negative impact will be maximum. The complete system will have to be restored to the previous state, at least doubling the downtime.
  - Minimum downtime patching; a subset of nodes is shut down. The patch is applied to this subset. Once verified, the patched subset is brought up again. Another subset is powered down. The patch is applied to this second subset, and so on, until the completion of full patching. The downtime is minimized, but the application of patches is more complicated and takes longer.
  - Rolling patching; the first node is shut down. The patch is applied to the first node. This first node is brought up. The second node is powered down. The patch is applied to the second node. The second node is brought up. The third node is shut down, and so on. The rolling patching strategy is the safest one, but it takes a longer time to complete than the previous strategies. This strategy produces the longest window of corporate exposure.
- Documenting and adding the patch to the baseline configuration. In many enterprises, there is a set of default baseline configurations that are deployed on new systems. Therefore, it is important that these baseline configurations be the most up-to-date ones. This will ensure that newly deployed systems are secure.

The criticality of the targets should also determine the order of deploying security patches. In order of decreasing criticality:

- Mission critical systems have to be permanently operational. Any downtime has severe business impact. These systems have to be patched in order of priority as the effects of an attack may be significant. Unfortunately, some patches may require a reboot or a network disconnection. This is not always possible. Thus, the patching is done first on a spare or backup system. Once this patched system is operational and thoroughly tested, it replaces the current operational system. The previous operational system can now be patched. This deployment presents no risk of downtime. Usually, to prevent any issue, patching of critical systems

is done manually. This human monitoring ensures a fast application of the back-out plan in case of failure. Automation of patching would not extend this safeguard.
- Business critical systems have to be permanently available. Nevertheless, short downtimes are acceptable provided they do not occur too often. These systems have to be patched in the second rank. Gradual patching (minimum downtime or rolling patching) should not produce any problem. The operational systems support a reboot without disrupting the business. In this case, automated tools can do the patching.
- Business operational systems do not need permanent availability. Disruption of service has no serious impact on business. Simultaneous patching is acceptable. In this case, patching is better served by automated patching tools.

Patch management becomes even more complex in two contexts: Supervisory Control and Data Acquisition (SCADA) and consumer electronics.

SCADA is a control system that acquires data from sensors and controls remote equipment in manufacturing and industrial control processes. In June 2010, the STUXNET malware targeting an Iranian SCADA system made headlines. Using four 0-day vulnerabilities, STUXNET aimed specifically at programmable logic controllers used by the Siemens Step 7 system. STUXNET compromised Iranian systems used in nuclear installations. STUXNET is often regarded as the first known cyberweapon. The novelty of this malware was its target: SCADA. Usually, SCADA equipment handles critical controls or continuously running processes. There are two main issues: the patch failing rate and the availability of the patch [286]. There is never a guarantee that a patch will not wreck the patched element or the entire system. These control systems are complex and interact with many heterogeneous elements. Thus, there is a risk that patching one part hurts other parts of the system. The patching failure rate of SCADA systems is estimated at 60 %. Usually, they do not have a mirror system on which to test the soundness of the patching of one element of the SCADA. Thus, patching one element corresponds to taking a risk. Often, for financial or practical reasons, the owners of the system evaluate this risk as being higher than the risk of being targeted by a cyberattack. In any case, it is not conceivable to patch at the same pace as the IT world. It is not possible to expect a monthly patching event. The ideal solution would be to shut down the full system and patch every element and test the completely patched system. Unfortunately, this scenario is not realistic as these systems usually need to minimize downtime and often run 24 × 7. The second issue is the age of the SCADA system. The investment analysis usually forecasts that the control equipment will last about 20 years. This extended period raises the question of availability of patches for old code. The original manufacturer may have gone out of business, and the availability of specialists still conversant in a 15-year-old technology is questionable. Then, there is no possibility to develop such a patch. Thus, the estimation is that 50 % of the known SCADA vulnerabilities have no corresponding patch. In some cases, the patching process of a given equipment is so complex that it requires the manufacturer to undertake it rather than the operator of the SCADA system. Given the complexity of this problem, it is recommended to harden the isolation of SCADA equipment from the external global Internet. This isolation mitigates the risk of an external attacker accessing the system. Unfortunately, this does not prevent attacks from insiders or through infected computers if the production network is not air-gapped from the corporate network.

In the past, most consumer electronic devices were not connected to the Internet. Therefore, there was no need, and nor was it possible to patch the device. Currently, more and more modern CE devices are connected. Furthermore, with the advent of the Internet of Things (IoT), patching will become unavoidable [287]. Unfortunately, patching CE devices has some peculiar problems.

The first issue is that many CE devices are not designed for an upgrade. Nevertheless, most network-connected devices such as gateways are designed to allow it.[4] The biggest issue is the maintenance of the software. Contrarily to general computers, which use many generic components, IoT and CE devices use very specific hardware and custom binary drivers. Therefore, the vulnerability patch should be tested with all variants of hardware and drivers. Failed patching is not acceptable for consumer of a security patch on his CE device ending up with a broken device is a nightmare for any manufacturer. Furthermore, as there is no maintenance contract, nor sale of new versions of software, CE manufacturers have no incentive to propose as it is expensive to design a patch. With the widespread use of these connected devices, unpatched vulnerabilities will flourish.

The second issue is the application of the patch. The consumer must be aware of the availability of a security patch. Pushing a security patch to a CE device is not common practice. For instance, Android does not implement a system to automatically download security patches of the OS. Google delivers the OS security patches to the manufacturers of the mobile devices. It is the responsibility of the manufacturers and carriers to push the security patches to the devices. Unfortunately, Google does not require the handset manufacturers to deploy them quickly. The estimated rate of security patches deployed for Android devices by telecom operators is about 1.26 upgrades per year [288]. Given the wealth of Android vulnerabilities, this rate is insufficient. Interestingly, Android can push updates to installed applications but not to itself, i.e., the OS.

We could have expected consumers to be afraid and to apply security patches. Unfortunately, this is not the case. Consumers are not proactive with regard to their security. For instance, only about 70 % of the vehicles subject to recall are fixed within 18 months [289] despite the recall being free. This means that about 30 % of the drivers did not apply for the free fix even if it concerned a critical element of their vehicle. The same behavior is seen in the digital world. Consumers do not perceive upgrading the system as a top priority for real security [283]. Thus, the only effective method is pushing the patch, or at least actively informing consumers

[4]There is even a protocol dedicated to remote management and updates: TR69.

of its availability. It is not realistic to expect consumers to look for available patches proactively. The application of the patch must also be smooth and transparent. If it is cumbersome or requests many actions, consumers will not use it.

The complexity of patch management may explain why the delay between the release of a security patch and its application to all targets can be lengthy in the professional world. In the consumer word, both the complexity of applying patches and the lack of awareness of this best security practice may explain the even longer delay. Unfortunately, it is in the best interest of all actors of the Internet to reduce the window of exposure, which is currently far too wide.

6.3.3 Take Care of Your Keys

In July 2014, Cisco issued a security advisory [290]. Some Cisco routers had a unique, global, default, private key for `SSH` that was reserved for the Cisco support team. Unfortunately, the developers did not properly obfuscate this private key in the binary code. An attacker successfully extracted this private key from the binary code after some reverse engineering efforts. With this `SSH` private key, an attacker could connect to the system by using the integrated support account without requiring any additional form of authentication. Once logged into the router as a support team member, the attacker gets access with the privileges of the root user, i.e., the attacker could gain the complete control of the router. As the key was global, the attacker could access all concerned routers deployed worldwide.

This example highlights the importance of properly securing keys. As stated by Auguste Kerckhoffs, protecting the secrecy of the key should be at the heart of any cryptosystems. If a system does not adequately protect its secret keys, then this system is doomed.

Rule 6.3: Protect Always Your Keys

This example also highlights the difficulty of implementing such protection. The lifetime of a key has four different phases: creation, storage, usage and deletion. They all require particular attention.

The first phase is the generation of the key. The creation has two requirements. Of course, the first requirement is confidentiality. The generated secret must not leak during its inception. Usually, "black boxes" produces the keys in factories. Black boxes are rugged, tamper-resistant computers that may even use HSM. These specialized devices are designed to resist many sophisticated attacks, even physical ones. FIPS 140-2 defines four levels of security [291].

- Level 1 provides the lowest level of security. It requires a cryptographic module with at least one approved algorithm. Level-1 devices do not require specific physical security mechanisms.

- Level 2 also requires tamper evidence mechanisms. Tamper evidence detects an attempt of tampering with the principal, whereas tamper resistance attempts to prevent the tampering of the principal. Tamper resistance is harder to realize. Tamper evidence includes the use of tamper-evident coatings or seals. Attackers need to break these seals to access the critical cryptographic material. In addition, a role-based authentication phase must trigger the operations of the HSM. FIPS 140 Level 2 must be evaluated at the CC EAL2 (or higher).
- Level 3 requires the addition of a tamper-resistant mechanism. The physical security mechanisms may include the use of secure enclosures and tamper detection/response circuitry that zeroes in all cryptographic material when the removable covers/doors of the cryptographic module are opened. An identity-based authentication phase replaces the role-based authentication phase. Identity-based authentication provides finer granularity than role-based authentication. FIPS 140 Level 3 must be evaluated at the CC EAL3 (or higher).
- Level 4 is the highest security level. The physical security means should cover the entire HSM. Any attempt to intrude in the device as well as any unexpected environmental context may trigger the zeroing of the embedded cryptographic material. For instance, the device includes light detectors, temperature sensors, and low power sensors. Whenever the internal component sees some light, measures temperature too high or too low, or detects erratic power supply, it erases the cryptographic material. FIPS 140 level 4 must be evaluated at the CC EAL4 (or higher).

The transfer of the key from the generating device to the final host is usually protected. At least, the transfer occurs in a secure environment so that an attacker has no way to eavesdrop on the communication between the two devices. Confidentiality is not sufficient. At this step, the most important characteristic is the randomness of the generated secret key. If an attacker can guess the key, then the key is useless. Producing true random numbers is a complex task. There are mainly two categories of Random Number Generators (RNGs).

- True Random Number Generators (TRNGs) use physical phenomena to generate a random stream of bits. The generator measures a physical parameter or a set of physical parameters that is supposed to be random. Many physical phenomena can be used as a source of entropy. One possible source of entropy is the decay of radioactive material. These nuclear decays are totally unpredictable. Atmospheric noise is another excellent source of entropy. In recent years, quantum number generators have appeared using the phase of photons issued by a laser beam [292]. Quantum phenomena are expected to be random in principle and fundamental in nature. Indeed, any source of white noise is suitable. Often, SoCs use the noise generated by an electronic component, such as the Schottky noise of a diode or the thermal noise of a resistor. Computers sometimes use human-generated “noise” as a TRNG, for instance, measuring precisely the delay between different keystrokes or the speed of a computer mouse.

- Pseudorandom Number Generators are mathematical algorithms that attempt to generate random streams of bits. Some algorithms are excellent random generators. One limitation of PRNGs is their periodicity. After a given period, the PRNG produces the same sequence. Ideally, this cycle must be as long as possible. One interesting characteristic of some PRNGs is that they use a seed. Thus, they can be *deterministic* while remaining *unpredictable*. Deterministic means that the same seed generates the same sequence of random bits.

When should a designer use a TRNG rather than a PRNG? Table 6.3 provides a rough comparison of the two types of RNGs. PRNGs are faster, cheaper, and simpler to implement than TRNGs. They are software algorithms, whereas TRNGs require hardware, i.e., some physical contraption. Furthermore, the throughput of PRNGs is higher than the throughput of TRNGs. As already mentioned, PRNGs are deterministic, whereas TRNGs are not deterministic by nature. Furthermore, by nature, TRNGs are aperiodic, i.e., any sequence of random bits will never repeat in a regular period. Thus, TRNGs are "more random" than PRNGs but less practical, as they require physical measurements. Nevertheless, many modern SoCs have embedded TRNGs using internal sources of entropy.

The second phase is the storage of the secret material while at rest. There are mainly three methods to store secret material.

- The secret material may be stored as clear text. In that case, its access must be strictly controlled and restricted to authorized principals. If ever an attacker gets access to this material, the system is doomed. The optimal protection is storing it with hardware protection, in other words, in a vault that only trusted principals can access. In chips, this storage uses a tamper-resistant, secure, hardware memory on a secure processor. Only the software executed by the secure processor can have logical access to this secure memory space. No other components than the secure processor can physically access this secure memory space. Usually, the access to this memory is also protected against external physical access such as microprobing. Storing secret material as clear text in a non-secure memory is usually a design flaw.
- The secret material may be stored as a ciphertext. In that case, a master key encrypts the secret material. The encrypted material can be stored safely in any memory, even publicly accessible. Without the master key, an attacker cannot use the encrypted secret material. Of course, the new problem is to protect the

Table 6.3 Comparison between TRNG and PRNG

	TRNG	PRNG
Efficiency	Low	High
Determinism	Non-deterministic	Deterministic
Periodicity	Aperiodic	Periodic

master key and to control which principal can initiate the decryption of the encrypted material. In other words, the master key is as in the previous case.

- If the secret material is a password, then there is a third method. The reference password should be stored as a salted hash value (The Devil Is in the Detail of Sect. 7.2.2). When a user presents his password, the system computes the corresponding salted hash value. Then, it compares it to the stored reference value and checks whether it matches. As hash functions are irreversible, it is not possible to retrieve the password from its salted hash value. The hash value does not need confidentiality, but it needs protection for integrity. The attacker should not be able to replace the stored hash with a hash of a known password; else she would gain access to the password-protected service.

The third phase corresponds to the use of the secret material. If the trust model specifies that the execution environment is trusted, then the use of secret material should not require special care. The execution environment guarantees its secrecy. The only exception may be perhaps the protection against side-channel attacks (Sect. 6.2.2). If the trust model specifies that the execution environment cannot be trusted, then the use of secret material should be handled very carefully. Necessarily, at some moment, the algorithm has to handle the secret material in clear. At this precise moment, an attacker observing the process may see the clear secret material. It is mandatory to hide and obfuscate as much as possible the handling of the key. For instance, the transfer of the clear key should never use a standard `memcpy` command with the consecutive bytes. The different bytes of the key should be spread all over the memory and should be handled in small blocks rather than in one block unit. Writing a secure implementation of a cryptographic algorithm for a non-trusted environment is an incredibly complex task that should be handled only by seasoned security experts.

The fourth phase corresponds to the deletion of the secret material. There are two scenarios of deletion. The first one is the removal of the temporary image of the secret material that the third phase used. Indeed, execution of a cryptographic algorithm requires temporarily storing one or multiple instances of the clear secret material in the RAM and the cache memory of the system. Once these instances are not anymore useful, they must be erased properly from the memory. Erasing means that a random string should overwrite the corresponding physical memory location. For instance, it is not sufficient to `free` one memory space that the software dynamically allocated using `malloc`.[5] The free command releases the reserved memory but does not erase it. Thus, an attacker may find it. The same precautions

[5] `malloc` and `free` are standard C language commands used to allocate dynamically some memory buffers and later free the allocated spaces.

are needed if the stack is used to pass a secret key. Furthermore, it would be wise to use a random string to overwrite rather than a static string such as of zeros. Were a static text string to be used for overwriting the key, a search for it could help the reverse engineer locate the potential locations of the key during processing, offering a valuable clue.

The Devil Is in the Detail Zeroing a buffer may seem a simple programming task. Unfortunately, this is a wrong belief if the operation has to be secure, i.e., an attacker should not be able to retrieve the values that were zeroed. Let us assume that the software uses a temporary memory space of 128 bits labeled `key` within a C function `do_something`. The usual approach would be to use the `memset` C command.

```
void do_something( void)  {
 uint8_t key[32];
…
/* erasing sensitive data */
memset(key, 0, sizeof(key));
}
```

The `memset` function is expected to set the 32 bytes of `key` to zero. Unfortunately, modern C compilers analyze the source code to optimize the binary code. When analyzing this source code, the compiler detects that at the conclusion of the function, the temporary memory `key` will not be anymore accessed as it is a local variable. Therefore, compilers decide not to implement the `memset` command, as it is functionally useless. This optimization does not modify the observable behavior of the function. The final outcome is that the value of the key remains in the physical memory of the system once the function has concluded because the compiler has ignored the `memset` command.

Therefore, the C language defines a `volatile` keyword. `volatile` indicates that the value of the parameter may change between successive accesses, even if it does not appear to be modified. This keyword prevents an optimizing compiler from optimizing away subsequent reads or writes and thus incorrectly reusing a state value or omitting writes. Thus, a better implementation would be:

```
static void * (* const volatile memset_ptr)(void *, int, size_t)
= memset;


static void secure_memzero(void * p, size_t length){
    (memset_ptr)(p, 0, length);
}
void do_something( void)  {
 uint8_t key[32];
 …
 /* erasing sensitive data */
 secure_memzero(key, sizeof(key));
}
```

The call to the function `memset` is replaced by a call to a pointer `memset_ptr` in a new routine, `secure_memzero`. The declaration of the pointer `memset_ptr` as a constant volatile pointer tells the compiler that at some moment, this pointer may point to another function than `memset`. Therefore, the compiler will not try to optimize it and remove the `memset` call.

Lesson: The handling of secret keys while using modern, sophisticated languages is a complex task. Too often, compilers perform optimization with unexpected effects. When developing a secure implementation of a cryptographic implementation, the designer should study the compiled binary code and pay particular attention to clearing the locations that hold sensitive information.

The second case is when secret material has to be erased definitively. There are mainly two scenarios: at the end of the life cycle of the secret material or when detecting an intrusion from an attacker.

For the first scenario, the objective is to erase the value from a persistent memory such as a hard drive, a removable recordable medium or a flash memory thumb. The deletion function of the file system is not appropriate. This deletion function frees the segments allocated to the file from the allocation table. Thus, the file system may reuse the freed segments. The result is that until the file system reuses the freed segments, the actual values are still present in them.[6] Rewriting a value in the space to free may not be sufficient. Indeed, even when a magnetic head overwrites one "bit," it indeed does not fully overwrite the previous value of the "bit" because the alignment of the magnetic head is not precise enough. Modern investigation technologies, such as magnetic force scanning tunneling microscopy, allow guessing the "overwritten" values by physical exploration of the magnetic support.

The solution is to use a secure deletion function. This function will not only deallocate the segments to free them, but it will overwrite the freed segments with specific data. In 1996, Peter Gutmann proposed a scheme using 35 data patterns to write sequentially on deleted memory [293]. These patterns addressed the different coding schemes utilized by the magnetic material. This solution addresses the magnetic supports but not solid-state based supports such as flash memory, EEPROM, or solid-state drives. The physical destruction or the demagnetization in the case of magnetic support is the only 100 % sure solution, provided that the destruction is thorough.

The second scenario erases keys in the case of an intrusion. This is usually present in HSM. This type of erasure employs physical components for the destruction.

6.3.4 Think Global

Security is global. A common mistake when designing is to concentrate on local or low-level elements of security without having a global view. Only the global picture can ensure proper security. A global view encompasses many dimensions.

The first dimension is the breadth of the context and the space of the analysis. Even if the assets to protect are usually within a limited known perimeter, the scope of the analysis must extend further than this perimeter. We already explained that the security perimeter is blurring if not disappearing. For instance, the analysis has to take into account the principals that it controls, but also the principals that it does not control but with which it has to interact with. Systems are increasingly interconnected to other systems. This dependence opens additional venues of attacks. The attacker may use the externally connected system to violate the perimeter, especially if this external system is weaker. Indeed, increased connectivity violates Rule 4.2: Minimize the Attack Surface. Unfortunately, this trend is unescapable.

Similarly, the analysis must take into account both the outsider attackers and the insider attackers. As part of the insiders, the analysis should encompass the insiders of principals the external system is connected to.

The second dimension is the breadth of technologies in use. Rule 1.4: Design In-depth Defense, states that defense should use several complementary solutions.

---

[6] This is why the undeleted function of an OS can retrieve some files from the recycle bin. If the segments are still free, the function just reinitiates their location in the allocation table.

The analysis has to enforce that these technologies cultivate synergy that ends up with increased security. It is essential to verify that one technology does not diminish the impact of the other protections used.

The third dimension is the depth of the protection. The protection has to encompass all the aspects of the assets to protect. It covers physical security, the security of hardware components, the security of the software elements, the security policies and their enforcement, and the people who interact with the assets. Once more, all these elements of the global scheme of security have to create a synergy rather than participate piecemeal.

Furthermore, only a global vision of the deployed security solutions can help identify the weakest element of its overall security. Without the knowledge of the weakest link, it is hard to optimize the defense. Once the weakest link is identified, the efforts should be on strengthening this link rather than enhancing the security of other elements that are already better protected than the weakest point.

6.4 Summary

Law 6: Security Is No Stronger Than Its Weakest Link
Security is the result of many elements that interact to build the appropriate defense. As a consequence, security cannot be stronger than its weakest element.

Rule 6.1: Know the Hardware Limitations In this digital world, most of the technical effort is put into developing software. The focus is often on protecting the executed piece of code. Nevertheless, the code executes on hardware. Hardware introduces constraints that are often unknown to contemporary software developers. Ignoring these constraints may lead to interesting attack surfaces that a seasoned attacker will, of course, use.

Rule 6.2: Patch, Patch, Patch Security is aging. New vulnerabilities are disclosed every week. As a result, manufacturers and publishers regularly issue patches. They are useless if they are not applied. Unfortunately, too many deployed systems are not properly patched. Attackers look first for unpatched targets.

Rule 6.3: Always Protect Your Keys Keys are probably the most precious security assets. Their protection should never be the weakest link. Ideally, they should represent the strongest link. They need protection not only at rest but also when in use. A software implementation of cryptographic algorithms has to be carefully crafted, especially when operating in a hostile environment. It is expert work.

Chapter 7
Law 7: You are the Weakest Link

Given a choice between dancing pigs and security, users will pick dancing pigs every time.

E. Felten, Securing Java [294]

7.1 Examples

Over the years, credit card and debit card companies have developed many enhancements in their security posture to secure users' accounts. The credit card and debit card numbers are confidential information. Nevertheless, the early ATMs displayed the complete card number on the printed receipt delivered after each transaction. Many people left this receipt at the ATM booth or dumped it into a nearby trash can. With the advent of e-commerce, this became a serious vulnerability. Malevolent people collected these receipts and used the printed card numbers. Thus, ATMs were updated not to print anymore the complete card number but only its last few digits. Furthermore, best practices in Web site design recommend also displaying only the four last digits of a credit card or debit card rather than the complete card number. In 2005, Visa created the Credit Verification Value (CVV)[1] to complement the previous evolutions. Soon, the other credit card providers adopted this new solution. CVV is a three-digit value (respectively, four digits in the case of American Express) that is printed on the back of the card (respectively, on the front of the card). Online merchants request the CVV together with the card number as proof that the user has physical access to the credit card used for the transaction. All these precautions were designed to prevent an eavesdropper from stealing the credit card credentials from the user. A strong assumption of banks is that the user will never share this information with untrusted persons. This assumption may seem reasonable. For many years, some people have taken a picture of their credit card and voluntarily shared them on social networks such as Twitter or Instagram. The Twitter account @needadebit card compiles such bad

[1]Many different acronyms seem to be used for CVV, such as CVV2 or CSC (Card Security Code).

E. Diehl, *Ten Laws for Security*, DOI 10.1007/978-3-319-42641-9_7

practices. Some people even share a picture of the back of their credit card, i.e., the CCV. Of course, the image of the front of the American Express card reveals both the credit card number and the CCV. Mitigating bad practice of users is hard. Fortunately, many banking institutions are using two-factor authentication to compensate for the possible weaknesses of their customers.

The previous example highlights that many people do not understand the risks they are taking with digital technologies. They do not understand how the digital technologies they employ everyday works. Thus, some people can be easily tricked into taking stupid actions "dictated" by their digital gadgets. In 2012, Trevor Harwell was sentenced to 1 year in jail and 5 years of probation [295]. He worked partly as friendly computer repair technician. When fixing computers, he also installed spyware on his victims' Mac computers. The spyware stealthily controlled the integrated webcam. The spyware displayed a strange error message.

> You should fix your internal sensor soon. If unsure what to do, try putting your laptop near hot steam for several minutes to clean the sensor.

According to the collection of pictures recovered from Harwell's computer by the police, indeed many victims brought their computers to the bathroom when taking a hot shower.

An underappreciation of the implications and the complexity of digital technologies can sometimes be lethal. In 2015, Air Force General Hawk Carlisle, head of the US Air Combat Command, disclosed that analysts at the 361st Intelligence, Surveillance, and Reconnaissance Group spotted a selfie snapped and posted online by an ISIS fighter standing outside an ISIS secret command-and-control building in the Middle East [296]. The fighter was bragging about the capabilities of this center on many open, public fora. Analyzing the published selfie, the analysts guessed the actual location of this command center. Twenty-four hours later, the US Air Force bombed and wiped out the center.[2]

According to a report from the UK Information Commissioner Office, during the first quarter of 2014, only 7 % of the data breaches were due to actual hacks. The remaining 93 % were due to human errors, inadequate processes and systems in place, and lack of care when handling data. People made mistakes, security practitioners did not design proper processes, or users did not accurately follow the defined processes [297]. Clearly, humans were the weak link. Humans include all segments of the population that deal with security: analysts, developers, practitioners, and end users. They put the system in danger.

Attackers often entice users to become this weakest link. Phishing and scams exploit the human weakness. These attacks become even more sinister if the attacker circumvents legitimate security mechanisms. In 2015, Symantec reported a new method used to steal the accounts of users [298]. Let us assume that Mallory wants to gain access to Alice's account. He knows Alice's email address and her

[2]The General did not disclose the techniques used to locate the command center. It may be possible that the fighter did not disable the geotag feature of his camera. See Sect. 8.1.

mobile phone number as well as her account number. Mallory contacts the service provider of Alice's account and requests a password reset. For the verification, he selects the method that sends a one-time digital code to Alice's phone. The service provider sends an SMS to Alice's mobile phone with this code. Simultaneously, Mallory sends an SMS to Alice impersonating the service provider. The SMS explains to Alice that there was some suspicious activity on her account. She must reply to this SMS with the code that was sent previously to her. Gullible Alice obeys. Mallory has now the one-time code that the service provider requests to reset Alice's password. Mallory gains complete access to Alice's account with the involuntary help of Alice.

Sometimes, the weakest link in the chain is the developer of the system. On May 24, 2008, Yahoo issued its new browser, AXIS, and its extensions for Chrome and Firefox browsers. As a standard best practice, Yahoo signed these extensions to authenticate its origin. Unfortunately, Yahoo made a mistake. Security researcher Nik Cubrilovic disclosed that the private key used to sign this applet was present in the applet [299]![3] As already explained several times in the book, private keys should never be published. Private keys have to remain secret. This condition is a non-negotiable security assumption. With the Yahoo private key, an attacker can sign any piece of software and pretend that Yahoo issued this signed software. Yahoo quickly released a new version of its extension using a different signing private key, published the certificate of this new signing key, and revoked the leaked private key. The developer(s) who wrote the applet seemed not to have understood what a signature is. The developer was the weakest link, as he introduced a major flaw in the software. A similar issue occurred in 2014 with the Belkin WeMo home automation system. As recommended by best practices, the firmware used by the embedded devices was signed, and the home automation system properly checked this signature. Unfortunately, the private signing key was distributed within the firmware! Furthermore, the download of the signed firmware used SSL. Unfortunately, the home automation system did not check the certificate of the server that distributed the signed firmware [300]. Creating a malicious rogue firmware and installing it on a remote machine was easy. With the private key distributed within the genuine firmware, the attacker can sign any arbitrary piece of code. She can impersonate the download server as the home automation system does not validate its certificate. Any properly formatted SSL certificate would work.

In January 2014, Daniel Wood reported one vulnerability in the Starbuck iOS application. The application stored the credentials used to access the Starbuck's account in the clear on the user's mobile device. If the attacker gained physical access to the victim's mobile device, she could access his credentials by reading a file (`session.clslog`). Indeed, this file contained the HTML code with the victim's OAuth credentials, i.e., the token and the signature, to access his account. The developer did not follow best practices, such as never to store credentials in files, or at least to encrypt them.

---

[3]There is no reason why the signing private key should be present in the signed application.

Unfortunately, often the weakest link is the person who operates a system. It is not always voluntary. This person may be induced to perform unauthorized actions. In a live performance at Defcon 2012 [301], Shane MacDougal demonstrated this. He called by phone a Walmart store manager of a small military town in Canada pretending to be a Walmart representative: Gary Darnel. He told the manager that Walmart had a multi-million-dollar opportunity to win a major government contract. He announced that his store was picked up as a likely pilot spot. For ten minutes, Darnell described his role and the outlines of the wondrous contract. He explained that he needed to understand the operating mode of the shop. Hence, he learned many details about logistics of the store. For instance, Darnell convinced the manager to disclose the type of computers used and the version numbers of the OS, browser, and antivirus. Finally, Darnell directed the manager to a malicious external Web site to fill out a survey to prepare for the upcoming visit. The manager obeyed. Fortunately, Walmart's IT security department blocked the access. Darnell said that he would have the IT department unlock it. The manager answered that he would try again in a few hours. Through a simple phone conversation, MacDougal collected confidential information and even succeeded in having the manager connect to a potentially dangerous Web site. This type of social engineering attack is extremely common. Phone-based social engineering attacks are even coined vishing. They are the first steps of many penetration attacks.

7.2 Analysis

7.2.1 Bring Your Own Cloud

For many reasons (convenience, economic, flexibility), BYOD has made its way into the corporate landscape. Authorizing the personal devices of users to connect to the corporate network increases risks as the owner of the device is controlling the device rather than the corporate IT team. Not only does BYOD increase the attack surface, but it also puts a part of the burden of security on the user who is not necessarily suitable for performing this task. Users bring into the corporate landscape a new threat.

Usually, the primary goal of users is to perform their task smoothly and efficiently. Users will reject or try to bypass everything that will impede this goal. Unfortunately, security is often perceived as an obstacle. Thus, users will circumvent it when security is an obstruction to the fulfillment of their task. In 2013, the cloud security alliance released "The Notorious Nine" threats for clouds [302]. Unfortunately, the most critical threat was missing: "Bring Your Own Cloud (BYOC)." BYOC occurs when an employee uses a cloud-based service without the blessing of his company for business purposes. The employee puts the company at risk. The employee might wittingly bypass all the security policies of the enterprise. He may as well circumvent the fences the company may have created to protect its intellectual properties or infrastructure (such as firewalls, antiviruses, and DLP tools).

BYOC is easy to perform and, unfortunately, is convenient for employees. An employee just needs to enroll in a free Software as a Service (SaaS) offer to use it immediately. Enrolling in a SaaS is sometimes faster than getting the company's IT team to launch a similar service. This enrollment in a cloud-based service often occurs without the company being aware. The employee will most probably not check whether the selected system is secure or approved by the company. The default settings of these services are not necessarily the ones that the company would use. Of course, the employee will likely not have read the Service License Agreement (SLA) when enrolling.

Employees can use the company credit card to open an account with an Infrastructure as a Service (IaaS) or a Platform as a Service (PaaS) provider. This is clearly faster than asking the IT team to purchase and install a bunch of servers in the DMZ. The security settings and the management of the security of these platforms are under the responsibility of the licensee and not of the IaaS and PaaS providers. Most probably, the BYOC user will not have the skill set to secure and maintain these platforms. The result is that some platforms operating on behalf of the company will not operate in compliance with the security policies of the enterprise and that too without the company being aware.

The fast and free or cheap enrollment in cloud services makes it extremely attractive for employees to use it as a BYOC. Moreover, they do not make this wrong choice maliciously. They have a strong rationale for their action.

BYOC can quickly become a nightmare for the company when things go wrong [303]. BYOC raises many issues.

- Security: As already mentioned, the default security settings of cloud-based services are not necessarily the ones that the company would use. Furthermore, the employee would not set them up correctly, and neither would he patch the systems in the case of IaaS or PaaS.
- Control of the account: When opening the BYOC service, the employee enrolls with his identity rather than the company's one. The employee may even use a personal email rather than the company's email. If ever the employee leaves the company, the company will have difficulty in retrieving the account if the employee did not transfer the ownership of this account to the company.
- Data privacy: Some countries, as in Europe, have regulatory restrictions on where the private data can be stored and how they must be protected. The employee will most probably not even be aware of these laws. Furthermore, the compliant privacy-aware implementation is complex and often needs the help of lawyers. For instance, not all cloud providers are compliant with the Safe Harbor doctrine.
- Contractual issues: When selecting a cloud provider, the SLA is important. An SLA defines the conditions of operation, such as guaranteed uptime, disaster plan, recovery plan, and liability issues. Only the legal department of the company will be in a position to negotiate more acceptable terms for the SLA. Without a proper SLA, business continuity cannot be guaranteed.

What can we do? Cloud is inevitable. Thus, we must anticipate the migration. A few possible actions are as follows:

- Provide a company-blessed solution in the cloud for the type of services that will be needed. This blessed solution can be fine-tuned to have the expected security requirements. The master account will be in the name of the company, thus manageable. Premium services often offer better security services, such as authentication using Active Directory, logging, and metering, than free services.
- Update the security policy to make it mandatory to use only the company-blessed solutions and a violation to use BYOC.
- Educate the employees so that they are aware of the risks of BYOC.
- Listen to their needs and offer an attractive list of company-blessed services.
- Make it convenient to enroll in the company-blessed services.

7.2.2 Authentication

Passwords are the most common form of authentication. Unfortunately, they are also the weakest. In 2010, a study of 2500 users highlighted some interesting practices about user passwords [304]. About 40 % of the respondents used the same password for logging into various Web sites. Forty-one percent of respondents had shared passwords with at least one person in the past year. Forty percent of the respondents never used special characters (e.g., ! ? & #) in their passwords. About 20 % respondents had used a significant date, such as a birth date, or publicly known information such as their pet's name, as a password. Sharing passwords with other people is, of course, dangerous practice. This issue is especially true if the user uses this shared password for other accounts. Special characters are useful to make the password more complex and thus harder to brute-force. In 2011, following the hack of two Web sites (Gawker and rootkit.com), Joseph Bonneau analyzed the disclosed passwords and their users to identify users who used the same password for both sites. He evaluated the ratio of password reuse to 30 %.[4] This value is rather worrisome as one of the two Web sites (rootkit.com) attracts a security-aware audience, from which we would expect good security practices. In 2014, Bonneau and his four colleagues analyzed a larger sample of passwords. They found that 43 % of accounts shared the same password, and 19 % of the others were substrings of more complex passwords [305]. These figures underscore that good practices are not natural for users when handling their passwords. Using the same password for multiple sites may become an issue whenever an attacker compromises one of these Web sites. An attacker may attempt to log into other Web sites using the credentials of the compromised Web site. Furthermore, lists of

---

[4]Indeed, he found a larger ratio of 43 % but interpolated it to a lower value. The 2014 study confirmed his initial finding.

hacked passwords are freely available on the Internet [306, 307], or even sold on the Deep Web. Thus, attackers may try these passwords on other sites. Furthermore, with a strategy that takes into account the password policy enforced by the Web site, it is possible to increase the guessing rate by 10–30 %. The attacker may use the already known passwords and apply transformations such as leet substitutions, deletions, insertions, or capitalizations [305]. For instance, the password `trust-noone` can be easily turned into `TrustNoOne`, `TrustN0One`, `TrustNo1`, or `Trust_No_1`.

The Devil Is in the Detail Usually, an OS stores each password as a hash. In theory, having only the hash of the password, it is not possible to retrieve the password (because the hash is the result of a one-way function). The only solution is a brute force attack, i.e., exploring all the possible passwords and checking whether their hash matches the targeted hash value. If the attacker has the hash value (and the hashing algorithm), she is not limited or throttled by the limit on failed password presentation.

The attacker's objective is to find one password P_0 such that $H_O = H(P_O)$, where H is the hash function and H_O is the targeted hash value. The password P_0 is not necessarily the password of the user but it will present the same hash value. Normally, the attacker should explore all possible values H_i until one fits the equation. If the attacker has to retrieve several passwords "protected" by the same method, for instance, in the case of an OS, one potential method to accelerate the process is to create precalculated tables with all the hashes. Unfortunately, this method requires too large databases of precalculated tables. Tools like `hashcat` use dictionary attacks with a large collection of passwords and passphrases. Nevertheless, if the password does not belong to the dictionary, then brute force is the ultimate solution. This means that if the attacker explores all the possible combinations with the expectation that user may employ all random values, assuming the use of SHA-1 as the hash function, i.e., 20 bytes per hash, the size of the database would be as follows:

- For six-character passwords using only alphabetic Roman characters, it means 52^6, i.e., about 20 billion, combinations, which mean a table larger than 500 GB.
- For eight-character passwords using only alphabetic Roman characters, it means 52^8, i.e., about 55 trillion, combinations, which mean a table larger than 1500 TB.
- For eight-character passwords using only alphanumeric characters, it means 62^8, i.e., about 220 trillion, combinations, which correspond to a table larger than 6000 PB.
- For eight-character passwords using alphanumeric characters and these twenty special characters & ' {([- \ @]) = + $ % ! :; <>, it means 82^8, i.e., about 2000 trillion, combinations, which correspond to a table larger than 60,000 PB.

In reality, these dictionaries would be smaller, as they would most probably focus on "reasonable," i.e., not fully random passwords. That space is smaller. Nevertheless, the size remains huge and the calculation time needed to compute the dictionary is extremely large.

In 2003, Philippe Oeschlin designed a method called rainbow tables that drastically accelerated the exploration of hashed passwords [308] by finding a trade-off between storage space and speed. The process uses two steps. The first step constructs the tables. It uses a hash function H and a set of reduction functions R_i. The reduction function R turns a hash value into a password. The hash value is usually longer than the password.[5] We create a chain of n hash values starting from an arbitrary password P_0 with the following process:

- Calculate its hash value $H_0 = H(P_0)$.
- Reduce with reduction function R_0 this hash value to a new password $P_1 = R_0(H_0) = R_0(H(P_0))$.
- Calculate its hash value $H_1 = H(P_1) = H(R_0(H(P_0)))$.
- Continue to the next hash value with reduction function R_1, which gives
- $H_2 = H(P_2) = H(R_1(H(R_0(H(P_0)))))$,
- until reaching the last password P_n of the chain.

With this data, we can build the rainbow table. The first item of the rainbow table is the first password P_0 and the last password P_n. We select a new password to generate the second chain. The second item of the rainbow table is this password and the last hash value of the generated second chain. Figure 7.1 illustrates this creation process with a chain of three elements.

Rainbow tables use different reduction functions at each stage rather than one unique reduction function for all stages. Indeed, as reduction functions generate data far smaller than the input data, there is a serious risk of collision, i.e., two or more hash values produce the same password. A collision means that two chains merge at one point, reducing the efficiency. Using different reduction functions drastically lowers the probability of collision, as collisions have to appear in precisely the same position in the chains.

Once the rainbow table created, it is used to find the password corresponding to the hash value H_{crack}.

- Calculate the first reduced value P_{crack_1} using the last reduction function R_n such that $P_{crack_1} = R_n(H_{crack})$.
- If this value is not present in the last column of the rainbow table, then calculate the next candidate P_{crack_2} such that $P_{crack_2} = R_{n-1}(H(R_n(H_{crack})))$.
- Check whether P_{crack_2} belongs to the last column of the rainbow table.
- Iterate until either we find a candidate or we have completed n iterations.

[5] It is rare to use a 160-byte password.

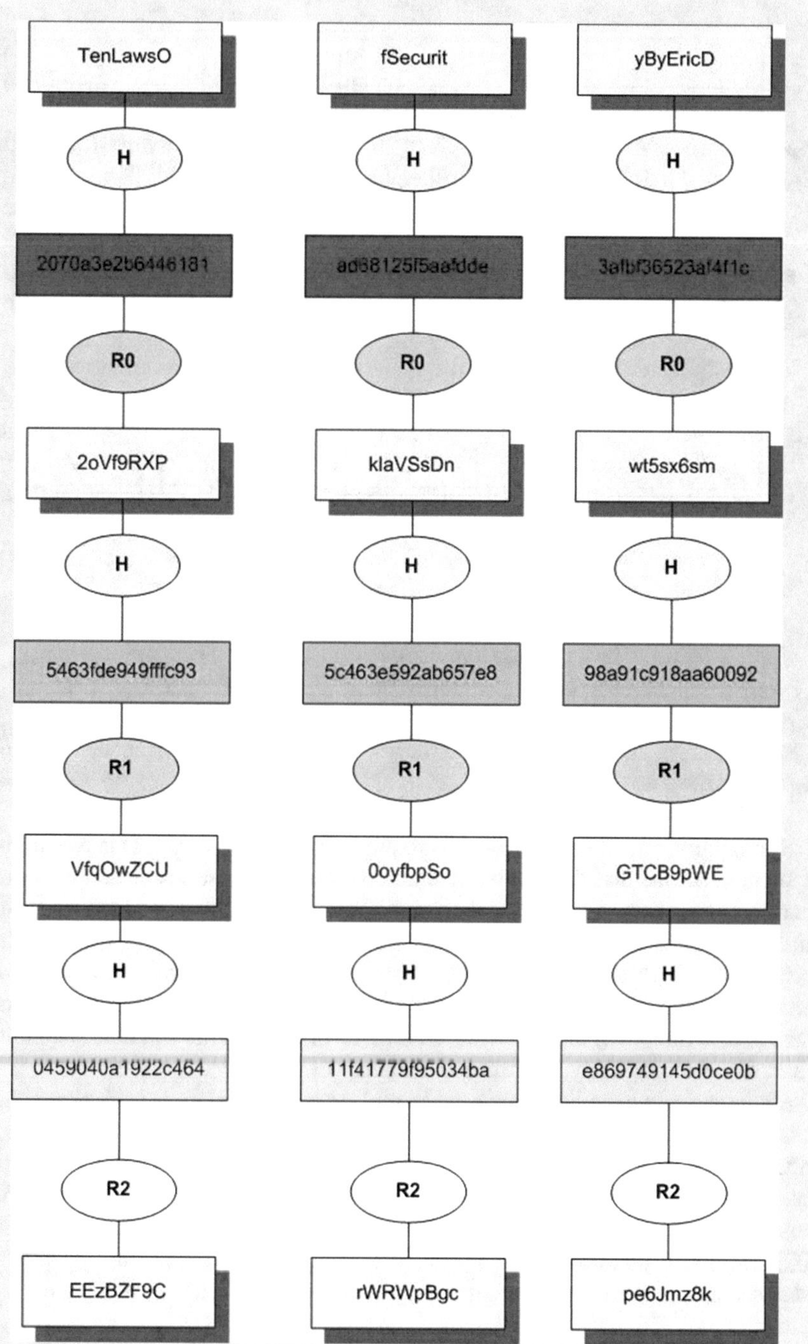

Fig. 7.1 Creating a rainbow table

If we found a candidate, then we know which chain of the rainbow table generated the corresponding end point and its position within the chain. Then, starting from the beginning of the corresponding chain, we calculate the actual value by regenerating the chain as when creating the rainbow table. If we did not find a candidate, then the attack failed.

The Swiss company Objectif Sécurité, founded by Philippe Oeschlin in 2004, [6] sells such rainbow tables for hash functions, used by some operating systems. For instance, the rainbow table that explores eight-character passwords (alphanumeric and special characters) has a size of 2 TB (compared to 60,000 PB).

Lesson: The first lesson is that the answer to attacks using rainbow tables is preventing the attackers from building such rainbow tables, i.e., by using "nonstandard" hash functions. This is why passwords are stored with salted hashes Hash (password|salt), where salt is a random number and Hash is a standard hash function such as SHA-3 or MD5. If the implementation uses many salts, then the attacker cannot build enough rainbow tables.

The second lesson is that often security is not always only the responsibility of one single person, but it may also be the responsibility of a chain of individuals. For instance, the security of the password relies on the end user's using a strong password and protecting it adequately. Security of the password depends also on the developer of the system that uses the password through his implementation of its proper management, for instance, using salted hashes and clearing the memory buffer used to temporarily store the password.

Strong authentication via password requires strong passwords. This requirement of strength means that the password should be long, and use lower and upper case characters, numbers, and special characters. These constraints put a heavy cognitive burden on the user's long-term memory. Therefore, users sometimes use some trick such as "`TrustNo_1`," "`Hell0`," or other trivial character substitutions. Hackers' dictionaries contain the most trivial substitutions. A solution for some users is to write down the password on a piece of paper. This solution is definitely weak, but, unfortunately, too often encountered [309].

Many variants have been explored to replace weak passwords such as graphical passwords or pattern-based passwords. In version 2.0, Android introduced a pattern-based lock screen. To unlock his screen, the user has to draw a pattern on a nine-digit grid. The user may select the complexity of the pattern. Do users have more secure practices with such pattern-based passwords? In 2015, Marte Løge disclosed that it was not the case [310]. Her study shows that at least 10 % of the users select a pattern representing a letter from among C, M, L, N, S, and O. Of

[6] The company's initial name was Ophcrack after the name of the first password cracker.

Table 7.1 Strength of pattern-based lock

Length of pattern	Number of possible patterns
4	1624
5	7152
6	26,016
7	72,912
8	140,704
9	140,704

course, these patterns are not more robust than the typical "123456" or QWERTY. Furthermore, a nine-digit grid offers a limited number of possible patterns. Table 7.1 displays the number of possibilities with the length of the pattern. For instance, a five-figure pattern is weaker than a four-digit PIN.

Furthermore, most people start their pattern from one of the four corners. About 50 % of the users even begin their pattern from the top left corner. About 75 % of the users start from either the top corners or the bottom left corner. This predictable behavior reduces the strength of this type of lock screen. When associated with smudge analysis (Sect. 6.2.2), the pattern-based lock screen becomes weak.

7.2.3 Social Engineering

For decades, Nigerian scams, also called 419 scams, have been around. These scams have continually evolved to adapt to the newest communication tools. Nigerian scams started with postal mail, switched to email once it became common, and now employ social networks. Nigerian scam is a generic term for a category of scams that always follow the same scheme. The target receives a message that always tells the following story. A widow/lawyer/son/exiled person has an enormous sum of money blocked in a foreign banking account. The emitter of the message needs the help of a trusted person to exfiltrate this money. The recipient of the message is this trusted person. Of course, the trusted person will be generously rewarded for his help. If the recipient offers to help, soon the scammer asks him for funds to complete the needed official paperwork or to bribe the proper official. Of course, in the end, no money is transferred to the gullible victim. The Nigerian scam is a perfect example of social engineering. It exploits human weaknesses such as greed and gullibility.

As the Nigerian scam is an old and well-known scam, a valid question is why do attackers still use such a known trivial ploy? The seemingly obvious answer, that the attackers may be stupid, is not correct. Herley Cormac, from Microsoft Research, provides a more convincing explanation [311]. Scammers try to reduce false positives, i.e., potential victims who start to respond but do not fall into the trap at the end. This type of scam needs a lengthy interaction with the target. Before asking for money, scammers exchange many custom, personalized messages with the "collaborating" potential victim to build trust. Although the communication channel is cheap when using emails, this interaction is costly in terms of time and effort. When starting the interaction, the scammer would rather avoid a false positive. For an optimal ROI, the attacker should only begin with viable targets, i.e., targets that will pursue the interaction through the successful skimming. Intuitively, we may guess that the more gullible the target is, the higher the chance of success is. Therefore, using such a worn-down ploy filters the initial respondents. It skims out only the most gullible persons. Thus, using absurd, non-credible scenarios lowers the rate of false positives.

Cormac analyzes the Receiver Operator Characteristic curves that are usually used to define the trade-off between true and false positives of classifiers. He checks for the optimal operating point. He analyzes the impact of density (i.e., the ratio of viable targets) and the quality of the classifier. Then, he applies the outcomes to the Nigerian scams. He demonstrates that the "dumbness" of the mail is a good classifier whenever the attackers try to operate with a better overall profit.

Nigerian scams exploit greed and gullibility. Scams use many other human weaknesses. In April 2014, scammers were able to trick Facebook Indian users to hack themselves by clever social engineering [312]. A video posted on the Internet outlined a method to ostensibly hack the Facebook accounts of friends. The method promised to disclose their friends' password. The alleged method required pasting a command into the browser. The command was available on Google Drive. Unfortunately, when pasting this code, gullible users launched a self-cross-site-scripting attack. It stealthily executed a set of actions in the name of the oblivious user, such as following new users or lists, and liking some pages or comments, all defined by the scammer. The scam seems to have been rather successful. Later, scammers extended this attack worldwide by using other infection vectors such as phishing emails or Facebook notes [313]. Social engineering often exploits the worst aspects of human beings. In this example, it was curiosity and a bit of mischief.

A good social engineer also exploits the current breaking news and the curiosity of users. With the advent of social networks and its vector of dissemination of the latest news, the breaking news has become a standard avenue for attackers. For a few years, every major disaster has seen spams mushrooming and fake sites pretending to collect charities for the victims of the catastrophe. In 2014, this even started to become a vector for the APT. On March 8, 2014, Malaysian authorities announced that they had no news of the flight MH370 to Beijing. It took several weeks before the confirmation that the Boeing 777 had crashed into the ocean. Meanwhile, the eagerness of learning fresh information on this topic was used for political spying. One day after the initial announcement, a specially crafted malware, carried by an attached, forged `pdf` file, infected 30 computers at the Malaysian Department of Civil Aviation. Once installed, the malware sent back confidential information about the catastrophe to a China-based server [314]. Two days after the disappearance of the plane, members of one undisclosed government of the Asian–Pacific region received a spear-phished email with an attachment titled "`Malaysian Airlines MH370.doc`." Of course, this document was empty but contained a Poison Ivy malware [315]. On March 14, 2014, the same attacking group sent a different spear-phished email to a US think tank with an attachment titled "`Malaysian Airlines MH370 5 m Video.exe`." Once more, the attachment was a malware. Many other malwares exploited the same catastrophe without participating in an APT, but, rather attempting generic random attacks [316]. Some phishing Web sites, mimicking the Facebook look, were used to collect data from infected users. The sites supposedly presented a video of the discovery of the missing plane.[7] Before users could view the video, the Web site proposed that the users share the video with their friends, thus facilitating the dissemination of the malware. Then the Web site asked the users to answer some personal questions, such as their age. In other words, the phishing Web sites scammed the curious tricked users. Current events and news updates have become the go-to social engineering bait of attackers. Therefore, some Security Operation Centers display in real-time breaking news feeds to be aware of the potential inspiration for new social engineering-based malwares.

All these attacks have in common that they are social engineering attacks. Social engineering is the art of using social interaction to influence people and convince them to perform actions that they are not expected to do, or to share information that they are not expected to disclose [317]. Social engineers manipulate their human targets using several human social norms for creating artificial contexts favorable to manipulation. Among these social norms are as follows:

- *Similarity*: Under the similarity norm, people are expected to align their behavior with that of other people and favor people who are similar to them. If Alice has the same beliefs as Bob, and if Eve has a different belief than Alice, then Alice may prefer Bob to Eve. A social engineer may use the similarity norm by impersonating a person who has the same problems as her target, or the same opinions. The sympathetic target will be more inclined to help the social engineer.
- *Reciprocity*: Under the reciprocity norm, people are expected to answer to an interaction in a similar way. If Alice gives a gift to Bob, she expects Bob to return the favor.[8] Similarly, if Alice is mean to Bob, most probably Bob will also be mean to her. A social engineer may use the reciprocity norm by initiating a positive mood, or offering something with the expectation of getting a favor in return. Humility is also a powerful tactic. Feigned ignorance is a form of reward that the target victim will appreciate. The target becomes the correct and knowledgeable one in the exchange. This feeling is highly uplifting, and the victim will be inclined to demonstrate his extensive knowledge.
- *Proximity*: Under the proximity norm, people are expected to avoid more a conflict with an antagonist who is physically close than with one who is remote. A social engineer may use the proximity norm when on-site. It is harder for Bob

[7]Only 16 months after the crash did the first debris of the aircraft land on a shore of the French island Réunion.

[8]This is one of the levers used in merchandising when offering free samples and free additional quantities.

to refuse something to Alice while she is in front of him than when she is on the phone.

- *Authority*: Under the authority norm, people are expected to obey to any form of identified authority. If Alice is higher in an organization's hierarchy than Bob, Bob will tend to obey Alice without questioning the validity of her commands. The Milgram experiments are an iconic example of this social norm.[9] A social engineer may use the authority norm when impersonating a top manager or a controller.
- *Contrast effect*: When offered the choice between one option and a less constraining, but dangerous alternative, people will naturally tend to select the second one. For instance, Alice wants to collect Bob's email address and his manuscript signature on the street. First, Alice asks Bob for a donation of $30. When Bob refuses, she suggests a smaller donation of $15. When Bob refuses again, she proposes him just to sign her petition. Given the context, the last option seems reasonable to Bob, who signs and gives his email.

Chapter 7 of the Open Source Software Testing Methodology Manual (OSSTMM) provides a list of tests to perform to assess the ability of an organization to defend against social engineering [112]. For instance, it recommends checking how easy it is to get illegal access to premises when:

- impersonating as a member of an "internal" support or "internal" delivery department without any credentials (similarity norm),
- impersonating as a member of the management team or other key person (authority norm),
- impersonating as a member of support or delivery service from outside the organization (similarity norm, proximity norm),
- inducing terror, or panic (authority norm, contrast norm),
- providing misinformation regarding security policy in an authoritative manner to circumvent or violate security policies (authority norm).

OSSTMM provides an impressive list of tests. Indeed, it represents a catalog of techniques that a social engineer may use.

The only defense against social engineering is increasing the awareness of the organization against this type of attack. It requires the potential target to do critical thinking. The target must detect the social engineer when attempting to create favorable contexts. Role-play-type training is the best tool to raise this kind of awareness. Scripted answers for people that may be more natural targets, such as customer service or reception desk employees, are an efficient countermeasure.

---

[9]In the 1970s, Maurice Milgram conducted a set of impressive experiments. Volunteer subjects participated in a fake research experiment. The objective was to study whether punishing a learner with electric shocks would increase his learning capabilities. Sixty-five percent of the subjects agreed to apply the allegedly lethal electric shocks to the learning individual because the researchers told them to do so. Of course, the learner was an accomplice of the experimenters [318].

7.2.4 Biometrics

Personal authentication is an important but difficult task. In these days of interconnectivity, being able to prove that an individual is the person she claims to be is of utmost importance. Login and passwords are the most deployed solutions. Unfortunately, this solution is notoriously insecure. Authentication can rely on four types of information [319].

- Something that you know: For instance, a password, a passphrase, a Personal Identity Number for credit cards, a secret key (in the case of a computer or software), or even the name of your pet (a rather weak authentication but used often for password recovery).
- Something that you physically own: For instance, the physical key that will match a given lock, an entry badge, a token such as a SecurID, or a mobile phone.
- Something that you are: This is a characteristic unique to your person, such as your voice or your fingerprint. This type of authentication has been in use for centuries, with handwritten signature. Unfortunately, a handwritten signature is a rather weak form of authentication as it is easy to forge and because it is difficult and cumbersome to validate the authenticity of a manuscript signature.
- Something that is part of your skills: For instance, the knowledge of a foreign language, or the capability to identify a given set of images. This field of authorization is rather an exploratory research. Captchas are an example of such a system using character recognition (Sect. 6.1). The aim of a captcha is not to authenticate a given person but to authenticate the "authenticated" principal as a real human being and not an automated software.

Depending on the number of types of information required, an authentication scheme is either mono-factor or multifactor. Login/password-based authentication is a one-factor scheme. Login with a one-time password sent via an SMS is two-factor authentication. The user has to know the password and thus he needs to have access to the mobile phone that will receive the SMS with the one-time password.

Passwords are widely deployed because they are easy to implement and rather user-friendly. A robust password needs to both be long and have random characters. Without random characters, specialized cracking tools, such as OPHCRACK or John the Ripper, can easily discover the password [320]. Unfortunately, remembering complex random passwords is difficult for most humans. Thus, people use simple passwords easy to remember or use one complex password for all Web sites. Of course, the first solution is inadequate as it is prone to simple attacks. The second solution is also risky if ever this robust password leaks out. Indeed, even strong passwords may leak in the case of a Web site hack [321]. Once the strong password in the wild, it may be used to access other Web sites illegally.

Using an intrinsic characteristic of a person seems to be more effective than using a password and, of course, it removes the burden from the human brain. You always carry your personal characteristics. The purpose of biometrics is to capture these characteristics. Any biometric-based authentication uses five steps.

1. During the first phase, the person presents his biometric characteristic. Many types of human characteristics can be used: face, voice, fingerprint, palm structure, retina, vernacular network, the way a user types on the keyboard [322], the writing style of a user (so-called stylometrics) [323], the way a user walks through gait recognition [324], and even butt authentication [325].
2. The corresponding biometric reader captures these characteristics. Depending on the biometric parameters captured, the biometrics reader can be more or less intrusive. Readers based on cameras (for face recognition, gait recognition), or microphones (voice recognition) are quasi-transparent to the user. Fingerprint, vein pattern, or palm readers are not intrusive. Many modern devices have embedded fingerprint readers. The most intrusive readers are the iris/retina-based ones. These readers require the person to place his eye in front of the device precisely. Many people do not readily accept iris/retina recognition [326]. They perceive it as being too intrusive.
3. The reader translates the captured biometric characteristic into a digital file. This file may be encrypted and signed to avoid forgery of this data before being transferred to the decision engine.
4. The decision engine compares the captured biometric characteristics to those in the reference database. During an enrollment phase, the reference database will have been populated with a set of initial characteristics of the person. This initial collection of samples is done with the individuals to be identified. The comparison is done either with one reference sample if the purpose is to identify one unique person or with a set of reference samples if the purpose is to authenticate a collection of individuals. The comparison is not a strict comparison of data but rather a measurement of the similarity of characteristics. The result of the comparison is a score of similarity.
5. Depending on the score of similarity, the system decides whether the user is authenticated. A threshold defines when the recognition is accepted. The value of this threshold determines the accuracy, the false positive, and true positive.

Biometrics is not the silver bullet of authentication. Biometrics has one major advantage compared to all other authentication systems: user friendliness. Compared to passwords, it does not put any burden on the user's long-term memory. Indeed, it requires no memory (if used alone). Unlike with physical tokens, you cannot forget to take biometrics with you; indeed, these credentials are permanently with you. In most cases, the user interface of biometrics is convenient. Indeed, it is easier and faster to present your finger to a reader or to just repeat word or sentence rather than to type a password. Where is the weakness? Biometric parameters cannot be revoked! When hacked, you cannot change your fingerprints, your voice, or your eyes. Replacing biometric parameters by new biometric parameters is impossible. You can try to modify them, but simple transformations should not fool an efficient biometric system.

Nevertheless, most biometrics systems can be fooled. Voice can be recorded and played back. Faces can be shot by a camera and displayed using a high-resolution screen. Even fingerprints can be reproduced. Sometimes, the manufacturers make the impersonation easy [327]. In some cases, the reference database is not encrypted and accessible to the world. In other cases, the interface to the sensor can be eavesdropped on, and the attacker can capture the biometric characteristics.

The Devil Is in the Detail In 2013, with its new iPhone 5s, Apple introduced a new feature: Touch ID. Touch ID is an integrated fingerprint recognition system. Once his fingerprint is enrolled in the phone, the user will be able to unlock the phone by pressing his finger on the home button. The home button encapsulates an 88 by 88 pixel 500 dpi capacitive sensor [328]. The sensor measures the difference in capacity between the edges and ridges of the fingerprint, the so-called subdermal ridge flow angle. This system advantageously replaces the mandatory PIN presentation that was introduced with iOS 7. Furthermore, biometric authentication may also serve as credentials for purchases on Apple Web stores or with Apple Pay. Unfortunately, in September 2013, the CCC announced it had cracked Touch ID [329].

For this version of its fingerprint detection, Apple used an already known and deployed technology. The only enhancement was the use of a fingerprint sensor with a higher resolution than that of the previous generation.[10] As the previous generation of Apple's fingerprint was already hacked once, the hacker had only to improve the performance of the known attack.

The first step of the attack is to acquire the victim's original fingerprint [330]. The potential sources are glasses, glossy paper, or even the victim's phone screen. Everybody leaves fingerprints as skin oil and sweat residue. To expose these fingerprints, the attacker will use techniques used by forensic agents. This usually implies using colored powders or cyanoacrylate. Once a fingerprint has been exposed and is visible, a flatbed scanner at 2400 dpi captures it. Then, the captured image is converted to black and white. This transformation reveals the ridges of the fingerprint. It may be useful to do some graphical cleansing and enhancement using tools such as Photoshop. The image is then inverted and mirrored. The result is the master fingerprint.

The second step is to create a fake rubber fingerprint. For that, the attacker uses a typical photographic process with the master fingerprint acting as the negative image. Thus, the master fingerprint is printed onto a transparent sheet at 1200 dpi. This printed mask is exposed on the photosensitive PCB material. The PCB material is developed, etched, and cleaned to create a mold. A thin coat of graphite spray is applied to improve the capacitive

[10]Indeed, there were some improvements that lowered the false positive ratio and made better use of the secure processor within the A7 System on Chip. Nevertheless, these enhancements did nothing against hacking by impersonation.

response. Finally, a thin film of white wood glue is smeared into the mold to make it opaque and create the fake finger.

Where is the mistake of Apple's designers? Their trust assumption was wrong. They assumed that the system checked the actual owner of the fingerprint. Indeed, this type of sensor checks an image of the fingerprint rather than the actual fingerprint. An image of the fingerprint can be stolen, as illustrated above. Then, the attacker can spoof the fake identity. The real fingerprint is a particular template of ridges attached to a *living* being. The missing element is the liveness that is absent. Thus, the verification cannot control whether or not the image is attached to its actual owner. This weakness is not unique to fingerprints. It is valid for any biometric. More sophisticated sensors check, in addition to the configuration of the ridges, for some form of liveness behind the captured image. This verification can be done through several measurements such as that of temperature or by checking whether there is some actual blood pulsing [331]. Unfortunately, such sensors are far more expensive and complicated than basic image sensors.

In September 2014, Apple introduced the new iPhone 6 with the new payment scheme, Apple Pay. Unfortunately, the new model was still vulnerable to fake fingerprints. In 2014, Samsung introduced its new Samsung Galaxy S5 smartphone with a similar fingerprint authentication mechanism. Three days after the commercial launch, researchers from SRLabs duplicated the same exploit for the Samsung phone as for the iPhone 5s [332]. It is worrying that companies do not learn from their mistakes or the mistakes of their competitors.

Lesson: Verify that your trust assumptions always remain valid when making decisions driven by financial considerations.

Learn from the mistakes of others. If a competitor made a mistake, do not reproduce it.

Of course, there will be a race between hackers and the designers of biometrics systems. The former will design new sophisticated lures that better mimic human characteristics whereas the latter will increase their ability to detect faked biometrics. Advanced research attempts to find some biometrics characteristics that cannot be easily spoofed. Vein pattern recognition seems to be such a characteristic. Since deoxyhemoglobin in the blood absorbs near-infrared lights, vein patterns appear as a series of unique dark lines. This pattern is complex and hard to mimic. The biometric reader is a near-infrared camera. Unfortunately, this type of secure biometrics is also expensive.

Nevertheless, according to Law 1, attackers will always find their way. For instance, in September 2015, the US Office of Personnel Management acknowledged that the fingerprints of 5.6 million US federal agents had been stolen following a massive breach of its system [333]. If biometric characteristics are accurately forged or genuinely stolen to impersonate one person, there is no way to revoke them, or to replace them with newer ones. The only solution is to ban the spoofed user from the system that may be lured by corresponding forged biometrics characteristics. This may be acceptable for a solution protecting one device, such as a smartphone. It may not be acceptable if the system controls access to business offices, or if it equips banks' ATMs. Once more, the human is the weakest link, as human biometric characteristics cannot be revoked.

Biometrics is promising and alluring. Biometrics is definitively a user-friendly method of authentication. Nevertheless, biometrics has limitations that the implementer has to be aware of. The designer has to reduce the corresponding risks and must not blindly trust biometrics. Multifactor authentication may mitigate this risk.

7.2.5 Do Users Care About Security Warnings?

This is an important question. The commonly accepted assumption is that most people are oblivious to security issues. They will not care. Several studies confirm this belief. Devdatta Akhawe (Berkeley) and Adrienne Felt (Google) launched an empirical research by observing more than 25 million real interactions following security warnings from Chrome and Firefox browsers [334]. This study occurred from May 2013 to June 2013. They collected information using the in-browser telemetry system. Telemetry measures and collects non-personal information, such as memory consumption, responsiveness timing, and feature usage. Users have to opt into activating this function.

The researchers studied phishing warnings, malware warnings, and SSL warnings. They measured the click-through ratio, i.e., the number of times that users bypassed the warning, i.e., when users did not take into account the browser's recommendations.

Table 7.2 provides the ratio of bypassed warnings by browsers. The good news is that most users take into account the security warnings in case of malware or phishing. As the detection mechanism uses Google's Safe Browsing List, the ideal ratio should be near 0 % as the ratio of false positive in the list is extremely small. For SSL warnings, the ratio is significantly higher. Of course, many legitimate sites generate such warnings (due to misconfiguration of the server or self-signed certificates). Thus, the ideal ratio cannot be null. Nevertheless, this ratio seems too high, especially given the latest issues with CAs.

Table 7.2 Click-through ratio on browsers' warnings

	Firefox (%)	Chrome (%)
Malware	7.2	23.2
Phishing	9.1	28.1
SSL	32.2	73.3

Interestingly, Chrome has a higher click-through ratio than Firefox. In other words, Chrome users care less of the warnings than Firefox users. In the case of SSL, several reasons can explain the huge difference (+40 %). The main reason is that Chrome users receive more warning than users of other browsers. For instance, by default, Firefox memorizes an accepted SSL warning whereas Chrome will repeatedly present the same warning, regardless of the first decision of the user.

The paper presents some interesting findings. Consistently, Linux users did have a higher click-through ratio than other operating systems' users. Two things may explain this difference. Usually, Linux users have good knowledge of information technologies. Linux is not a mainstream OS. Most probably, these users feel more confident in their skill set than average users. As they better master the system, they may have lower risk aversion than average users. Usually, it is admitted that Linux is less prone to malware than Windows, for instance. Thus, Linux users may feel that they are more secure. Unfortunately, this assumption is wrong for the threats addressed by this paper. No OS is safe from phishing attacks or forged SSL certificates.

Furthermore, users who disregarded the warnings spend less time on the page (1.5 s) compared to users who took into account the warnings (3.5 s). This response time means that they made their decision extremely fast. Therefore, we may question whether this decision is the result of a rational analysis. This study highlights that we are a long way from helping average users make rational security decisions. In other words, where possible, the security designers should take critical decisions on behalf of users.

Rule 7.1: Where Possible, Make Decisions on Behalf of the End User

Most users are not enough security literate to make adequately informed decisions. Unfortunately, often the system requires end users to make decisions that may impact the risk factor. The security model of Android is an excellent illustration of this problem. Android's security model is well defined and designed. From its inception, it has provided an impressive level of granularity of 143 security controls. This model is a dream for any security engineer. He can fine-tune every element of the system. An application has to request the permission before getting access to each required functionality. When installing a new application, the end user has to approve the requested permissions, else the installation fails. For that purpose, the installation displays a screen with the requested permissions to the user. The user has to accept them or cancel the installation. Unfortunately, this model is not a dream for the end user. Users may understand what is requested by the permission `SEND_SMS`, but do they understand the consequences of allowing the permission `DUMP`? Unfortunately, many users will accept the requests without checking what they agreed to permit. Furthermore, nobody educates these end users on what the consequences are of authorizing access to the file system or the MMS. According to a recent study, only 17 % of users paid attention to permissions during the installation of an Android application. Fewer than 25 % of users demonstrated reasonably, albeit not perfect,[11] comprehension of the requested permissions [335]. In other words, the security system of Android is not safe for end users as it relies on end users taking decisions that they are not able to analyze rationally.

To properly manage permissions, there are mainly four possible strategies.

- The designer automatically grants permissions without involving the end user. This strategy is wise if the developer makes the right decision and if no application abuses the end user. Thus, it does not work if malicious applications can be installed. In any case, the end user should be able to reverse the decision.
- The designer integrates the decision process into the task that the end user fulfills, and that will require a new permission. This is what happens when the user decides which directories a friend may access on his cloud-based repository or social network. This same strategy mandates that the user presses a button before sending a message. Usually, the end user is not even aware that he is taking a security decision. The end user is not distracted from his primary goal, i.e., performing the task (allowing a friend to access his files, or sending a message to somebody). Furthermore, the user is making his decision in context, and thus it is more informed.
- The designer has a dialog box launched when a decision has to be taken. The end user is distracted from his primary goal. Therefore, these dialog boxes should be rare and restricted to decisions that would have severe adverse consequences.
- The designer gives the user the option to select all permissions at the installation time. The initial Android versions employed this strategy.[12]

For the two previous scenarios, the user should be helped with explanations that will highlight the potential risks he takes when making a decision. An ideal product would mix the four approaches [336].

Many studies have emphasized that users do not respond appropriately to security warnings for mainly four reasons.

- Users do not understand the warning. The message may not be explicit enough, providing inadequate information (for instance, giving just a warning reference number rather than an explanation). The warning message may not be adapted to the audience. The message should be crafted for the targeted audience by using proper vocabulary and appropriate level of technical details.
- Users do not understand the offered options. Without a clear understanding of the different choices at hand, users cannot make the right decision. The user should be informed of the impact each choice will have.
- Users do not comprehend the corresponding risk. They may not be aware of the risk or may underestimate it. The cure is better security awareness.

[11]Indeed, only 3 % of the participants of this study demonstrated a perfect comprehension.

[12]After the M development version, Android applications installed without asking for permissions. Permissions were requested the first time needed (i.e., the third strategy).

- Users' mental models do not match the real model (Sect. 7.3.1, Understand Your Users).

Explaining to the user the problem is a common difficulty. Usually, best user interface practices recommend that the warning messages for a broad audience be simple, non-technical, brief, and specific [337]. This is a hard challenge. Microsoft released some interesting rules for deciding when and what to display to users with regard to security warning [338]. For that purpose, Microsoft proposed two nice acronyms. A security warning should be Necessary, Explainable, Actionable and Tested (NEAT). In other words, the designer should only present a security warning to the user if the user is needed to make a decision and if the issue can be precisely described to him. Unfortunately, explaining a security warning to non-savvy users is a difficult task. Thus, Microsoft proposed another acronym. The explanation should clearly explain the Source of the issue, define the Process that the user can follow to address it, describe the Risk, be Unique to the user (with his or her context), offer some Choices and give Evidence (SPRUCE).

Another factor that may help the user in his decision is the possibility of reversing a security decision. Since Android Marshmallow (6.0), users have been able to revert to a security decision previously taken. A user may be less afraid to take a decision if he can change his mind later. This was not the case with earlier versions of the Android OS.

7.3 Takeaway

For most users, security is not their most important concern. In most cases, security is even not a concern for them. Sometimes, security is even perceived as being an obstacle to fulfilling the primary goals of the user. In that case, the user will do everything that is needed and possible to succeed in his primary goals, even if it means violating security rules.

There are several ways of avoiding this problem.

- Security has to be as transparent as possible to end users.
- The user should better understand the importance and need for security.
- Security objectives should, when possible, not compromise the primary goals of the user.

7.3.1 Understand Your Users

In any case, ideal security should be transparent to users. For security to become transparent as much as possible, it has to follow the best practices required for user interface design. In designing the user interface, the first step is to define the user's goals when he fulfills a given task [339]. Unfortunately, as already mentioned, most of the time, security is not part of the user's direct goals.

For most businesses, security is mandatory for preserving their bottom line. For instance, a security breach may corrupt sensitive data or block critical infrastructures. Such breach impacts the resilience of the business. Unfortunately, business people will reject blindly imposed security features that will either slow down, increase the cost, or reduce the flexibility of business. As business people do not grasp security concepts, any constraint is perceived as unnecessary and nefarious for good business. The business team and the security team do not a share common vocabulary and handle different concepts. This difference broadens the gap between their two worlds. The security team should better understand the business organization and its constraints. Hence, security personnel should attend business courses to grab the basic concepts handled by the business team. Once educated on business, security practitioners should spend time with their business peers. This way, they learn about the processes and methods that the business team uses. Then, they can demonstrate an understanding of business objectives and the drivers [340]. With this newly acquired knowledge, the security staff can design security procedures that disrupt as minimally as possible the current business processes. Ideally, security should totally, or at least partly, blend into the existing business process. Furthermore, the security practitioners can then explain the benefits that these security measures will bring to the business of the enterprise. Also, the explanations and procedures can employ the vocabulary of the business team, making them more understandable and acceptable. This approach is not limited to business units. It is suitable for any department of an enterprise.

People make a decision following a mental model that they have of how a system operates. Security is not different from other fields. Experts or technically well-informed people may have mental models that are reasonably accurate, i.e., the mental model fits reasonably with the real-world operation. For ordinary users, the problem is different, as their mental models are not accurate. Rick Wash identified several mental models used by ordinary users when handling security issues [341]. He extracted four mental models dealing with the issues of virus and malware.

- *Viruses are bad*: According to this model, viruses are nefarious programs that the user will never encounter. People using this mental model have little knowledge about viruses and, thus, believe they are not affected. They have thought themselves to be immune to viruses.
- *Viruses are buggy software*: According to this model, viruses are ordinary, poorly written software. Their bugs may crash the computer or produce strange behavior. People believed that to be infected they had to download and voluntarily install such viruses. Thus, their protection solution was to install exclusively trusted software.
- *Viruses cause mischief*: According to this model, viruses are pieces of software that are intentionally annoying. They disrupt the normal behavior of the computer. People do not understand the geneses of viruses. They believe that the

infection comes exclusively from clicking on applications or visiting dangerous Web sites. Their suggested defense is being careful.

- *Viruses support crime*: According to this model, the end goal of viruses is to identity theft or the sifting of personal and banking information. As such, people believe that viruses are stealthy and do not impair the behavior of the computer to remain unnoticed. Their suggested protection is the use of antivirus software.

Walsh also extracted four mental models used to understand hackers.

- *Hackers are digital graffiti artists*: According to this model, hackers are skilled individuals who breach computers purely for mischief and to show off. They are often young geeks with poor ethics. Their victims are random.
- *Hackers are burglars*: According to this model, hackers treat computers like burglars treat physical property. The goal is financial gain. These hackers choose their victims opportunistically.
- *Hackers are criminals targeting big fishes*: According to this model, these hackers are similar to the previous ones, but their victims are either organizations or rich people. The victims are targeted.
- *Hackers are contractors who support criminals*: According to this model, these hackers are similar to the graffiti hackers, but they are henchmen paid by criminal organizations. Their victims are mostly large enterprises.

When using these mental models, it is obvious that some users will never apply best practices, regardless of their pertinence. These users cannot understand the value of these best practices, as they do not fit their mental model. Alternatively, some users may feel these best practices do not concern them. For instance, the users who believe that viruses are bad or buggy software cannot understand the benefit of installing an antivirus on their computer. The users who associate hackers to contractors cannot believe that hackers may attack their home computers. Better understanding the mental model of a user highlights where awareness is needed to adjust his mental model to reality. This understanding helps us also to design efficient, secure solutions that may seem to fit the user's wrong mental model, although the solutions operate differently in the actual world model.

Rule 7.2: Define Secure Defaults

One of the consequences of Rule 4.3: Provide Minimal Access is that the designers should always set the default values of the system to the most restrictive conditions. Often these default values are set to the settings those the most user-friendly environment. Convenience often outweighs security. When it comes to security settings, this is often a bad strategy. The following is a list of such weak strategy-based choices.

- Servers, network appliances, network devices, and connected devices are delivered to the customer with an initial default password for the administrator account. Usually, this default password is the same for all the devices.[13] This default password is printed in the user manual, which can often be found online. This information is a valuable resource for hackers. Unfortunately, many administrators do not change this password. When scanning the Internet, once an attacker has profiled her target, she can easily attempt logging into the administrator account of the targeted machine using the default password. This simple attack too often works [342, 343]. A simple modification of this default password would defeat this attack.

- Sometimes, manufacturers poorly manage this administrator password. For instance, the baseboard management controllers (BMCs) provided by Supermicro had their admin password stored in the clear in a password file named `PSBlock.txt`. Furthermore, this file was accessible via open port 49152 [344]. A BMC is a system that monitors and controls the physical state of the computer. It uses sensors to measure temperature, humidity, fan speed, and other physical parameters of the computer. Servers have such BMCs to control their environment. With the retrieved administrator password, an attacker could modify the actual environment parameters. For example, she could allow the server to overheat by increasing the threshold for the fans, thus sabotaging it.

The Devil Is in the Detail Despite the fact that having a serialized administrator password for each machine would be better for security, why do manufacturers define the same default password for all the devices? The reasons are mainly financial and practical. Let us assume that a manufacturer decides to define a different administrator default password for each device. This insert explores this example to highlight manufacturing and maintenance constraints that make best security practices unpractical or, at least, expensive.

The first issue is the value of the default password.

1. The most logical and most secure solution is to generate a different random default password for each device. Once generated, this default administrator password is stored in the memory of the device during the personalization phase, together with other unique information such as the serial number, or the MAC address. Usually, for reasons that this section will explain later, these random numbers have to be stored in a secure database for later use. Thus, this database becomes a valuable, sensitive asset and requires adequate protection.
2. The second solution is to generate the default password using a "secret" function fed with parameters unique to the device. The factory may calculate the default password and then store it in the memory of the device. An alternative solution implements the "secret" function in the software of

[13]Too often, the administrator account is admin and the default password is admin.

the device that will use it whenever it has to check the default password. Both solutions are weak as the disclosure of the "secret" function annihilates the secrecy of the default password [345]. Cracking the first implementation may require the collaboration of an insider who will reveal the function. The embedded solution is even weaker as the attacker has potential access to the "secret" function by studying the binary code. Indeed, reverse engineering of the device may disclose this secret function. Nevertheless, deriving the secret password from device's parameters has the advantage of being simple and cost-efficient.

The second issue is how to convey the serialized default password to the customer. There are mainly two solutions.

1. Print the serialized default password in the user manual; unfortunately, from the logistics point of view, this is difficult. The booklet is printed independently of the actual manufacturing of the physical device. The booklet and the device are packaged together only at the end of the manufacturing chain. Therefore, there is a need to match a booklet with its counterpart device. This pairing increases the complexity and cost of manufacturing. Furthermore, it is prone to error.
2. Print the serialized password on the sticker with the serial number that will be glued on the device. If the device is a Wi-fi router, the same label also displays the SSID and the WPA2 key.

Of course, the password must be available at the printing time of the booklet or the sticker. This availability requires access to either the database containing the random passwords or the secret function and the associated parameters. An additional difficulty is that the booklet or sticker has to be packaged with the proper device. If there were to be a mistake, then the customer would not be able to access his administrator account.

The third issue is the user. What happens if the end user has lost his booklet, or its sticker is not readable anymore? The hotline support must be able to communicate the default password to the customer. If the default administrator password is the same password for all devices, then this is straightforward. If the default administrator password is individualized, the hotline should be able to retrieve this default password. The hotline team needs access to either the database or the secret function. A necessary condition is that the hotline has remote access to the device, which is a kind of authentication of the caller. Furthermore, the problem is worse if the user has changed the default password. This modification is highly recommended as a best practice. How can the user recover the account if he forgot his new administrator password?

Usually, a procedure allows restoring the factory settings of the device, including the administrator password. Customer service teams already use this procedure. Maintenance teams need to access the administrator account even if the user has changed it. For this reason, this procedure requires physical access to the device. If it were to be possible to reset the settings remotely, then this procedure would be a real trapdoor for attackers.

Lesson: When implementing a security feature in a device, it is essential to investigate its impact on the usability and maintenance of the device once it is in the field. If not well thought through, the security feature may become a showstopper and create massive financial problems and generate frustration from the end users.

7.3.2 Align the Interests of All Actors

Sometimes, fast-food restaurants display a sign asking the customer to verify that the cashier has given him a receipt. If the cashier has not given a receipt, then the customer is entitled to claim the meal for free. Why should the owner of the restaurant want to offer free lunch to his customer? A common fraud is one in which the cashier takes the money from the client without generating a receipt. In that case, the clerk might keep the money without its being traced in the daily account balance, as there is no record of the sale. It is important for the restaurant's owner to prevent the cashiers from cheating. The owner cannot monitor in real time all the cashiers' transactions. Thus, with this incentive, the owner delegates this monitoring task to the customer. The owner incentivizes the customer to undertake the monitoring on his behalf by offering a free lunch for denouncing a potential fraud. With this delegation, the owner aligns his customer's security interest with his own interest. Furthermore, the owner does not lose money. When a customer reports such a fraudulent transaction, the owner will reimburse the customer. However, he will get back the money from the employee.

Many security systems employ more than one party. In that case, often the trust model assumes that they collaborate and work together in good faith. The optimal case is when the interests of all actors are aligned. If the interests diverge then the assumption weakens.

Rule 7.3: Align the Interests of All Actors

If the entity that is in charge of ensuring the security of a system is not going to be impacted whenever a hack occurs, then there is the risk that the entity may not be making its best effort. This first type of misalignment may seem contradictory. Nevertheless, there are real examples to illustrate it. If ever a DVD player or Blu-ray player is compromised, the primary entity that will lose money is not the manufacturer of the compromised player but rather the content owners whose content will be ripped. Why should the manufacturer put a more expensive secure implementation for a feature that will not be a selling point for the customer? A similar situation occurs with CE gateways and home routers. For instance, owners of home routers infected by malware that enroll them in a botnet used to mount a DDoS attack will only slightly suffer from the attack [343]. The same goes for infected, home computers. It is the target of the DDoS that will be impaired not the owner of the infected, home computer, nor the manufacturer of the infected devices. Why should the owner of a home computer pay a subscription for an anti-malware service that would not harm him directly? Why should he bear the cost? Of course, viruses may also impact the owner of the computer, thus enticing the owner to install the protection. This requires the education of the owner. Another solution is to introduce some legal liability. Through legal action, the entity suffering the loss may hold the lax entity liable. Of course, legal answer is not the most satisfactory response. Nevertheless, the fear of legal action may strongly motivate an entity to do a proper job in securing its part.

Common sense dictated that the optimal solution is one in which the interests of all actors are aligned. In that case, we may expect all actors to collaborate to get better security. This is not necessarily the case and not necessarily required. As with reliability [346], there are three types of situation for security.

- *Total effort*: The security of the system relies on the global effort of all actors. It requires a cumulative effort to secure the system. These are some examples.
  - Reviewing the source code of a security component, the ratio of detection of vulnerabilities depends on the total number of reviewers. The higher the number of reviewers, the greater the likelihood of discovering vulnerabilities.[14]
  - If users do not change the default administrator password on their routers, a hacker can take control of them. If the attacker controls enough routers, she can build a botnet to launch DDoS attacks [343]. To defeat this type of attack, it requires a majority of users to change the default administrator password. Defeating botnets is usually a collective effort of the owners of the infected principals of applying better security practices, thus reducing the likelihood of infection.
- *Worst link*: The security of the system depends only on the weakest actor. This is, for example, the case with the infection of a corporate network through malware or spear phishing attack. An APT needs only to breach one strategic entry point. It is a typical illustration of Law 6: Security Is No Stronger Than its Weakest Link.
- *Best effort*: The security relies on the maximum effort of one actor.

The two first situations require all players to collaborate for better security. In the total effort situation, the more the players are aligned, the better the overall security is. In the worst link case, the weakest actor is not known usually. The more the actors will align, the lower the likelihood will be that one actor is the weakest actor. In the best effort situation, the principal player is usually known. Therefore, it is easier to strengthen the security of the system by focusing on this known actor.

[14]The ratio is a logarithmic function of the number of reviewers.

7.3.3 Awareness

People perform tasks better when they either agree to do it or, at least, when they understand the purpose of the task. Therefore, it is paramount that users understand the importance of security. Once they are convinced of the interest, most probably, they will accept performing the additional tasks mandated by security. Committed people are far more responsive to threats and mischief than non-committed ones. In 1972, Thomas Moriarty conducted a field experiment. Fifty-six subjects were confronted with the theft of a possession of an unknown confederate victim [347]. The experiment used two sets of subjects. In the first set, subjects had no interaction with the future victim. In the second set, subjects had interaction when the victim asked them to keep an eye on his belongings. Only 20 % of the subjects of the non-committed first set responded to the theft. Ninety-five percent of the subjects of the second, committed set attempted to stop the thief.[15] Once people understand the risks and potential losses, they will be engaged and committed, even if they will not suffer directly from the loss or attack. Thus, they may react more wisely to a potential ongoing attack.

Many companies train their IT and security teams and encourage them to stay up-to-date in the security arena. Sometimes, they do not nurture enough the basic security awareness of their employees. Unfortunately, many attacks, and often also APTs, start with an infection by malware by one of their employees. If every employee is not vigilant, the security of the company can be compromised through such an employee. Basic security awareness is critical. IBM recommends the following [349]:

- Train all employees on security best practices and how to report suspicious activity. It is important that the reporting be easy to perform and well documented. If reporting is too cumbersome, some employees may not make the necessary effort to report suspicious activity. Unfortunately, the earlier a suspicious activity is detected, the easier it will be to identify an ongoing attack, and the less adverse will be the impact of the attack may have. Swift reporting is an essential element of the defense.
- Consider conducting periodic, mock phishing exercises where employees receive emails or attachments that simulate malicious behavior. This recommendation is rather unusual. Nevertheless, it is one of the most efficient methods of increasing the awareness. Such mock exercises demonstrate the reality of such attacks. People learn better from real examples than from abstract

[15]In 2015, a French study reproduced a similar field experiment on a larger scale, employing 150 subjects rather than 56 subjects [348]. The outcomes were similar. Eighty-eight percent of the committed subjects intervened whereas only 18–34 % of the non-committed set reacted. In the latter case, the lower rate occurred when two bystanding confederates did not react to the theft. This variation does highlight the potential impact of conformism. People tend to align their behavior with the mobs.

descriptions.[16] Extrapolating the mock phishing exercises to the personal domain may also increase the efficiency, as people may be more receptive to the message if it concerns their personal information. Of course, the mock phishing exercises must be innocuous for the security of the organization.

- Provide regular reminders to employees on phishing and spam. Awareness fades with time. Thus, regular reminders reinforce the message.

Not every employee needs the same amount of knowledge about security. Therefore, the NIST introduced a model to describe the so-called security learning continuum [350]: SP 800-16. This model addresses the level of knowledge and the targeted audience. The model defines three levels of learning programs within an organization.

- *Awareness* focuses on the attention of the audience on security. Awareness presentations should attempt to educate individuals in understanding the security concerns and acting correspondingly. The audience of an awareness program is global and thus often broad, diverse with regard to knowledge and background. Thus, it should be attractive and simple to understand. The NIST published guidelines for designing a security awareness program: SP 800-50 [351]. The awareness program is successful whenever the audience accepts the message: Security is everyone's responsibility. In 2015, Absolute Software published a report based on the interview of 762 US adult employees using at least one employer-owned mobile device [352]. Most employees acknowledged that corporate data had enormous value and that their mobile devices hosted such data. Nevertheless, half of the respondents believed that IT security was not part of their responsibility.
- *Training* focuses on producing security skills and competencies for non-security specialists. The audience of this training should be smaller than the audience for the awareness program. The content of the training should be customized to address real security issues that the audience may encounter. SP 800-50 provides guidelines for designing training programs. Training is successful whenever the user understands the importance of security and applies the security procedures consistently and uses the right tools.
- *Education* focuses on building the knowledge pool needed by the security practitioners. The audience of the education programs is small. Education programs are highly specialized and dedicated to specific topics. Education is successful whenever practitioners can apply a proactive defense. Only continuous training allows security staff to stay aware of the newest threats and latest attacks. Without this knowledge, they cannot prepare the defenses against these threats.

Figure 7.2 illustrates the three types of learning programs and where they apply within an organization. Security awareness is for all the members of the

[16]As such, mock phishing exercises are similar to regular mock evacuation exercises.

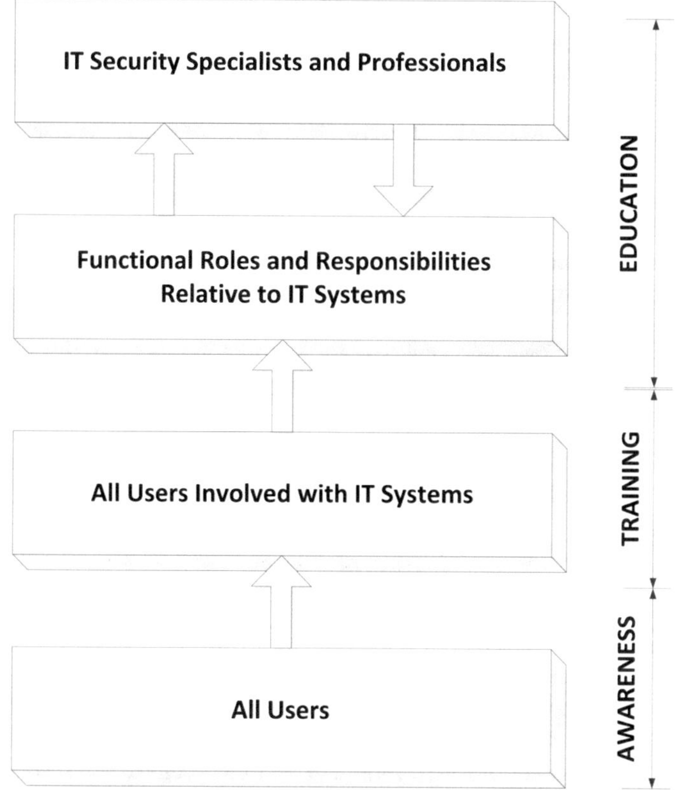

Fig. 7.2 The NIST-proposed learning continuum

organization. Training provides security basics to all the users of the IT infrastructure. Furthermore, training provides the skill set needed to handle security issues to all the IT team members. Education increases the knowledge and experience of the security practitioners. All three types of learning programs are necessary for an organization to stay safe.

Rule 7.4: Educate Your Employees

Educating the staff has an additional advantage. An educated person is less prone to falling into the traps of attackers. For instance, education is the only efficient defense against social engineering attacks. Only aware staff can detect such deception and be able to detect an ongoing social engineering attack, or identify anomalous behavior. Training increases the vigilance of users. The human brain is good at identifying known patterns and is poor at identifying unknown schemes. Thus, the brain must be exposed to mock social engineering attacks.

Obviously, it is key that IT security practitioners be trained in the latest defense tools, as well as in the most recent threats and attack techniques. This training may be challenging, as the security landscape is evolving incredibly fast. Every month, new attack vectors appear. Everyday, new variants of malware flood the cyber-space. Thus, the training has to be routine to be sensitive to the latest trends. It has a cost in human effort.

Rule 7.5: Train Your Security Staff

The ideal training for security staff should expose it to real hacking exploits. If the defenders are not aware of the skill set of their attackers, then they will usually underestimate the corresponding risks. Hacking conferences such as CCC or DefCon are interesting educational events. Regular security assessments by independent evaluators are also excellent training opportunities. Not only do such security assessments lead to a more secure, patched system, but also educate the defending team at least about the discovered vulnerabilities and methods used by the penetration team.

7.4 Summary

Law 7: You are the Weakest Link
Law 6 states that the security of a system is no stronger than the security of its weakest link. Human beings are often part of an overall system. They may operate the system. They may feed information into the system. Unfortunately, in many of these systems, humans are the weakest link. The primary threat is a social engineer inducing a person to perform an action that will compromise the security or will leak valuable information. The second threat is users making wrong security decisions or not applying security policies and procedures.

Rule 7.1: Where Possible, Make Decisions on Behalf of the End User As the end users are not necessarily able to make rational decisions on security issues, the designer should make the decisions when possible. Whenever the user has to decide, the consequences of his decision should be made clear to him to guide his decision.

Rule 7.2: Define Secure Defaults The default value should always be set to that for the highest or, at least, an acceptable security level. User friendliness should not drive the default value, but rather security should.

Rule 7.3: Align the Interests of All Actors Only if the interests of all actors are aligned can we expect that all actors will do their best to secure the system. If one actor will not suffer from a breach (directly or indirectly), then its incentive to do a proper job is reduced. It is likely that it may be the weakest link.

Rule 7.4: Educate Your Employees The best answer to social engineering is enabling employees to identify an ongoing social engineering attack. This detection is only possible by educating the employees about this kind of attack. Training employees increases their security awareness and thus raises their engagement.

Rule 7.5: Train Your Security Staff The landscape of security threats and defense tools is changing quickly. Skilled attackers will use the latest exploits. Therefore, it is imperative that the security personnel be aware of the latest techniques. Operational security staff should have a significant part of their work time dedicated to continuous training.

Chapter 8
Law 8: If You Watch the Internet, the Internet Is Watching You

Une sortie, c'est une entrée qu'on prend dans l'autre sens.
Boris Vian, Traité de civisme

8.1 Examples

In 2011, the Web site, `youhavedownloaded.com` was launched.[1] It quickly started to generate some buzz. The Web site collected information about pieces of content downloaded using the P2P BitTorrent protocol. Indeed, it stored for each IP address the list of content that this address allegedly downloaded. The Web site gathered more than 55 million unique IP addresses. The Web site displayed the allegedly downloaded content for the currently presented IP address of the visitor. Furthermore, the visitor could check the available records for any IP address.[2] The Web site Torrent Freak reported that some people at Fox, Sony Pictures, Google, and NBCU (or, at least, IP addresses attributed to these companies) downloaded copyrighted content [353]. According to the Web site's authors, the announced purpose of this site was not mischief. They announced:

> Don't take it seriously
>
> The privacy policy, the contact us page—it's all a joke. We came up with the idea of building a crawler like this and keeping the maintenance price under $300 a month. There was only one way to prove our theory worked—to implement it in practice. So we did. Now, we find ourselves with a big crawler. We knew what it did but we didn't know how to use it. So we decided to make a joke out of it. That's the beauty of jokes—you can make them out of anything.

The Web site demonstrated that it was neither difficult nor expensive to collect such potentially embarrassing information. Thus, malevolent people could use this

---

An exit is an entry that you take from the other direction.

---

[1]The Web site stopped collecting data in 2012.
[2]Some experiments with my team seem to indicate that at least some of the data were true.

E. Diehl, *Ten Laws for Security*, DOI 10.1007/978-3-319-42641-9_8

technique to ransom downloaders of copyright infringing content. This kind of ransomware is already deployed in the wild [354, 355], but usually it targets random users without taking advantage of real infringement information. The blackmailers expect that the targeted victim may have participated in such questionable activity, and thus may be more gullible. If the attacker were to collect real data beforehand during the discovery phase and target exclusively victims who had engaged in such activity, the likelihood of success in ransoming would be far higher. Such a scenario could also be used as an entry vector for an APT.

In November 2014, the UK-based Data Watchdog announced that the Russian Web site www.insecam.org featured a database listing about 73,000 live streaming IP webcams or Closed Circuit TVs [356]. The Web site allowed watching each of these video cameras live. The video camera's owners were probably not aware that their webcam was broadcasting the live video to the world.[3] The webcams were spread all over the world. They were used for office surveillance, baby monitoring, shop surveillance, pub surveillance, and so on. All the major manufacturers were represented among the exploited webcams. The procedure involved two steps. First, a thorough scan discovered the webcams connected to the Internet. Then, the exploring system attempted to log into the webcam using the default manufacturer password. If successful, it recorded the information related to the vulnerable webcam. The Web site claimed that its purpose was educational. The Web site displayed the following statement:

> Sometimes administrator (possibly you too) forgets to set the default password on security surveillance system, online camera or DVR. This site now contains access only to cameras without a password and it is fully legal. Such online cameras are available for all internet users. To browse cameras just select the country or camera type.
>
> This site has been designed in order to show the importance of the security settings. To remove your public camera from this site and make it private the only thing you need to do is to change your camera default password.

In these previous examples, the victim may have been participating in questionable activities or may have been negligent with security. What if the attacker could misappropriate, and, worse, exploit legitimate security-related activities of her victim? What if security could be the vector of attack? For a few years, the market has offered wireless IP cameras. These cameras can stream live video to any connected computer or mobile device. They are convenient for use in video surveillance. Thus, many systems of personal video surveillance have begun to use them. It becomes easy to verify whether everything is OK at home while outside. Unfortunately, if the owner can watch the camera's stream remotely, then attackers can observe the same stream. At the conference "Hack In The Box 2013," researchers at Qualys demonstrated this kind of attack [357]. They showed how to

[3]Meanwhile, the site has filtered out cameras that might have breached individual privacy, for instance, inside a home. They also removed any camera when requested by its owner. Unfortunately, the problem is still around and most of these webcams are still accessible from the Internet.

install a backdoor on a commercially available camera from manufacturer Foscam. For that purpose, they used several known vulnerabilities and the default password of the administrator account (about 20 % of the deployed cameras still had "admin" as the default password, as illustrated by the previous example). Once the backdoor is installed, the attacker can watch the video issued by the surveillance camera. Furthermore, once the attacker gains access to the firmware, she can modify it to add stealthy features. The authors concluded that these cameras should never have been exposed to the Internet. If the exposure to the Internet were to be necessary, then strict firewall rules should isolate the cameras, for instance, by white listing the authorized connecting IP addresses. Nevertheless, is it sufficient to have properly managed surveillance cameras and an efficient firewall isolation? Between 2008 and 2010, the British GCHQ randomly collected snapshots of webcams that were used by Yahoo video chat services [358]. In 2014, Edward Snowden revealed this questionable activity. The alleged purpose was to create an enormous database of pictures to experiment and train facial recognition software. For many years, the GCHQ stealthily spied on several million accounts.

The scenario becomes even more nightmarish when an everyday consumer device turns into a spying tool. In 2014, Benjamin Michele and Andrew Karpow hacked Samsung Smart TVs [160]. Their attack illustrates such a scenario. Their exploit allowed them to capture the video of the camera integrated into the TV set and the ambient sound with the embedded microphone. The captured information could then be sent to a remote server. The consumer was not aware that his smart TV set was infected and that it had spied on him.

The scenario turns even scarier when the attack can endanger the life of the user. In July 2015, a few days before their presentation at Black Hat 2015, Charlie Miller and Chris Valasek demonstrated to a journalist that they could remotely take the control of the journalist's Jeep [359]. Using an Internet connection, they broke into the control system of a confederate journalist's vehicle, of course, without his assistance. Once in control, they could initiate benign actions such as modifying the volume of the radio or turning on the wipers. Unfortunately, they could also initiate critical actions. For instance, they slowed down the automobile of the willing journalist as he entered a highway. However, they could also have forced the vehicle to move promptly to the left, forcing it to crash. Chrysler equips its Jeeps with an Internet connection called UConnect. UConnect enhances the automotive entertainment system with Internet features and smartphone features. After 1 year of analysis, the security experts discovered a vulnerability in UConnect that allowed them to breach the car's control system. A few days before the Black Hat presentation, the parent company of Jeep, Fiat Chrysler, issued a patch and informed 1.4 million customers to upgrade their vehicle's software. This is a good illustration of the concept of Return on Non-loss (Sect. 2.2.3). Had Jeep known about the vulnerability before shipping its cars, it would have saved the cost of this recall and avoided a tarnished reputation. Furthermore, the ISP handling the connections to the UConnect system now filters out the attack [360], protecting the drivers who had not patched their car.

A communication link on the Internet is always bidirectional. In some cases, this communication channel may be used to strike back. In 2014, security researcher Raashid Bhat published an interesting countermeasure against attackers and spammers [361]. He reverse-engineered the Zeus Banking Trojan, which he received via an infected email. The Trojan was self-protected. Its payload had to be encrypted with RC4 for a successful installation. He discovered the corresponding encryption key and reverse engineered the Trojan. Then, he used a known vulnerability of some Zeus versions to create a shell on the computer of the C&C that controlled this instance of the Zeus Trojan. Using the open source penetration tool Metasploit, he launched a Meterpreter shell on the spamming server. With the Meterpreter script `webcam_snap`, he took pictures of the operator in front of the spamming computer using the webcam. Obviously, as he was root user of the private server, he could have performed other actions, even some destructive ones. The sprinkler was sprinkled.

Spying on people's computers no longer requires the skill set of a blackhat expert. Powerful hacking tools are available either as open source software or for sale on hacking forums. Among those tools are Remote Administration Tools (RAT) such as PoisonIvy, XtremeRat, and Blackshades. The use of RAT is legitimate in a professional environment. The IT department may use an RAT to manage and fix the employees' computers remotely. In that case, the IT department may install it on a user's workstation with the knowledge of the user.[4] Nevertheless, RAT can be perverted to nefarious use if its installation and activation are performed without the consent of the computer's user. BlackShades is a very powerful RAT and is available for a few hundred euros. Once the RAT is purchased, the buyer needs to infect her victim's computer with Blackshades. To facilitate this, Blackshades comes with a spreader that enables its dissemination on other machines. Once the RAT is installed, the hacker controls the infected remote computer. Blackshades provides many features, such as logging of keystrokes, accessing of the file system, encrypting of files for ransomware, posting of notes on the victim's Facebook site, setting of an alarm for when a specified window title or keyword is present on the victim system, turning on of the microphone, and even controlling of the webcam stealthily [362].

The combined use of the Internet and new technologies may create dangerous situations. Most modern mobile devices are equipped with an embedded Global Positioning System (GPS) receiver and camera. Similarly, most digital cameras have GPS features. Many services use geolocation with GPS to provide enhanced features to users. Nevertheless, the combination of GPS and camera may turn into a privacy issue. Currently, most devices featuring GPS and camera embed a geotag, i.e., the precise location defined by the GPS, in the metadata of pictures shot by the camera. Off-the-shelf, standard tools can extract this metadata. Many people post personal pictures on social networks such as Instagram, Flickr, Facebook, or

---

[4]Good ethical practice requires that the user agrees to the activation of a remote session of the RAT.

Craigslist. If Bob took a picture from his home, and then posted it on a social network or Flickr, Eve can easily extract the longitude and latitude coordinates of the shooting location by reading the associated metadata. With this information, she can locate the actual address of Bob's home, for instance, by using Google Street View.[5] In other words, publishing a picture of a house indirectly may disclose the address of this house. Disabling the geotagging feature on every mobile device is possible. Unfortunately, many people are not aware of the existence of this feature and do not understand the corresponding risk. For convenience, the manufacturer enables geotagging by default (Rule 7.2: Define Secure Defaults). Security would require this option to be opt out by default.

In 2012, Keaton Mowery and Hovav Shacham proposed a new original method to fingerprint a browser using HTML5 [364]. The method uses one of the new features of HTML5: canvas area. The markup `<canvas>` defines a borderless empty rectangular area on the screen. A JavaScript module can draw within this canvas area. To fingerprint the machine, they wrote a text, a pangram,[6] into a defined canvas area. They retrieved the rendered bitmap of the canvas area (using command `toDataURL`) and calculated a digest from this image. Web browsers use different image processing engines, export options, and compression levels. Operating systems use different algorithms and settings for anti-aliasing and sub-pixel rendering. Furthermore, the graphic card and the version of its driver also influence the final rendering. All these variations impact the actual rendering of the same text, thus generating different digests. They succeeded in identifying the browser used. Furthermore, experimentation demonstrated that this fingerprinting canvas may differentiate between users. When combined with other fingerprinting parameters such as the HTTP agent or the set of available fonts on the computer, the uniqueness of the fingerprint is high. In July 2014, Gunes Acar and his five colleagues studied different tracking methods used by the top 100,000 Web sites (ranking by Alexa) [365]. They discovered that 5.5 % of these sites used a fingerprinted canvas. Most of these Web sites employed the "www.AddThis.com" system. By reverse engineering the `AddThis` code, they discovered that AddThis improved the technique described in the Mowery and Shacham seminal paper. For instance, the AddThis developers used a perfect pangram or drew two rectangles and checked whether a particular point was part of the overlapping area. Without storing cookies on the user's computer, and without installing any spyware on it, the fingerprinted canvas uniquely identified every visitor. Canvas fingerprinting does not require the collaboration of the user. It does not exploit any vulnerability but rather uses a standard feature of the browser. It bypasses any cookies protection

[5] For instance, a Web site discloses the location of pictures of cats that were posted on Internet to demonstrate this threat [363]. Cats are less likely to be identified than people. Thus, the project does not endanger people. Nevertheless, the demonstration is impressive.

[6] A pangram is a sentence that uses all the letters of the alphabet at least once. A perfect pangram is a sentence that uses all the letters of the alphabet only once. A well-known English pangram is "The quick brown fox jumps over a lazy dog".

and even the browser's private mode. Furthermore, it is compatible with any modern browser, as they all support HTML5. Thus, the Internet may identify you even if you are cautious.

8.2 Analysis

8.2.1 Protect Your Corporate LAN

The previous section showed that if Bob's network is connected to the Internet, then Bob may become the subject of serious mischief. Today, using the Internet is unavoidable. Therefore, we have to mitigate these risks. Fortunately, decades of network security have led to the development of efficient protective tools and strategies. Network security has mainly two goals.

- Isolate the LAN from the Internet
- Control the data transferred between the LAN and the Internet

An essential component to reach these goals is the firewall. A firewall is a system that isolates two networks or sections of the network.[7] At a minimum, the firewall hosts packet-filtering services. Packet filtering inspects the packets of the network layer (layer 3 of the Open Systems Interconnection model) and the transport layer (layer 4). As networks rely on IP addresses, a firewall mainly inspects the IP (for the network layer) and TCP/UDP (Transport Control Protocol/User Datagram Protocol) headers (for the transport layer). Then it decides to drop or pass the packets depending on multiple criteria such as their source IP address, their destination IP address, the source port, the destination port, the protocol used, and the flags describing the direction of transfer (inbound or outbound). An IP address has 2^{16} possible ports. Therefore, the firewall has to manage about 65,000 doors (the approximate number of ports) and has to decide which remote server or application is authorized to employ each of these individual ports. Firewalls also usually implement packet inspection. Stateful packet inspection checks whether the corresponding packet is legitimate in the context of the ongoing communication called a network session. This inspection can be done even for stateless protocols like UDP. Firewalls start to implement packet normalization as an additional layer of filtering. This inspection can spot abnormal sequences and discard them. A new generation of firewalls acts also at the application layer (layer 7). Web Application Firewalls are the most common ones. They handle HTTP transactions.

A firewall is the first line of defense against the Internet. Figure 8.1 illustrates the simplest architecture to isolate a corporate LAN from the Internet. The firewall is inserted between the Internet and the LAN. It intercepts all incoming and outgoing

[7]The term firewall comes from the construction industry. A firewall is a wall built with fire-retardant materials to stop or at least slow down the propagation of fire.

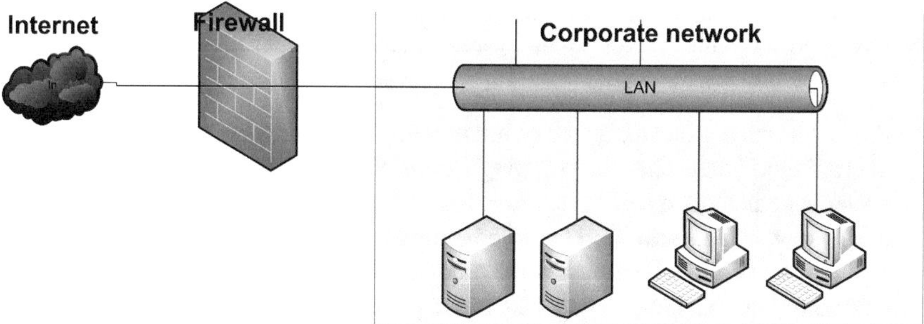

Fig. 8.1 Simple isolation

data exchange. The firewall should allow only access to the ports needed by the computers inside the LAN. The firewall filters out the unexpected or unauthorized packets of data. This minimal architecture has many disadvantages. The most significant drawback is that all computers on the LAN share the same network and are isolated from the Internet just by one single firewall. The public servers that are handling Internet services such as Simple Mail Transfer Protocol (SMTP) servers, DNS servers, or HTTP servers, are the most at risk of being compromised. As these servers are on the same network as the corporate computers, whenever an attacker compromises one of these public servers, she has immediate access to the other computers of the company. The likelihood of such an intrusion is high even if the firewall is correctly configured and the Operating System of the servers is hardened. Indeed, public servers run applications that always have new vulnerabilities that can be exploited by the attackers. This risk is not acceptable in a corporate environment. Despite the drawbacks, simple isolation is usually found to protect personal home networks and some small companies' networks. Implementing it is rather simple, and thus, it is attractive. Usually, the Internet gateway implements the firewall features.

The previous architecture lacks real separation of workstations and corporate public servers. The DMZ provides such isolation. Historically, a DMZ is a geographical area in which international treaties forbid the settlement of military, industrial, and personal installations. Usually, the DMZ resides between the boundaries of two antagonistic militaries. Currently, there is such a DMZ along the 38th parallel separating North and South Koreas. Antarctica is another example of a DMZ. In this case, the purpose is not to protect against armies invading Antarctica but to preserve it for scientific research, allowing any country to send its researchers without having to seek the authorization of any other country.

In the computing world, DMZ means a physical or logical subnetwork that is an airlock between two networks, e.g., the Internet and the LAN or between two LANs. A DMZ is a subnetwork inserted between a protected network and an external network to provide an additional layer of security [366]. No server or computer on the LAN should have direct access to the Internet or should be accessible directly from the Internet without passing through a firewall. Furthermore, the DMZ should not authorize to initiate any connection toward the corporate LAN. Thus, the DMZ hosts all the Internet-facing public services or servers such as Web servers, email servers, DNS servers, or FTP servers [367]. Usually, the DMZ hosts some specialized servers.

- Proxy servers interact on behalf of internal hosts with the Internet. For instance, an internal workstation has to go through a proxy to access the Internet. Each proxy server is dedicated to a given service, for instance, Web, email, or FTP. Thus, this proxy server has a profound knowledge of the expected syntax of the exchanged data for the service. It performs deep analysis of these exchanged data to detect non-compliant packets, to flag suspicious packets or to filter out specific packets. For instance, a proxy server may use a white list of authorized requests or a blacklist of forbidden requests. The white list strategy is more secure than the blacklist. This strategy exhaustively defines what it authorizes and denies every request outside of this list. The blacklist strategy is less constraining for day-to-day management than the white list one. A proxy server may also filter out given content. Furthermore, a proxy server may require the principals from the LAN to authenticate themselves prior to serving them. In this case, the proxy server may apply a personalized security policy unique to the authenticated principal.
  Some Web proxy servers may also serve as cache servers. In that case, the Web proxy server keeps in its memory the most frequently accessed Web pages. Caching frequent Web pages enhances the response time and decreases the needed bandwidth.
- Reverse proxy servers interact on behalf of external principals on the Internet with Web servers internal to the corporate network. They can also handle other protocols such as POP3 and IMAP. They play the inverse role of proxy servers. They isolate the internal servers from the Internet. A reverse server may only serve authenticated Internet principals. Some reverse proxy servers may also balance the load between different internal Web servers. They can even handle TLS connections and forward decrypted data to protected internal servers.
- A server accelerator is a reverse proxy server that caches the most frequently accessed pages, decreasing the load on an internal Web server.

Figure 8.2 depicts the "three-legged" DMZ architecture. The DMZ (in the light gray area of the figure) hosts all the services and servers facing the Internet. The servers and computers on the LAN (in the dark gray area of the figure) never have a direct connection to the Internet. All their connections pass via the DMZ's servers. The firewall enforces this policy by banning all direct communication between the LAN and the Internet. It only authorizes communication from the LAN to the DMZ. The firewall also filters out the unexpected and non-compliant connections between the DMZ and the Internet. With this simple architecture, there is now an airlock between the Internet and the corporate computers in the LAN. If the firewall is correctly configured, an attacker cannot have direct access to employees' computers and internal servers. She can only compromise the machines in the DMZ. Of course, once inside the DMZ, she will attempt to breach the corporate LAN. The firewall should prevent it.

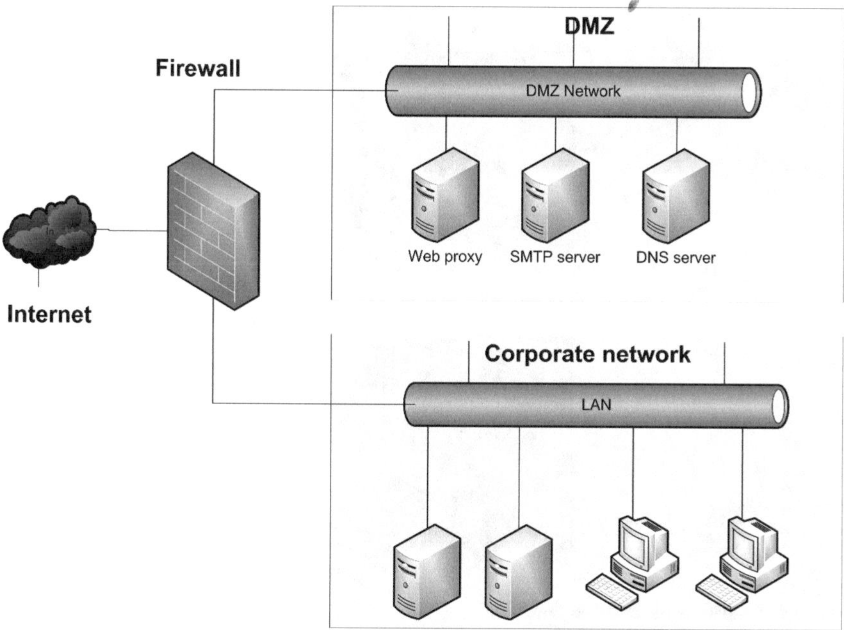

Fig. 8.2 Three-legged DMZ architecture

Figure 8.3 presents an enhancement of the "three-legged" architecture. Instead of using one single firewall, this architecture employs two firewalls. The first firewall manages and authorizes only connections between the LAN and the DMZ. The second firewall handles and permits only connections between the DMZ and the Internet. Each firewall handles only rules dedicated to one interface. Hence, the configurations of both firewalls become simpler and thus less prone to errors (Rule 4.4: Keep It Simple). This architecture also mitigates the risk of inconsistent rules as each firewall manages rules devoted to its domain. In this configuration, the attacker has to breach two firewalls to reach the LAN. This second firewall should have no rules allowing direct connection from the DMZ to the corporate network. To be even more resistant, the two firewalls should use different technologies or suppliers. Thus, if an attacker were to exploit a vulnerability of the Internet-facing firewall, the likelihood that of her exploiting the same vulnerability in the second firewall would be small. This diversification is an example of Rule 1.4: Design In-depth Defense.

Starting from this dual firewall DMZ architecture, it is possible to elaborate on more complex and sophisticated isolations by multiplying specialized DMZs. Figure 8.4 presents such an enhanced architecture. For instance, DMZ1 may host the public Internet servers that anybody may access while DMZ2 hosts servers that require authenticated connections from the Internet. Of course, the rules of the ingress filtering from DMZ1 are different from the rules of the ingress filtering from DMZ2.

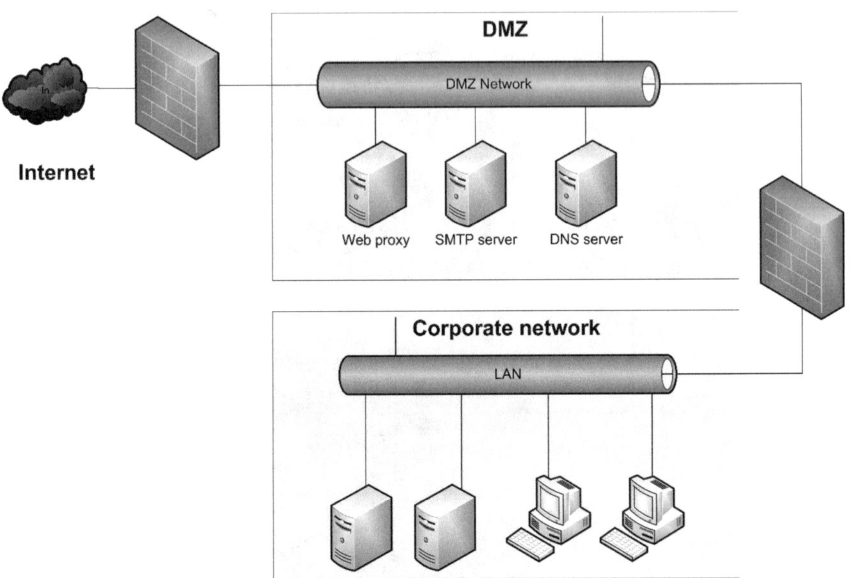

Fig. 8.3 Dual firewall DMZ architecture

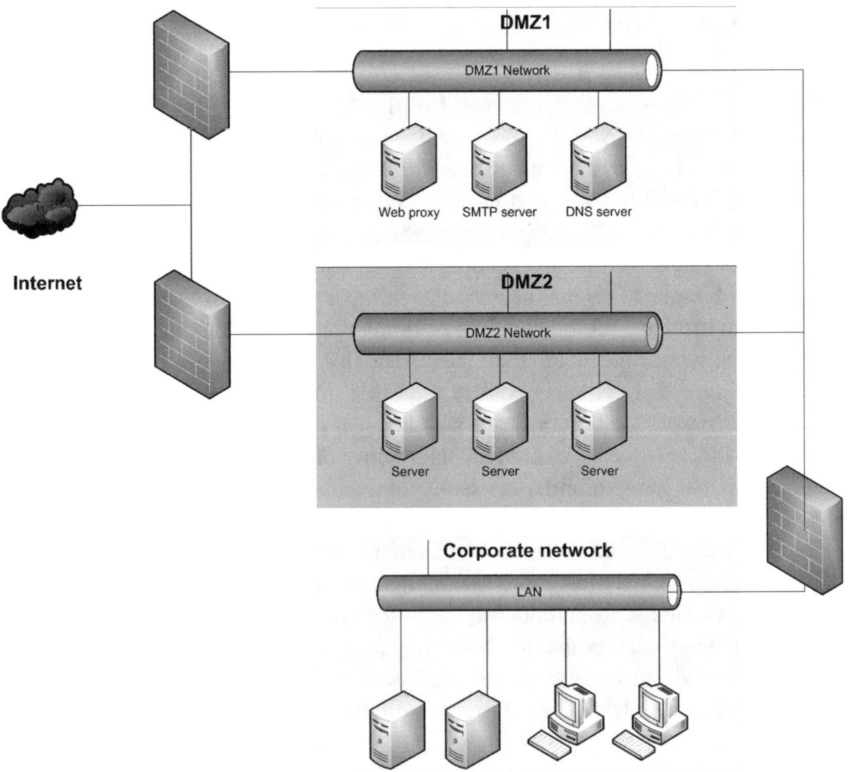

Fig. 8.4 More complex DMZ architecture

The DMZ operates in a more aggressive environment than the corporate network. It is outside of the trusted perimeter of the company. The DMZ may be the first fire line. Thus, its servers should be hardened to their maximum. These servers are sometimes called bastion servers when they are hardened for such hostile environments. The bastion servers should host only the minimum set of services needed for their operation. These services should run with the lowest privilege possible. Each service should also run with its own permissions. Only the required accounts should be opened on the server. They should avoid default accounts and settings as well as default admin passwords. Each administrator should have his own account, for accountability and monitoring. The bastion servers should always run the latest security-supported OS, and all security patches should be installed. Their logfiles should regularly be checked to detect potential intrusions and accordingly responded promptly (Sect. 9.3.2).

In the previous figures, the corporate network was designed as a flat entity. Best security practices attempt to define logical subnets of the LAN and isolate them with intranet firewalls. This internal segregation provides additional obstacles to an intruder in reaching her target. Furthermore, if these subnets are organized to mimic the organizational structure of the enterprise, then they provide some segregation between entities. Such isolation is good practice to protect against potential attacks from insiders.

8.3 Takeaway

8.3.1 Assume External Systems Are Insecure

A sound security assumption is that all external systems are insecure and already compromised. External systems mean all systems that are not inside the security perimeter or that the owner of the internal system does not control. As such, a DMZ should ideally be considered as an external system. DMZs are clearly in a hostile environment, and they cannot be treated as trusted as they may be compromised. Many APTs or simple cyber-attacks used a third party, i.e., an external system, to breach the target system. For instance, the APT against the US retailer Target infected first the supplier of the air conditioning system. The supplier had a connection to Target's corporate network. From this position, the attackers breached Target's corporate network to infect the POS.

Segregating these external third parties is paramount. They should not be given direct access to the corporate network. Ideally, this access should be managed by a dedicated server in a DMZ. This server should handle only the services dedicated to this third-party supplier or partner. The dedicated server in the DMZ would act as a proxy for the corporate system. This server should be allowed to access within the corporate LAN only the services that the third-party supplier should legitimately access. Furthermore, it should only be granted access to data needed for the normal operation of the third-party supplier. If possible, the DMZ should handle in its perimeter a mirror of the required data. The mirrored data should be stripped down to this strict necessary minimum rather than the full dataset being stored on the corporate LAN (Rule 2.1: Give Access Only to Those Who Need to Know). The firewall between the DMZ and the corporate network should enforce this strict isolation. As the goal is well defined, the configuration of the firewall can be tight down and adequate monitoring can be instantiated. With such architecture, the risk of an unsecured or breached external system is mitigated. If an attacker were to access the third-party supplier's system, she would first access only the dedicated server in the DMZ. This server has no more information than what the third-party supplier already has in its own system. Breaching within the dedicated server would not provide additional information. The attacker must now penetrate the corporate system through the second firewall to gain some benefit.

Rule 8.1: Do Not Connect Directly to the Internet

A device or a local network should never be directly connected to the Internet. Not everybody may need to install a DMZ or can install a DMZ, especially at home. However, everybody should install a firewall between his network and the Internet. In a consumer environment, the firewall should by default ban every ingress communication. For egress, the firewall should only enable the ports of the most often used protocols such as 80/TCP (HTTP), 443/TCP (HTTPS), 25/TCP (SMTP), 53/TCP/UDP (DNS), 143/TCP (IMAP4), and 110/TCP (POP3). Power users or applications may need to authorize other ports, but they should be skilled enough to manage the creation of corresponding firewall rules for these additional ports. The Universal Plug and Play protocol (UPnP) may also handle the addition of necessary rules when installing new applications on devices. Ideally, the gateway should host anti-virus software (or use a cloud-based anti-virus service). This gateway-hosted anti-virus does not preclude the use of anti-virus software on individual computers. It is just an additional defense.

Many critics estimate that such perimetric defense is outdated. Their argument is that the perimeter is becoming fuzzier and porous. Once inside the perimeter, malware will continue to spread, or attackers will navigate laterally. It is true that perimetric defense is not anymore sufficient. Nevertheless, it remains mandatory. It is a proper application of Rule 1.4: Design In-depth Defense. Perimetric defense is the first barrier that the attacker must bypass to enter the system. Without such a barrier, the attacker may too easily breach a network. Network security would become one of the weakest links. Perimetric defense is complementary to other defenses.

The world has become increasingly connected. With the advent of cloud computing and cloud storage, companies and individuals increasingly use these convenient remote external services. The promise of the cloud is to turn computing and storage into a commodity such as power supply or telecommunications [368]. A cloud can be viewed as a special case of an external system. Will the cloud modify the trust model and require extending trust to these external service providers? In the case of cloud storage, this change of trust model may not be necessary. If the function of the cloud is just storage, then it should be possible to store only encrypted content in the cloud. The local computer could encrypt data prior to storing them on the cloud. Similarly, the local computer could decrypt data after retrieving them from the cloud. The keys remain under the control of the user. In this configuration, the storage provider does not need to be trusted, at least for the confidentiality and integrity properties. Even if an attacker compromised the storage provider's service, the attacker would not gain access to information. Nevertheless, the storage provider needs to be trusted for the availability of the data. This is especially true if the storage provider is also in charge of the backup of data. In the case of cloud processing (such as SaaS), the service provider has to be trusted (Sect. 4.2.4) Nevertheless, the security analysis should still consider the possibility that an attacker may breach the cloud. The cloud is both an external and an internal system. When using cloud computing, the same techniques used to protect corporate computers have to be applied to the cloud infrastructure. Most cloud providers offer network security tools such as firewalls, DMZs, or load balancers.

8.3.2 Privacy

Snapchat is a photo and video messaging service that allows users to take pictures, record videos, append text, and drawings and send them to selected recipients. The sender defines a lifetime for a piece of content called a snap. The lifetime begins once the recipient starts to watch the snap. After expiration, the recipient can no longer watch this content. Most users believed that Snapchat had no memory: "What is posted on Snapchat will be erased after a while, and no instance of the posted snap will exist anywhere." This mental model is inaccurate. In October 2014, over 100,000 Snapchat photos were leaked, surprising the believers of the ephemerality and privacy enforcement of Snapchat [369]. The 13.6 GB file containing the stolen pictures was first posted on the Web site www.viralpop.com. Soon, copies of the file were widely spreading over the Internet. The Snapchat system was not breached, but rather a breach occurred at some undisclosed, third-party application provider who serviced Snapchat. The promise of ephemeral messages was wrong. Applications such as `snapsave` allow users to back up the snaps that they have received, regardless of its user-defined lifetime. In other words, these applications bypass the promises of the ephemerality of Snapchat. Thus, these applications give way to breached privacy. Obviously, security-aware people were cautious and challenged the claimed ephemerality of snaps. Unfortunately, this was not the case of most end users. On February 6, 2015, a 16-year-old teenager was charged with the murder of a classmate [370]. The teenager took a selfie with the corpse of his dead victim in the background. He sent this picture to a friend via Snapchat. The recipient saved the picture and showed it to his mother who called the police. In 2014, the US Federal Trade Commission (FTC) had already complained that Snapchat misled users about the self-deleting capabilities. Snapchat agreed to a settlement with the FTC that included a regular monitoring of their practices for the next 20 years [371].[8]

Rule 8.2: Thou Will Be Traced

Service connected to the Web may identify the host through many means, even if it did not identify itself. Many services or Web sites would like to infer the identity of the connected host, or, at least, know whenever the host returns. In some cases, services may even attempt to collect connections from a given host across various Web sites to profile the behavior of the user. This data collection is a serious privacy breach. Currently, there is an arms race between the user profiling services and the privacy conserving solutions proposed by different browsers, often via extensions or plugins.

How can a service uniquely identify the connected host? The first trivial solution is to use the IP address of the host. However, the IP address does not uniquely identify the host for all possible scenarios. If a system uses a static IP address, profiling becomes straightforward. If a system uses a dynamic IP address, then, at each new connection, the network may attribute a new IP address for the host. In that case, profiling via IP address is thwarted. If the system is part of a subnet within a LAN, then the service sees the IP address of the Internet gateway rather than the device's.[9] Furthermore, the gateway address will be the same for all devices connected to the LAN. Once more, profiling via IP address is thwarted. Furthermore, by using proxy servers, anonymizing proxy servers, or an anonymizing network such as Tor, the host can proactively hide its IP address. Once more, profiling via IP address is thwarted. Profiling of users must take place at the application layer.

In 1994, Netscape engineers invented a second solution. The profiling service could store a proprietary identifier of the host locally in a small 4 KB text file, called HTTP cookie. When connected to the host, the service looks for the presence of its persistent cookie. The HTTP cookie may contain information such as triggered hyperlinks and parameters.[10] An HTTP cookie is in a readable format and may have an expiration date. As the HTTP cookies are text files stored in a known directory unique to the browser, utility software can delete all the cookies. Browsers provide a similar cleansing feature via the cleaning cache command. Such cleaning defeats user profiling based on HTTP cookies.

In 2004, a third solution appeared. The new solution could store locally more information than the 4 KB available to the sole HTTP cookie. This information would be more persistent than HTTP cookies. Furthermore, several Web domains could share the information, allowing cross-site profiling. Macromedia's Flash's locally

---

[8]In September 2015, Snapchat proposed a paid feature allowing the recipient to replay a Snapchat. Each replay was monetized. Ephemerality was not anymore a dogma.

[9]This statement is true for IPv4. With IPv6, this statement is wrong, as the IP address encompasses the main address and the address within subnets.

[10]The Web site cookie keeps this information in order to change the status of a hyperlink from active to already visited.

shared objects, also known as flash cookies, present these interesting characteristics. They are stored in a non-trivial location, can handle up to 100 KB of binary data, and do not require JavaScript, which users sometimes disable. The Macromedia Flash Player plugin manages Flash cookies via a flash application (`.swf`). With the prevalence of Flash format, most of the deployed browsers currently have the Flash Player plugin installed. Thus, Web sites can use it to store flash cookies stealthily. More interestingly, advertising networks may use them to track the browsing history of Web sites using the same advertising network, independently of the browser used. Deleting Flash cookies is a tidy task that requires some expertise out of the reach of most users.[11] Nevertheless, the deletion is still possible. In 2015, some browsers, e.g., Chrome, started to deprecate the Flash Player plugin.

In 2005, United Virtualities designed a new, unusual fourth method to increase the persistence of HTTP cookies. Flash cookies could restore deleted HTTP cookies. This operation is called cookie respawning. The Flash cookie stores the information that was stored in the HTTP cookie. At the beginning of a new session, the Web site checks in its Flash cookie whether a recorded HTTP cookie was deleted. If so, the Web site restores the HTTP cookie using the backup information stored in the Flash cookie. Cookie respawning is a widely deployed method [372].

The advent of social networks raised a flurry of privacy issues. The goal of social networks is to share information with other people. Social networks handle different types of data. Bruce Schneier proposed a taxonomy of this kind of data [373]. The taxonomy embraces the point of view of the account owner.

- Service data is the data used to manage a service, such as the name of the account owner. Usually, the account owner has control over the creation of this type of data. Nevertheless, some social networks require the account's owner to give real data and forbid pseudonyms.
- Disclosed data represents the information that the account owner posts on the social network. The information is published visible to the world or a limited set of viewers. The account owner has full control of this type of data.
- Entrusted data represents the information that the account owner posts on other people's pages. The account owner has control only over the creation. Once posted, he loses control as the data is on another user's page. The account owner of this page may delete or alter the entrusted data. Bob may post information on Alice's wall. She can delete this posted information as she controls her wall.
- Incidental data represents information related to the account owner, but that other person posted. When Alice tags pictures with the name of Bob, or Alice publishes a picture featuring Bob, she creates incidental data related to Bob. The account owner has little control over its creation, or on its life. Entrusted data created by Alice are incidental data for other people.
- Behavioral data represents information that the site collects about the account owner's habits by recording what he does and with whom he interacts. This collection is the "raison d'être" of many social networks. Most business models

[11]For instance, under Windows, Flash cookies are stored in hidden directories.

of social networks are based on monetizing their users' profiles. The primary revenue stream of such social networks is from targeted advertising. The account owner has little control over behavioral data. People rarely fake their behavior just to misdirect behavioral analysis.

- Derived data is data about the user that is inferred from the five previous types of data. Derived data allow the social networks to refine a user's profile and thus increase its value. The more the social network knows about a given user, the more relevant are the advertisement or personalized services the social network offers. Derived data is the most valuable asset of a social network. The account owner has definitively no control on derived data.

Users can control only the second and third types of data. However, to adequately control this data, a user must understand the consequences of publishing information on social networks. Awareness programs on the cyber risks related to social networks is essential. "*Je publie, je* réfléchis"[12] is the name of a French Internet site that aims at sensitizing people (mainly young audiences) on the risks of publishing information on social networks. The CNIL (French authority for IT and Liberty) designed this site. It provided ten recommendations to follow before publishing, such as "ask yourself if you would do the same in real life" and "Don't publish content that may harm the reputation of somebody else."

Information about people is not leaked exclusively through social networks. Web sites and Web services are collecting information continuously. Marketers aggregate them and analyze them, via so-called big data mining, to better profile the user. For marketing purposes, there are four sources of information.

- First-party data are collected by the marketer itself, usually behavioral and derived data.
- Second-party data are received from partners, for instance, brands.
- Third-party data are delivered, often purchased, from other sources, such as data services providers.
- Fourth-party data originate from IoT-connected objects that can identify their user, usually service data.

Users have little control over the three first types of data. They may have control over the last type of data. Unfortunately, users may not be aware of the consequences of authorizing the sharing or publication of this fourth type of data. Furthermore, when blocking this kind of data, the utility of the connected device may decrease. It is not certain that many people will trade some utility and convenience for increased privacy.

Privacy is the ability of an individual or group to exclude itself, or information about itself, and thereby express itself selectively. The scope of this information differs for each culture. All the previous examples show that privacy is in danger in the digital world. The US privacy law has defined the concept of Personally

[12]"I publish, I think" in French.

Identifiable Information (PII). PII is information that may identify an individual with a high level of certainty. The association of several PIIs identifies an individual without any doubt. The NIST classified the following information as PII: Full name, home address, date of birth, birthplace, private email address, telephone number, national identification number (such as the Social Security Number), passport number, IP address, vehicle registration plate number, driver's license number, face, fingerprints, handwriting, credit card numbers, digital identity, genetic information, login name, screen name, nickname, and handle. Currently, enterprises are extremely careful when handling PIIs. They may be liable in case of leakage of PII due to negligence. Many corporate data policies encompass one particular category for PII. Best security practices would require PII to be encrypted and their access strictly controlled.

Unfortunately, in the past years, numerous breaches have leaked tons of PIIs. The Darknet is a flourishing marketplace for this type of stolen information [374]. Nevertheless, the digital world continues to collect a massive amount of information about its users. As the business model of many Web sites or services relies on the monetization of the gathered information or the monetization of their analysis (profiling), this trend does not change [375]. For instance, most of the revenues of Google are derived from advertisements. A targeted advertisement is presented only to a precisely defined, targeted audience. The impact of targeted advertising on consumers is more efficient than the impact of regular advertising. Thus, advertisement agencies pay a premium for such targeted advertising. Google's profiling of users enables the optimal delivery of targeted advertisements. Anonymity is not an offering by such businesses unless the user is ready to pay for it. Until tools enable automatically better privacy, privacy will erode.

8.3.3 Anonymity

As already mentioned, the connected digital world collects massive datasets. Datasets relate to many domains and forms: data such as traces and streams of information, data related to movement and localization, time series, biomedical data, and data related to Web usage and social networks. The collected datasets may serve two primary purposes.

- User profiling: when focusing on one given target, the analysis of the dataset attempts to guess better the behavior and preferences of one particular individual. It also exploits data from other users to find similarities that may enhance the profiling.
- Statistical analysis: The datasets can be useful to discover trends or compute statistics about population, transportation, diseases, and so on. This activity is often called data mining or knowledge discovery [216]. It is different from user profiling, although it may use the same initial dataset. There is no doubt that its benefits will be decisive for academics, governments, companies, and end users.

Nevertheless, even with this seemingly benign usage, there are serious related privacy issues.

In 2006, in a famous paper, Michael Barbaro and Tom Zeller, two journalists from The New York Times, disclosed that it was possible to identify users from an AOL anonymized query logfile [376]. AOL published a collection of 20 million anonymized Web search queries that it had collected over time. AOL assigned a number to each searcher to protect his or her anonymity. The aim of AOL was to help the research community. In their study, the two journalists focused on the user with number 4417749. Number 4417749 conducted hundreds of searches over a 3-month period on topics such as "numb fingers," "60 single men," and "dog that urinates on everything." By exploring the information that each search revealed, the journalists found the actual identity and address of number 4417749 without too much difficulty.

Thus, in the second type of usage, the objective is to anonymize the dataset, i.e., transform the dataset into a sanitized dataset that contains no more personally identifiable information. Two fundamental concepts are used to anonymize a dataset: randomization and indistinguishability. Randomization modifies sensitive data by combining them with random values, thus decreasing their accuracy. Indistinguishability alters sensitive data either by masking them or by using synthetic techniques that generate new data that replace the sensitive data. k-anonymity and l-diversity are two major properties of anonymized datasets. A released sanitized dataset provides k-anonymity protection if the information for each person contained in the released dataset cannot be distinguished from at least k-1 individuals whose information also appears in this released dataset [377]. A block of data is l-diverse for sensitive attribute S if it contains, at least, l "well-represented" values for this sensitive attribute [378].

Many techniques are used to anonymize datasets.

- Condensation is one such method used to preserve privacy. Condensation regroups close data within larger groups. The output groups still maintain some statistical properties of the original dataset (for instance, first-order and second-order statistics). Condensation enhances indistinguishability.
- Suppressing samples to remove vulnerable items that may be too easily identifiable, thus increasing indistinguishability.
- Reaching k-anonymity by co-clustering, at least, k similar data, thus once more increasing indistinguishability.
- Perturbation techniques, adding noises to the dataset. The choice of the type of noise is crucial. By employing techniques inherited from signal processing, such as wavelets and digital filters, an attacker may extract sensitive information. For instance, adding white noise is not sufficient to sanitize a dataset. A simple compression algorithm followed by the decompression would desanitize the sanitized dataset. Compressed information usually has reduced noise compared to the original data. Perturbation techniques increase randomization.

The challenge of anonymizing a dataset is to find the right equilibrium between privacy preservation and the utility of the sanitized dataset. The larger the perturbations are, the higher the preservation of privacy is. Conversely, the larger the perturbations are, the lower the accuracy of the statistical properties is. Thus, its utility diminishes. Finding the ideal equilibrium is the task of the data scientist. Currently, there are no available techniques that cannot be defeated if the sanitized dataset keeps a reasonable amount of utility.

The ability to communicate anonymously is another significant challenge in our digitally connected world. Many users seeking anonymity employ HTTPS anonymizing proxy servers or commercial VPN operators. A proxy server is a server that acts as an intermediary. The Web server interacts with the proxy server rather than the actual browsing device. The user configures his browser to forward all communications to the anonymizing proxy server using TLS. The anonymizing proxy server forwards the received communication to the final target, thus hiding the IP address of the issuer from the receiving site. Many anonymizing proxy server services offer the possibility of selecting the country from which the communication may seem to originate. This trick efficiently thwarts geofiltering, as geolocation can only detect the issuing address of the proxy server. Nevertheless, using anonymizing proxy servers has several drawbacks. Network sniffing may reveal a large amount of encrypted data exchanged with one service and thus signal the utilization of a remote proxy server. Filtering out the corresponding ingoing address blocks this communication. Similarly, filtering out the corresponding outgoing addresses kills the service. Therefore, anonymizing proxy servers are not efficient against state censorship or when stealth is needed. The second drawback is that the anonymizing proxy server operator knows the IP address of the user. If this service provider logs all the transactions, then there is a risk that it may hand them over upon receiving a judicial request. Hence, the anonymizing service provider has to be trusted. All service providers do not offer the same level of anonymity [379].

For strong anonymity, a better solution is necessary. The Onion Router (TOR) provides such a solution. TOR is the most deployed software for anonymous communication over TCP. The project started in 1999 under the code name Free Heaven Project. In 2004, it switched to a new project with enhanced features. The goal of TOR is to protect against traffic analysis and censorship. The aim of traffic analysis is to find out who is talking to whom over the surveyed network. The aim of censorship is to block the transfer of some information or to prevent some people from communicating. TOR attempts to hide the IP address of the originator as well as the IP address of the recipient from spying ears that eavesdrop on used network.

TOR uses a network of volunteer-operated servers called onion routers. Each onion router has a unique public/private key pair. If Alice wants to communicate anonymously with Bob, TOR selects a random path from Alice's computer to Bob's computer by randomly choosing n onion routers, so-called a circuit. The secure communication of the message m requires multiple successive encryptions, called onion routing [380].

Following is a simplified version of onion routing. The public key of the ith onion router of the circuit is K_{pub_i}. First, Alice encrypts the message m with Bob's public key $K_{pub_{Bob}}$. The encrypted message is m_I.

$$m_1 = E_{K_{pub_{Bob}}}(m)$$

Then, Alice encrypts the previously encrypted message m_I with the public key of the last onion router of the selected circuit.

$$m_2 = E_{K_{pub_n}}(m_1) = E_{K_{pub_n}}\left(E_{K_{pub_{Bob}}}(m)\right)$$

And so on recursively until the first onion Router.

$$m_n = E_{K_{pub_1}}(m_n) = E_{K_{pub_1}}\left(\ldots E_{K_{pub_n}}\left(mE_{K_{pub_{Bob}}}(m)\right)\right)$$

The encrypted message m_n is sent to the first onion router that decrypts it with its private key pair. While the encrypted message is passed from one onion router to the next onion router, each onion router peels away a layer by decrypting the message with its private key. Figure 8.5 illustrates the decryption process with three hops. Each tube represents the same encrypted data, with a different color for each key. Each onion router removes a layer of encryption.

TOR uses a more complex encryption scheme than the previously described one for enhanced efficiency. Using asymmetric cryptography to encrypt the actual message would be too expensive in terms of computation and too slow. The message m is encrypted using symmetric cryptography, and an asymmetric cryptographic scheme protects these symmetric keys.

Thus, an attacker who eavesdrops on the communication cannot decrypt the message as there is always, at least, one remaining encryption layer. This layering ensures confidentiality of the message. The protocol establishing a random circuit ensures that each used onion router knows only the IP address of its direct predecessor and the IP address of its immediate follower. The protocol ensures also that an onion router cannot discover the IP address of any other onion router of the circuit than its immediate neighbors. An attacker may own an onion router. Indeed, as the onion routers are volunteered, an attacker can propose her server as an onion router. Hence, according to some revelations of Edward Snowden, the NSA

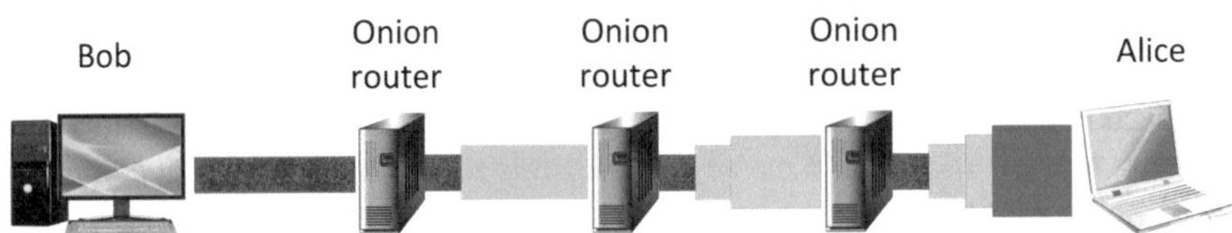

Fig. 8.5 Onion routing

controls several onion routers. Nevertheless, the owner of this onion router does not gain any significant information, unless it is the last onion router of the circuit. This last exit point knows the final destination, which may be interesting for an attacker. Nevertheless, TOR users do not need to trust one single onion router. They trust the fact that one individual malicious entity does not control the majority of the onion routers of the TOR network. In 2015, there were more than 6000 onion routers deployed all over the world.

Recent information leaked by Edward Snowden seems to indicate that TOR is actually private and secure. The NSA acknowledged its robustness in an internal memo. Nevertheless, TOR has some weaknesses. For instance, it is possible to list all the exit nodes. A government can thus block them to enforce censorship.

8.4 Summary

Law 8: If You Watch the Internet, the Internet is Watching You
Most connections are bidirectional. The consequence is that information flows both ways. Controlling what is exchanged, and monitoring who is using the connection, is the role of network security. Fortunately, network security is a rather mature science.

Rule 8.1: Do Not Connect Directly to the Internet The access to the Internet should be carefully controlled. It should have at least a firewall and anti-malware filtering. When possible, implement a DMZ to create an isolation buffer between the Internet and your network that may discard attackers to intrude it.

Rule 8.2: Thou Will Be Traced The digital world increasingly keeps records of all the activities of users. Many Web enterprises build their business model on monetizing the results of this data collection. This constant monitoring is a threat to privacy and also a potential mine of information for attackers. Some tools, such as TOR, help in preserving anonymity on the Internet.

Chapter 9
Law 9: Quis Custodiet Ipsos Custodes?

> ***Muffley***: *General Turgidson, I find this very difficult to understand. I was under the impression that I was the only one in authority to order the use of nuclear weapons.*
> ***Turgidson***: *That's right sir. You are the only person authorized to do so. And although I hate to judge before all the facts are in, it's beginning to look like General Ripper exceeded his authority.*
>
> Dr. Strangelove [381]

9.1 Examples

The procedure that prevents the launch of American nuclear weapons without presidential authorization is claimed to be "fail-safe." It is a combination of manual processes, redundant controls, and technical means. Among these technical means, the Permissive Action Link is the most well-known one (Hollywood often illustrated it in its movies). A Permissive Action Link is a lock that requires an officer to enter an eight-digit code before authorizing the firing of a missile. In the 1960s, such a physical Permissive Action Link prevented an unwanted launching of the American Minuteman nuclear missile. Unfortunately, the Strategic Air Command, which is in charge of controlling nuclear weapons, pointed out that this safeguard would be dangerous in a time of crisis. The high command feared that under critical crisis conditions, either the nuclear code could be lost or that the lock could fail to unlock when presented the correct nuclear code. According to the high command, the safeguard during peacetime would become a risk during wartime. Thus, the Strategic Air Command decided, without consulting the White House, to set the code of all Permissive Action Links to 00000000 [382]. Furthermore, the nuclear launch checklist instructed the officers to verify that the digits of the locks were all set to 0 at any time. In other words, the Strategic Air Command circumvented the Permissive Action Link. Although a strict, fail-safe procedure was defined, documented, and approved, a high authority decided to bypass it. Nobody verified whether operators applied the written procedure, or, at least, no military officer with sufficient authority did this verification. Once this information was disclosed to the

E. Diehl, *Ten Laws for Security*, DOI 10.1007/978-3-319-42641-9_9

public, this circumvention was stopped, and the procedure was properly enforced. The current Permissive Action Links are supposed to be set to a non-null value.

When developers attempt to detect software pirates and denounce the infringers, things may also go wrong. The Japanese software editor Enfour published online dictionaries for iOS, such as the Oxford Deluxe. It was alleged that 75 % of the deployed dictionaries were pirated versions. In November 2012, the editor decided to employ a surprising strategy to fight back piracy. Enfour's developers appended an anti-piracy module to the latest version of their dictionaries [383]. At the time of installation of the new version, the anti-piracy module requested access to the user's Twitter account. Whenever the anti-piracy module detected a pirated version, it automatically posted to the user's Twitter account without his consent the following tweet:

> How about we all stop using pirated iOS apps? I promise to stop. I really will. #softwarepirateconfession.

The editor decided to shame publicly the alleged pirates. Besides the ethical issues, there was another technical problem. Rather than verifying from a remote database whether the user had acquired a legitimate license, the module used other heuristics for its shaming. Unfortunately, these heuristics were not reliable. For instance, whenever the module detected a jailbroken device, it decided that the version of the dictionary was necessarily pirated. Thus, many legitimate users, having paid for their licenses, but using a jailbroken device, were publicly shamed for a crime they did not commit.

In 2013, Verizon disclosed an interesting case [384]. The reviewing of the VPN logfiles of an undisclosed America critical infrastructure company revealed a worrisome fact. The auditors spotted an open and active VPN connection between their American corporate network and a computer located in Shenyang, China. To the best of their knowledge, there should been no such connection. This connection had been active during American business hours for many months. Furthermore, the VPN connection used the credentials of an American employee who was currently located on the American site, rather than in China. Uncannily, the VPN utilized a two-factor authentication based on a physical SecurID token from RSA Ltd. First, the investigators suspected that a 0-day malware had infected the computer of the US employee and that Chinese hackers had taken control of it. The hypothesis was legitimate as the case concerned critical state infrastructures. This was not the case. In the end, the forensics investigation highlighted a less technical solution. No cyberattack or foreign spying was going on. The American employee was a software developer who had subcontracted his software development work to a Chinese consulting company. To enable the remote connection, he had shipped his SecurID token to Shenyang. There, the Chinese subcontractor connected as the legitimate developer into the enterprise's system. As the subcontracted development work was irreproachable, the American developer's employer never spotted this deceit. Furthermore, the American employee was prized for being an excellent software developer. Only a careful analysis of the VPN logfiles highlighted the evidence of the unexpected activity.

This book has already cited Edward Snowden many times. The Snowden case is exemplary for illustrating Law 9. The security procedures of the NSA are known to be strict, with excellent cybersecurity practices. For instance, regular NSA employees cannot transfer confidential files to USB drives or any other type of external storage. Tools such as DLP enforce this policy. Snowden was a contractor working for Booz Allen Hamilton in Hawaii. Nevertheless, Snowden was granted system administrator rights [385]. A system administrator has exceptional privileges such as for handling any files or for accessing any accounts. He may even get access to the accounts of employees with higher security clearance than himself. A contractor should never have been granted such elevated privileges. Regular reviewing should have verified that all the 1000 NSA system administrators had the highest clearance. One thousand system administrators seem a far too large number. It violates Law 4: Trust No One. Such a large number dilutes the trust assumption. Control is impossible with such a large pool of system administrators. At least, there should be a segregation of systems with segregated system administrators. Since the leak, the NSA reduced by 90 % the number of its system administrators [386].

Credit scores are an important element of the financial lives of US residents. Specialized companies monitor and evaluate the credit scores of individuals on behalf of companies who want to assess the solvency of their customers. Experian is one of the leaders in this field. It is expected to be irreproachable. On September 2015, the American Telecom operator T-Mobile announced that personal data of 15 million customers had leaked [387]. The leaked personal information included name, address, birthdate, SSN, and ID number. The leak occurred not at T-Mobile. Instead, it happened at its credit monitoring service provider: Experian. For 2 years, attackers had obtained access to a server containing this unencrypted dataset. Only the SSNs and the ID numbers were encrypted. Furthermore, Experian determined that the encryption might have been compromised. The credit warden was not at the forefront of cybersecurity.

Bitcoin offers an example where neither a central authority nor a group of trusted auditors performs the monitoring of transactions. Every user validates Bitcoin transactions. No Bitcoin user has a particular role. Every few minutes, all the passed transactions are recorded in a block B_i. Then, the users have to validate this block. For the block B_i to be validated, there are two conditions.

- All the transactions in the block must be valid. All past transactions are recorded in a ledger that is a chain of all past blocks. This ledger is called the *Block Chain*. Every user has a copy of the block chain. Thus, users check the validity of the new transactions within their instance of the block chain.
- The block must be associated with a valid *proof of work*. To create a proof of work, the users attempt to solve a challenge. They have to find x such that the following equation is valid: $Challenge \geq Hash(B_i|x)$, where $Challenge$ is a value defined by Bitcoin. This value adapts the average time needed to solve the challenge. It depends on the number of users, called miners, and the total current calculation power. Solving the proof of work is costly and time-consuming, whereas verifying the proof of work is cheap and quick.

The miner who solves the challenge is rewarded a certain amount in digital money by Bitcoin.[1] This reward is an illustration of Rule 7.3: Align the Interests of All Actors. The combination of the common interests of all actors and the use of costly computation allowed the design of a system whose users are also the wardens of the system. This system of ledger is one of the most promising outcomes of Bitcoin technology.

9.2 Analysis

9.2.1 CobiT

Since 1996, the Control Objectives for Information and related Technology (CobiT) has provided best practices across a domain and process framework and presented security-related activities in a manageable and logical structure [388]. Its current version is 5.0. CobiT defines four interrelated domains, as illustrated by Fig. 9.1.

- *Domain Plan and Organize* describes the strategy of the organization, identifies the business units' needs, and verifies whether the risks are adequately addressed. Moreover, it provides guidance to domain "Acquire and Implement" and domain "Deliver and Support."
- *Domain Acquire and Implement* provides the solutions to fulfill the requirements defined by domain "Plan and Organize" and passes them to domain "Deliver and Support"; this domain selects the proper solutions and implements them in the organization. This domain handles the research and development part of security.
- *Domain Deliver and Support* receives the solutions and makes them usable for end users; this domain operates the solutions. This domain handles the operational part of security.
- *Domain Monitor and Evaluate* monitors all processes to ensure that the direction defined by domain "Plan and Organize" is followed. This domain addresses performance management, monitoring of internal control, regulatory compliance, and governance. It typically addresses the following management questions:
  - Is IT's performance measured to detect problems before they may occur?
  - Does the management team ensure that internal controls are effective and efficient?
  - Can IT performance be linked back to business goals?
  - Are adequate confidentiality, integrity, and availability of controls in place for information security?

[1] The current reward is 25 bitcoins. This reward is halved about every 2 years. Nevertheless, there is a fee associated with each validated transaction.

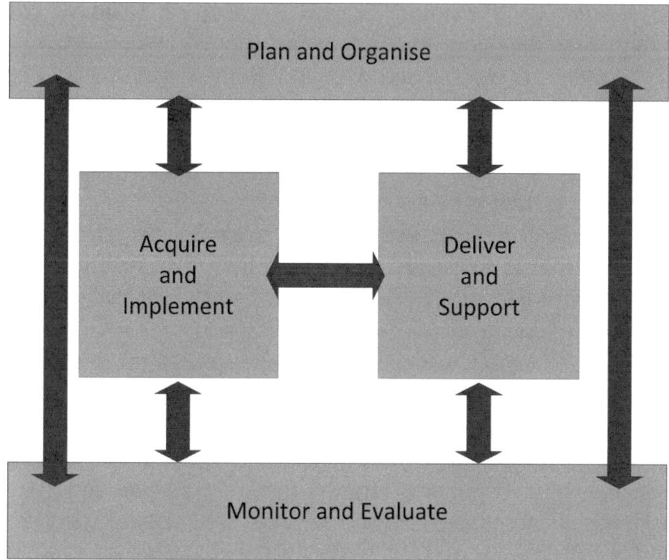

Fig. 9.1 The four domains of CobiT

For the last domain "Monitor and Evaluate," CobiT defines four control points.

- ME1: Monitor and evaluate IT performance: This control point includes identifying relevant performance indicators, systematic and timely reporting of performance, and prompt response to deviations. Monitoring is mandatory to ensure that the proper actions are taken and are in line with the directions and policies set by domain "Plan and Organize."
- ME2: Monitor and evaluate internal control: This control point includes the monitoring and reporting of control exceptions, results of self-assessments, and third-party reviews. A key benefit of internal control monitoring is to provide assurance regarding effective and efficient operations and compliance with applicable laws and regulations.
- ME3: Ensure compliance with external requirements: This control point enforces compliance with laws, regulations, and contractual requirements. This control includes identifying compliance requirements, optimizing and evaluating the responses, obtaining assurance that the requirements have been complied with and, finally, integrating IT's compliance reporting with that of the rest of the business.
- ME4: Provide IT governance: This control point includes defining organizational structures, processes, leadership, and roles and responsibilities to ensure that the enterprise's IT investments are aligned and delivered in accordance with the enterprise's strategies and objectives.

This chapter focuses on the control point ME2. CobiT defines seven associated control points.

- *ME2.1: Monitoring of Internal Control Framework*: Requires continuously monitoring, benchmarking, and improving the IT control environment and control framework to meet organizational objectives. The aim is to be sure that the control framework is up-to-date and adapted to the organization. The auditor verifies whether the proposed controls are adequate. This control point does not validate the results of the controls themselves, but rather assesses the implementation and usefulness of the controls.
- *ME2.2: Supervisory Review*: Monitors and evaluates the efficiency and effectiveness of internal IT managerial review controls. The auditor checks whether the management team actually reviews the reports and acts in line with the outcomes of these reviews.
- *ME2.3: Control Exceptions*: Identify control exceptions and analyze and identify their underlying root causes. Control exceptions should be rare and have valid business rationales. The auditor escalates the control exceptions and reports them to stakeholders appropriately. The management team must, at least, be aware of these exemptions. Where possible, the controller should institute necessary corrective action, and ideally, remove the need for a control exception.
- *ME2.4: Control Self-assessment*: Evaluates the completeness and effectiveness of management's control over IT processes, policies, and contracts through a continuing program of self-assessment.
- *ME2.5: Assurance of Internal Control*: Obtains, as needed, further assurance of the completeness and effectiveness of internal controls through third-party reviews. Internal auditors have to be reviewed. Third-party reviewing of internal auditors and internal reviews is a deterrent against insiders among the internal auditors. Without such an external control, the internal auditors would have too much power.
- *ME2.6: Internal Control at Third Parties*: Assesses the status of external service providers' internal controls. This control point confirms that the external service providers comply with legal and regulatory requirements and contractual obligations. This control mitigates partly the risks highlighted in Sect. 8.3.1.
- *ME2.7: Remedial Actions*: Identify, initiate, track, and implement remedial actions arising from control assessments and reporting. This is the usual remediation loop that fixes the problems identified by previous control points.

9.3 Takeaway

9.3.1 Separation of Duties

In January 2008, Jérome Kerviel became famous. This second-rate French trader lost five billion euros for his bank, Société Générale. In December 2007, Jérome Kerviel realized 1.5 billion euros of profit for this bank. The corresponding investments were toxic and more due to frauds than to real financial success. In January 2008, these dubious risky investments turned into an unstoppable five-billion-euro loss (about

5.5 billion dollars). The latest report highlights that the controlling mechanism of the bank did not work properly [389]. A second-rate trader like Kerviel was not authorized to invest such large amounts of money. Kerviel violated many rules. Kerviel stole or guessed the passwords of his colleagues. He used these stolen passwords to launch his instructions and to approve them. Before being a trader, Kerviel was on the technical team. There, he learned the mechanisms needed to bypass some controls with proper credentials. Typically, a controller should have spotted and signaled these violations. The question was whether the management team was encouraging auditors to close their eyes. It seems that his direct manager received several automatic warning messages, but he did not react. It is difficult to determine why the manager did not take into account the alerts. What would be the behavior of a controller who detects a violation that produces such enormous financial benefits? Would his management team retaliate for his obstructing such huge revenues? Should the auditor act ethically regardless of the consequences? Investigations highlighted a higher sensitivity of the audit process to the proper completion of trade rather than to the proper evaluation and management of financial risk. In other words, the control favored benefits over financial risks.

During the 1980s, while negotiating the reduction of nuclear weapons, President Jimmy Carter regularly told President Gorbachev: "Trust, but verify." This is an old traditional Russian saying. It clearly highlights the need for control in the enforcement of trust.

To cope with such conflicts, the usual solution is the separation of duties. As it reduces the power of one individual, the separation of duties limits the risk of fraud and decreases the risk of conflicts of interest. Several individuals need to collude for defeating adequately defined the separation of duties. Separation of duties is why, at the theater, one person sells the ticket and another employee tears the ticket in half. The sale of the ticket and the access control to the theater are separated to avoid fraud. The cashier cannot sell a seat without delivering the ticket, thus ensuring that each transaction is accounted for.

Rule 9.1: Separate the Roles

"Divide and conquer" is an old adage that was known in ancient Greece.[2] Rule 9.1 is its equivalent in the field of security. There are mainly two axes defining the mandatory separation of duties.

- Action and control
- Breadth of the functions

The first axis requires separating clearly the roles that are acting from the roles that are controlling the compliance of the actions. If there is not such separation, then the effectiveness of the control is questionable. Controllers and auditors must be independent of what or whom they control or audit. The independence is at two levels. They must be, as much as possible, hierarchically independent from the

[2]This maxim is attributed to Philip II of Macedon (third century BC).

controlled principal. This independence ensures that they fear no potential retaliation whenever the controlled principals or their direct managers do not appreciate their findings. It is difficult for an auditor to be entirely objective when a negative control may impact his career path or his bonus. The SOX regulation strictly requires this separation for the accounting world. The second level of independence is about the direct interest. Controllers and auditors must not have a vested interest in the controlled principal. It is difficult for an auditor to be entirely objective when a negative control may impact his financial investments or those of his relatives.[3]

The second axis attempts to fragment the responsibilities that one single person may have in the same system. Separating the roles has three benefits.

- It mitigates the risk of human error. If an operation requires two actions by separate individuals, then there is an inherent cross-control of the sanity of the actions. If the first operator did make a mistake, there is some likelihood that the second operator will spot it.
- It mitigates the risk due to an insider. If an operation requires two actions by separate individuals, then the insider has to collude with the second operator to perform its attack.
- It mitigates the risk due to an external attacker. If an operation requires two actions by separate individuals, then compromising the account of one single operator is insufficient. The attacker has to compromise the accounts of two operators.

Separation of duties may have a financial impact. Separation of duties requires more operators and fragments the operations. The efficiency may decrease. The overall cost may increase. Careful definition of the roles may mitigate this additional cost. Separation of duties may also have an emotional impact. Some individuals may resent the fact that they have limited "power." Some individuals may feel that restricting their role is a sign of lack of trust. Explanation and education may water down the emotional impact.

Rule 9.1 has many similarities with Rule 2.1: Give Access Only to Those Who Need to Know. Rule 2.1 concerns access control, whereas Rule 9.1 concerns functionality and the organization. Both rules derive from the same rationale: reduce the risk by reducing the potential scope of a successful compromise.

9.3.2 Logfiles Are to Be Reviewed

Logfiles are major elements of the security of any infrastructure. A logfile is a record of the events occurring within an organization's systems, computers, and

[3]For instance, the value of the financial grades delivered by credit rating agencies such as Fitch Ratings, Moody's and Standard and Poor's are sometimes questioned. The financial institutions that a credit rating agency grades also pay for the grading.

networks. Logfiles are composed of log entries. Each entry contains information related to a particular event that has occurred within a system, computer, or network. Each log entry has a time stamp in addition to the details of the recorded event. There are many sources of security-related logfiles. For instance:

- Anti-virus and anti-malware software record the detected malware, the cured files, and the quarantined files.
- IDS and IPS record detailed information on suspicious behavior and detected attacks and any action performed to stop malicious activities.
- VPN systems record the successful and failed attempts to connect and the volume of exchanged data.
- Web proxies record the URLs accessed through them and the principals accessing them.
- Vulnerability Management Software records the known vulnerabilities of each host and the applied patches.
- Authentication servers record the successful and failed attempts to authenticate with the submitted username and the origin, i.e., device and IP address.
- Routers and firewalls record all the details of blocked traffic.
- OS system events, log-failed events, and the most significant successful events. Many OSs permit system administrators to specify which types of events to log.
- OS audit records failed authentication attempts, file accesses, security policy changes, account changes, and the use of privileges. OSs permit system administrators to specify which types of events to audit.
- Applications may generate their own logfiles with security-related information.

Routinely reviewing and analyzing logfiles is beneficial for identifying security incidents, policy violations, fraudulent activities, and operational problems shortly after they have occurred. They provide useful information for resolving such problems. Once a security incident is identified, logfiles provide useful information for investigation and forensics analysis. However, logfiles are useful only if they are analyzed, else they are entirely useless.

Rule 9.2: Read the Logs

Besides the inherent benefits of log management, some laws and regulations require organizations to store and review certain types of logfiles. The Gramm-Leach-Bliley Act, HIPAA, SOX, and PCI-DSS are examples of such controls.

Logs contain sensitive data. Their integrity should be guaranteed. An attacker should not be able to modify a logfile, else the analysis may drive wrong conclusions, and potential forensics information may be repudiatable. The second expected characteristic is availability. An attacker must not be able to destroy any logfile. Some logfiles may record PII such as authentication credentials and a personal IP address. In that case, confidentiality is a required characteristic. For instance, the access to such logfiles should be strictly controlled. Users should not be able to modify the log records. Principals creating the log entries should only be allowed to append information to an existing logfile and be restricted from removing information. Furthermore, at a minimum, the logfile should be tamper-evident, and ideally, it should be tamper-resistant. The tamper evidence detects any modification, whereas tamper resistance attempts to prevent any change.

One important property of logfiles is their retention time. Logfiles have to be kept for some time. There are two reasons to keep logfiles. The first reason is that national regulations may require keeping some logfiles for a minimum retention time (HIPAA, Patriot Act, PCI-DSS…). The second reason is that the analysis of an attack needs to go back to the past to find the intrusion point of the attack as well as the modus operandi of the attacker. If it is not defined by regulations, the retention time is the decision of the Chief Information Security Officer. As some attacks are discovered only after many months or even years, the retention time should not be too short. Some care may be needed when destroying logfiles containing PII after their retention time has passed. Their deletion must be guaranteed as they contain confidential information.

The analysis of logfiles is a tough, tedious task. Log analysis, sometimes called audit trail analysis, is the act of extracting meaningful information from logfiles. The analyst exploits this meaningful information to draw conclusions about the actual and past security of the monitored system. Log analysis is a black art. Fortunately, tools help the practitioner. Ultimately, log analysis relies a lot on the experience of the analyst, his intuition, and a pinch of luck [390].

The first difficulty of this work is the heterogeneity of logfiles. The format of logfiles is not consistent. For instance, each OS uses its own format for OS-related logfiles. If the system uses many OSs, the corresponding logfiles will have different formats. All logfiles are not all in a human readable format. Many logfiles are binary. Of course, the syntax and structure of the logfiles vary for each software vendor. The representation of the same parameter may vary among logfiles. Tools need to harmonize and normalize the collected data. Furthermore, the time stamp is linked to the internal clock of the host on which the log system runs. All hosts may not be synchronized. It is then harder to analyze the temporal thread between logs generated by different hosts. Tools must temporally align the logs to one unique reference timeline before any analysis. Security information management (SIM) systems automate these tedious operations. They collect the logfiles from different sources. Then, they harmonize the data in a common format with common reference points.

The second difficulty is that the amount of generated data to analyze is enormous. SIM aggregates the data and reduces the size of the data. The last phase is the representation of the aggregated data and the extraction of any irregularities or deviations outside the limits defined by standard correlations. The human visual system is a remarkable pattern seeker and a powerful anomaly detector. Hence, visualization tools are excellent to highlight anomalies in one glance [391]. The tools map the log records into visual representations. There may be several types of visualization depending on the kind of analysis to perform.

- *Report analysis*: A report is a collection of events that occurred during the reported period. It is often reported in the form of a dashboard. A dashboard is a static synthetic view of all aggregated information. It must display in one single view the most significant information and be understandable to a broad audience. The choice of the data to display and the type of employed graphical charts is essential to the usefulness of the dashboard. Usually, it presents also the evolution of the trend compared to previous reference periods.
- *Time-series visualization*: A report presents a static view of one period, whereas a time-series visualization represents the evolution of a set of data over time. The aim is either to find past issues or to predict the future. When looking for clues, there are different types of searches. The first one is to detect new patterns, the variance of patterns, or gaps in an expected pattern. This kind of analysis looks for anomalies that may indicate an ongoing attack. The second one is to detect a correlation between events. This is extremely useful for forensics analysis to discover the operating modes of the attacker. Correlation may find out that event B is a consequence of the occurrence of event A. The third type of search tries to forecast the future. Trendlines facilitate this kind of analysis. The fourth type looks for potential links between seemingly unrelated data. Correlation graphs highlight the potential relationship between two data dimensions. The data dimensions may belong to the same logfile or may belong to different logfiles.
- *Real-time monitoring*: They display the current state of systems and ongoing tasks. They usually employ dashboards. Live feeds update the dashboard. The most salient current information appears in a single synthetic view with clear alert and warning indicators.

For all these visualization modes, interactive graphs are powerful tools that enable a drill down into particular parts of the reported information to fine-tune the analysis. Usually, they work in four steps. The first step displays the global view. The second step changes the attributes of the graphs and even the type of representation to find a potential issue. The third step zooms in or out the selected chart to adjust the right amount of information to the adequate scale. The final step examines the details of a particular part of the chart.

Many tools help the security analyst in his task to analyze the logfiles. Tools often issue warnings when they detect potential anomalies. Nevertheless, the security analyst must analyze the logfiles even in the absence of reported warnings. The automatic tools will not spot all problems. Sophisticated attacks attempt to fly under the radar. The thresholds of the tools may not be set at the proper value. The security analyst must proactively review the reports and dashboards looking for potential ongoing attacks, or questionable evolution of trends. They may be symptomatic of illegal activities. The analysis may also allow him to fine-tune the settings and thresholds of alarms. Regular analysis of logfiles is an application of Rule 5.1: Be Proactive.

9.4 Summary

Law 9: Quis Custodiet Ipsos Custodes?
Any security process should always have one last phase that monitors the efficiency of the implemented practices. This phase creates the feedback loop that regulates any deficiency or inefficiency of the security process. The quality and probity of this last phase have a strong influence on the overall robustness of the security.

Rule 9.1: Separate the Roles The scope of controlling and managing roles should be kept as small as possible. This reduces the impact of a malicious insider or the influence of an attacker who hijacked an administrator or controller account. Where possible, the scope of roles should partly overlap or be redundant between several individuals. This separation increases the chances of detecting an error or an act of mischief from an insider.

Rule 9.2: Read the Logs Logfiles are an important element for monitoring and auditing the efficiency of the security. They will be useful for detecting and understanding security incidents. Nevertheless, their full efficiency is reached when they are regularly analyzed to detect anomalies. Ideally, they have to be proactively analyzed.

Chapter 10
Law 10: Security Is Not a Product, Security Is a Process

If you think cryptography is the solution to your problem, you don't know what your problem is.

Neumann P.

10.1 Examples

In December 2013, the American retail chain Target suffered an enormous disclosure of confidential data. Information for 40 million valid credit cards was stolen. Attackers compromised Target's points of sale, collected stealthily the skimmed credit card information, and later sent the collected data to Russian servers. The attackers used credentials from a heating, venting, and air-conditioning third-party vendor who had access to Target's network. With the illegal access, the attackers compromised servers and installed a customized malware on many American POSs. The customized malware was not a sophisticated one. They also compromised three servers that the exfiltration process used later to transfer the stolen data to Russia. The POS sent the skimmed data to these three servers which forwarded it later to Russian servers. The exfiltration started on December 2, 2013. On December 12, 2013, the FBI informed Target of a massive data breach. Three days later, the malware was eradicated from all points of sale.

After the discovery of the breach, Target launched a thorough investigation to understand why this data breach occurred. In addition to the details of the exploit, an interesting fact popped up [392]. On November 30, 2013, a sophisticated, commercial, anti-malware system FireEye detected the spreading of an unknown malware within Target's IT system. It spotted the customized malware that was installing itself onto the POSs to collect the credit card numbers before sending them to three compromised Target servers. This malware called BlackPOS and designed by Antikiller, scraps the RAM of the POS. The malware analyzes the dumped memory to detect temporary storage of credit card information. The malware is

E. Diehl, *Ten Laws for Security*, DOI 10.1007/978-3-319-42641-9_10

designed for a given type of POS as it knows the location to search and the expected format of the data. BlackPOS collects the credit card information and forwards them to a remote server. Target's security experts based in Bangalore (India) reported the spreading malware to the American Security Operation Center in Minneapolis. FireEye graded the alert level to the most critical one. The Security Operation Center did not respond to this notification. On December 2, 2013, the Indian team sent a second alert to the Security Operation Center, which still failed to respond.

The exfiltration of the stolen data begun only after December 2, 2013. If the Security Operation Center had responded to these alerts, although it may not have stopped the collection of credit card information, it would, at least, have stopped their exfiltration to the Russian servers. As the details on the daily volume of critical alerts that the Indian team reported to the Security Operation Center are not publicly available, it is hard to blame anybody. Nevertheless, two top critical alerts were not treated properly. Despite powerful tools being in place, the process did not address the outcome appropriately. A few months later, Target's CEO had to resign. The financial impact of the data breach was severe. The breach costed the company $252 million in gross expenses in 2013 and 2014 [393]. Furthermore, Target was hit with more than 90 class action lawsuits by both customers and financial institutions, whose cost the previous figure does not reflect.

In February 2010, more than hundred Austin (Texas, USA) drivers found their cars disabled or with their horns honking intermittently [394]. All these cars were leased from the Texas Auto Center. The WebTeck Plus system from PayTeck technologies had equipped all these cars. The WebTeck controller installed in the vehicle allows disabling starting the engine or honking. WebTeck can be activated remotely in the case of delinquent payments. The investigation showed that the installed system was not faulty. A disgruntled employee of Texas Auto Center disabled the cars 1 month after he had been terminated. He kept passwords, which had not been changed after his firing. Once all passwords were reset, all operations were back to normal and the cars worked properly again. Texas Auto Center did not have a proper policy about departing employees or an appropriate enforcement of such a policy.

These examples highlight the importance of security policies and processes. The enforcement of compliance is of equal importance.

10.2 Analysis

10.2.1 The McCumber Cube

In 1991, John McCumber introduced a model of IT security. Figure 10.1 illustrates this Rubik's Cube-based model.

The McCumber Cube presents a synthetic view of many different aspects of IT security [395]. The first face displays three potential security properties of digital

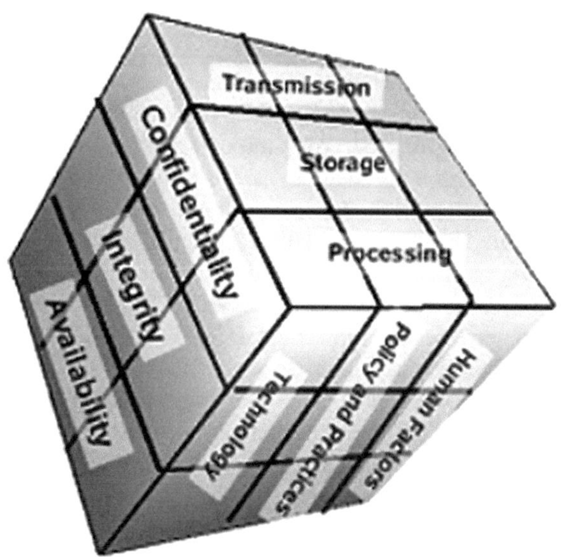

Fig. 10.1 The McCumber Cube

assets: confidentiality, integrity, and availability. The second face represents the corresponding enforcement of those properties while data is stored, while data is transferred, and while data is processed.

For this chapter's topic, the most interesting face of this cube is the last one. It puts at the same level three rows: people, policies, and technology. If one of these rows is not acting properly, security will fail. Too often, security is reduced to its technological aspect. For instance, a large portion of this book is about technology. The cube highlights that technology alone is insufficient. Policies help the security practitioners and the users to apply the adequate security practices and postures (Sect. 10.3.3). The first two rows (technology, policy, and practices) of this face of the cube represent the design and application of a security-enhanced information system. The last row of this model represents the knowledge necessary to protect information. This part of the model is about people. People are an important element of security. Human security experts defend a system. Human users employ the system (Chap. 7). Human attackers attempt to breach the system, even if they use automated attack tools. A global vision of security has to take into account the human factor. Humans need proper training and awareness. The smooth integration of people, policies, and technology is paramount for successful security.

10.2.2 Security Mindset

The first attribute that a security practitioner should display is a security mindset. What is a security mindset?

The first element of this mindset is a good degree of paranoia. Law 1 is the "raison d'être" of this paranoia. Perfectly securing a system is an intractable problem. It has no possible solution. Attackers are all around, and they will probe the system for any exploitable vulnerabilities. The current landscape of cybersecurity is turning everyday more into a cyberwar. The enemies are numerous and range from script kiddies, garage hackers who are curious or seeking attention to organized, well-funded teams. In the end, they will win and defeat the defender's system. A security practitioner should never be overconfident. Thus, he should always look for what may go wrong, even if it is highly unlikely. Only after having analyzed a threat should he discard it. The rationales for rejecting a threat are either that this threat is impractical as its likelihood is near impossible or that the cost to the attacker is too high.

The second element is an inextinguishable curiosity. It is about hacking. Hacking might be characterized as "an appropriate application of ingenuity." Regardless of whether one is defending or attacking, security mandates broad technological knowledge. The security expert must understand how systems are working. This understanding should range from the overall top-view architecture down to the finest details of hardware or low-level software drivers. When defending a system, this knowledge helps in detecting that something is going wrong, helps in stopping an ongoing attack, and helps in designing an efficient countermeasure. When attacking a system, this knowledge helps in finding ways to derail the system. The more secure a system is, the more skilled the attacker will need to be.

The third element is persistence. Security is not easy. It requires continuous effort. If vigilance goes down at any moment, the attackers may succeed. Security requires continuous training and learning to stay at the level of attackers. It requires tough, tedious tasks such as audit trail.

The fourth element is rigor. Defending requires following strict policies and procedures. Any exception to a policy or a procedure is a potential security hole. Implementing security mandates rigor down to the lowest details. This book has presented many examples where a tiny variation of the implementation defeated the overall security.

10.2.3 ISO 27005

Security is about risk management. Security practitioners spend their time at mitigating risks. What is a risk? The notion of risk encompasses two concepts. Risk means that there is some uncertainty that an adverse event will occur. Uncertainty is a significant element of this definition.[1] For safety, the adverse event may be a natural catastrophe or an accident, often referred to as a hazard. For security, the

[1]If there is no uncertainty but only certainty, then there is no risk. It is an actual attack.

adverse event may be an attack. The notion of threat encompasses the concept of the uncertainty of occurrence of a given set of attacks. Risk also means that there may be a loss or damage whenever the adverse event occurs. The risk is the "possibility of loss or injury" and the "degree of probability of such loss" [396].

Therefore, risk management is the art of identifying risks and bringing these risks to an acceptable level. Reducing the risk means either reducing the uncertainty or decreasing the potential loss. ISO 27005 defines an interesting framework for managing risks related to IT. Figure 10.2 illustrates this cyclic framework. It has mainly five phases.

- The first phase establishes the context and the environment in which this information security risk management operates. During this phase, all information about the organization is collected. It also describes the precise scope of this risk management and its boundaries. In addition, it designates a group that is in charge of establishing and maintaining this work.
- The second phase performs the actual risk assessment. It identifies the risks. It lists the potential associated threats. Finally, it evaluates the associated potential losses. Chapter 2 has already tackled in depth this phase.
- After the second phase, there is the first risk decision point. This checkpoint verifies whether the risk assessment is complete and whether the outcomes are exploitable. If this is not the case, the process requires enhancing the risk assessment and returns to the first phase.
- The third phase treats the risks. For each risk, there are four possible decisions and actions.

  - The risk has to be reduced. Reduction requires designing a countermeasure or a safeguard. The purpose of the countermeasure is to diminish the likelihood of the corresponding threats to an acceptable level at a limited cost. The space of possible solutions for protection is vast and encompasses the following:

    Correction: The solution modifies the system to reduce the risk. Once the correction is applied, the threats remain, but the risk becomes acceptable.
    Elimination: The solution modifies the system to eliminate all threats. Once the solution is implemented, the corresponding risk should be null.
    Prevention: Some modifications allow us to prevent the successful realization of the threats. The solution usually applies at the level of operational procedures or modification of policies.
    Impact minimization: If the impact can be lowered, then the risk may become acceptable as its associated costs fall below a defined threshold.
    Detection: Sometimes, it is sufficient just to detect in time the occurrence of the attack to stop it. If the attack is blocked at its beginning, then the impact will be null or small.

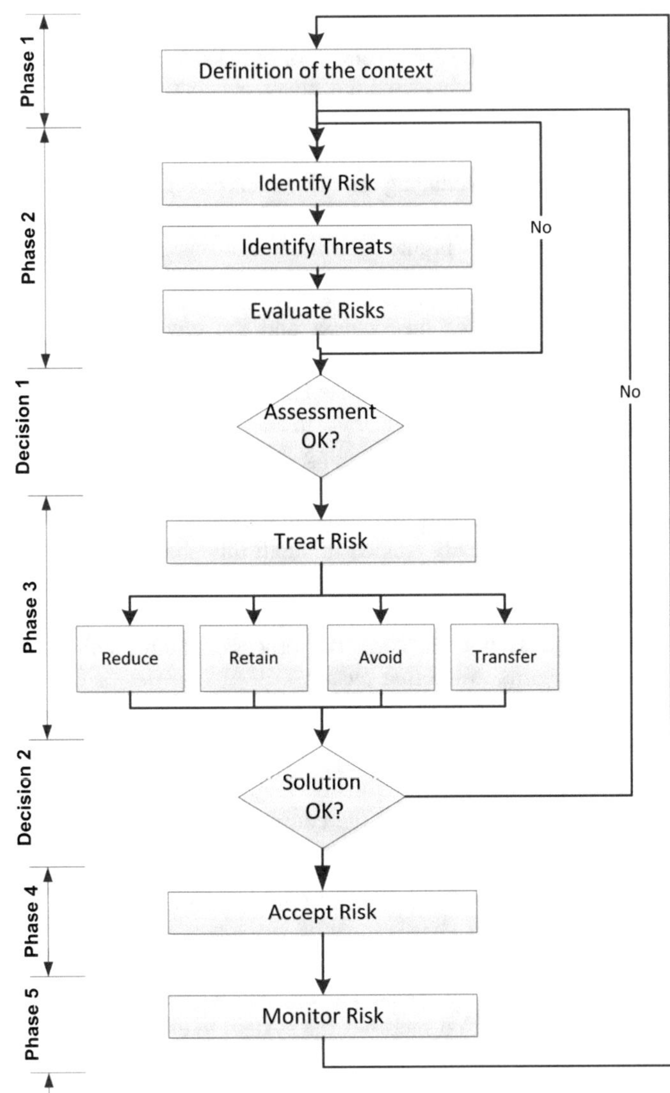

Fig. 10.2 A framework to manage risk

Awareness: Sometimes, just being aware that the risk exists and having good practices can annihilate the risk or, at least, bring it to an acceptable level. This is usually the solution for risks related to social engineering.
The countermeasure may use and mix all these techniques.

– The risk is retained. If the risk is under an acceptable level, there is no need to take any action.

  - The risk is avoided. If the risk is too high and the costs to cope with this risk are too high, then the conditions should be modified for the occurrence of this risk to disappear.
  - The risk is transferred to another party. The management of a set of risks can be outsourced to a third party (for instance, by delegating the physical security to a specialized company). Another solution is to transfer the risk to an insurance company. In the case of successful attack, the insurance company will bear the financial consequences.

- After the third phase, the second verification point checks whether all risks have reached an acceptable level once the corresponding countermeasures have been applied. The decision is taken by the top management team. If it is not the case, then the countermeasures or safeguards need to be redesigned, and the process returns to the third phase. This verification ensures that the organization's management team makes a deliberate, informed decision and that the decision criteria are documented [397].
- The fourth phase formally accepts the residual risks. The organization's management team verifies that the acceptance criteria are met. If this is not the case, the management team may nevertheless agree to take the risk. For instance, if the expected benefits of a risky new service are extremely high, the organization may take the risk. Another example is when the company cannot bear the cost of reducing a risk but needs the potentially hazardous service. In that case, the decision to accept the evaluated risk has to be recorded, documented, and communicated.
- The fifth phase continuously monitors the risk factor to check that new risks do not appear or that the level of the known risks is not growing too high. This phase is paramount. Without such a monitoring loop, security would decay inevitably. In the case of changes, the process returns to the first phase.

ISO 27005 is a high-level framework. It does not propose a detailed implementation of the security management. Each step needs to rely on a complete process. There are many other risk management methodologies, such as Mehari, EBIOS, and Corras [398]. Nevertheless, the ISO 27005 framework highlights that security is a continuous cyclic process.

10.3 Takeaway

10.3.1 What Makes a Great Hacker?

This qucstion is legitimate. Would a security defender also be a good hacker? We may expect that a security practitioner's skills should match some of the hacker's skills and complement some other ones. This section studies the skill set of Class II and Class III attackers of IBM's classification. It does not concern script kiddies or copycats, who have lower skill sets but are also far less dangerous.

As mentioned in Sect. 10.2.2, the obvious mandatory skill is deep knowledge of computer science. The hacker must master every aspect of computing, such as network administration, virtualization, database management, programming, reverse engineering, forensics, and basic cryptography. To break a system, the hacker must understand how the system operates to identify the elements that pose a challenge. To attack a system, the hacker must install and master sophisticated tools. She must script automated procedures of attack. She may even have to write her own hacking tools. To escape detection and to exfiltrate data, the hacker must know what information the system logfiles record and what information their monitoring may reveal. Ideally, the defender should master the same skill set. As highlighted in Sect. 7.3.1, training the security team is mandatory to ensure an up-to-date defense.

A hacker must be good at problem-solving. When attacking a well-defended system, the hacker encounters new seemingly unsolvable problems. Thus, she must think analytically. Hacking requires diagnosing which parts of the system may be the weakest ones and then breaking the problem down into smaller isolated challenges. The defender needs the same skill. He uses problem-solving skills when designing and deploying a secure system. Furthermore, he uses analytical skills when responding to or analyzing an attack. He must understand how the attack operates and find out how to counterstrike.

A hacker must be persistent. She must relentlessly try new attacks and explore all opportunities until one of them succeeds. This is especially true when attacking a given target rather than easy preys. On the other side, the defender must be vigilant.

In the end, the utmost quality of a hacker is creativity. Creativity is the impulse that makes the hacker think out of the box. When thinking out of the box, the hacker finds conditions that the designer of the system did not envisage. This is where the most devastating attacks come from. Creativity allows the attacker to jump over the fences. Should the defender be creative? For most defenders, the answer is negative. Security requires strict compliance with rules, procedures, and processes. Strict compliance is the enemy of creativity. The expected quality of a security defender is rigor and precision. For instance, proper cryptography requires extremely strict conditions of use and precise parameters. Often, one seemingly benign modification can totally ruin the efficiency of a secure protocol. Nevertheless, two categories of defenders may need creativity.

- Some security defenders need to create new solutions. These designers need to be creative to find new methods, but when implementing them, rigor prevails once more. Once well studied and understood, the newly created solution will have to be implemented. Its implementation must be rigorous to avoid bugs and vulnerabilities. Once deployed, its operation has to follow strict, defined compliance rules to ensure that the solution operates in the expected conditions.
- "It takes a thief to catch a thief." This old saying describes the second category of defenders, i.e., white hats. They may act like attackers, but their goal is to evaluate the defense and identify vulnerabilities that the practitioners must fix

before the black hats find them and exploit them. White hackers have to be creative like their nefarious alter egos.

Security practitioners, being white hats or black hats, should share the same similar skill set. This is why many black hats have turned later into powerful white hats.

10.3.2 Tools

Tools are necessary. It is impossible to defend efficiently without having automated tools to undertake the most tedious tasks. The perimeter must be protected against intruders. Thus, firewalls, proxies, and DMZs have to be installed and set up. IDSs have to be installed to detect anomalies that may be signs of an ongoing intrusion. DLPs may detect exfiltration of information by an insider or an external attacker who succeeded in penetrating the system. Perimetric defense is not anymore sufficient as this perimeter becomes more and more porous. Legitimate users may become the entry points of infection either through visits to infected websites, malware infected data, or contaminated personal devices that are allowed within the corporate network (BYOD). Antivirus software is mandatory for detecting known viruses and malware. Behavioral analysis tools may detect unknown threats or 0-day vulnerabilities as well as the suspicious behavior of insiders. Vulnerability scanning tools may identify unpatched resources or misconfigured firewalls. However, tools alone do not ensure an adequate defense.

Rule 10.1: Tools Are Not Autonomous

As attacks become sophisticated, defensive tools become equally complex. Properly configuring tools is a complex task. A minor error in the configuration may result in an open door that an attacker may exploit without triggering the warnings of the defending tool. The defenders may believe they are safe as the tool remains silent while the attack succeeds. Furthermore, a defender can configure his tools only against attacks that he knows. He cannot set the parameters to prevent unknown or unforeseen events. With the defense in depth strategy, many tools have to collaborate and complement their defenses. This synergy of tools makes the configuration task even harder. The configuration has to be fine-tuned to the current context.

Unfortunately, even if the tools were properly configured, the tools would not be sufficient to set a proper defense. Section 2.2.1 introduced three types of threats: regular threats, irregular threats, and unexampled threats. Regular threats are known threats with a high likelihood of occurrence. Defense tools must be configured to prevent them. As the threats are well known, and the strategy of defense is well documented, tools should be able to trigger the counterstrike automatically. The security practitioner should receive warnings only of the ongoing attack and the triggered automatic defense. Nevertheless, he should regularly check the logfiles to detect potential failures and holes in the defense. Irregular threats have a smaller likelihood and are usually more complex than regular ones. The defense tools may perhaps recognize some of them. Nevertheless, the response to irregular attacks cannot be delegated to the tools. The tools must forward alarms to security practitioners. The security professionals then analyze the reported alerts, cross-check data from various tools and logfiles. At the end of the analysis, the security practitioners must decide whether an attack is currently happening or has already succeeded. They adapt the response to the identified attack. Furthermore, many irregular threats may circumvent the defending tools. In this case, the defense tools will not alert the security practitioner. He must preventively analyze the logfiles and data of these tools to detect such an ongoing threat. The skill set of the security specialists is paramount in this case. Against unexampled threats, most probably the defense tools are useless, at least in their function to raise alarms. Their data collection is the source that the security practitioners will use preventively or a posteriori against these rare attacks.

The efficiency of reactive defense depends on the ability of the security practitioner to identify an ongoing attack. It is impossible to react to an ongoing attack without the awareness that the attack is occurring. The same goes for responding to a past assault. The security practitioner has first to detect the attack. For that, he usually relies on the warnings and alerts generated by his protection tools. He has to extract from many warnings and alerts the ones that portray a real attack. The configuration of the conditions that will trigger an alert or warning from the protection tools is critical. If the threshold is too low, then the tools will generate too many false positive alerts. A false positive alert is an alert that does not correspond to a real attack. The security practitioner may be flooded with too many false events to analyze. He may miss an important alert in this mass of false events. Furthermore, attackers sometimes generate voluntarily a bunch of false positive alerts to hide their actions. For example, attackers can construct a malware to mimick conditions nearing the threshold of the defense tool. Under these conditions, this malware generates much deceptive noise flooding the expert. The attacker may use this decoy to direct the scrutiny of the defender away from the actual attack, which will be far more discreet. Level III attackers may identify during their exploration phase the tools used by their target. Then, they can study the commercially available tools to create such diverting decoys. If the threshold is too high, then the tools will have too many false negatives. A false negative or alert is an absence of warning, although a real attack is currently happening. The security practitioner may miss the attack because no protection tools warned him. The adjustment of these configurations is a task that only a human expert can undertake. It is often the result of careful iterative trial and error. Once an attack has been detected, and all information has been collected, it is important to check whether the protection tools could have done a better job with different settings. This can be done by replaying the attack on a simulated system with the tools' parameters set to new values. If the tools would have detected the attack, then it is important to run a non-regression test on previous attacks to verify that the new parameters do not create new blind spots. The optimal experimented parameters replace the previous ones. This iterative adjustment may be time-consuming.

The administration of the tools is not necessarily the most expensive part of the human effort. Often, the operations following the discovery of an issue reported by a tool are far costlier. For instance, a vulnerability scanner reports all the unpatched systems. Patching the systems may have a high operational cost. It is a waste of time and money to run a vulnerability scan if there is no budget or manpower to apply the missing security patches. Logging all events is a best security practice. Nevertheless, the analysis of these logfiles will incur an important cost. When evaluating the cost of a new tool, it is key not to forget the extra cost. It encompasses human operators for the administration of the tool, the analysis of its outcomes, and the response to the reported issues. The total cost should be presented to the management team, rather than just the cost of acquisition and operation.

10.3.3 Written Policies

There is no proper corporate security without proper written policies.[2] Policies are the spine that architectures the defense and security processes. The security policy is the set of documents that state in written how a company plans to protect the company's physical and information technology assets.

Rule 10.2: Write Your Policies and Enforce Them

Unfortunately, there is no one-size-fits-all policy. Each security policy has to be fine-tuned for the targeted enterprise.

A good security policy should present all the following characteristics:

- *Be in touch with reality*: The security policy has to be anchored in the real world. As such, it must be relevant to the majority of the people impacted by it. It must reflect the actual context and environment of the enterprise and fit into its actual configuration.
- *Accessible*: Users must be able to locate the security policy easily and access quickly the parts that are relevant to their current problem. This characteristic has two requirements. The security policies must be accessible from the corporate intranet without our having to dive deep into the site. Ideally, the home page of the intranet should display a direct link to the repository of all security policies. The second requirement is that the security policy be structured to find information quickly and easily. If a security policy is not accessible, most users will not struggle to look for it, thus making it useless.
- *Well-marketed*: Security policies should promote themselves. Users must grasp the benefits that the application of the security policy would bring. Free and

[2]This section targets written policies. Security tools may also employ policies that are logical rules driving the behavior of the tools or system. This section does not tackle these types of policies.

willing users adhere more easily to the policy than reluctant users. Adhesion to policy is a good indicator of its actual use.

- *Informative*: The security policy must contain all information needed to understand it. Users should not have to look elsewhere for additional information. Nevertheless, the information should not be broader than required. It should be focused. The security policy should educate the impacted people when defining the rules.
- *Short*: If a document is too long, users will not read it. Thus, it is interesting to use a hierarchical approach. Policies should define high-level statements common to all parts of the company. Security standards should describe specific details of the previous policies. Security procedures should describe details of very specific security tasks. For instance, the policy may state that all accesses to the corporate network should require authentication. A corresponding security standard defines the way passwords are managed and their constraints. A security procedure defines the process of requesting a password reset. Such a hierarchical approach limits the size of the documents and enables faster access to the relevant information. Documents should contain a link to the related lower-level or upper-level documentation when suitable. It is often useful to define guidelines. A guideline is typically a collection of system-specific or procedural-specific suggestions for best practice. They are not requirements to be met. However, it is strongly recommended to follow them. Effective security policies make frequent references to standards and guidelines that exist within an organization.
- *Targeted*: Security policies must be specialized, where possible, to a targeted population. Once more, this enables faster access to more relevant information. Users are interested in reading the policy that concerns them. They do not care about policies that are irrelevant to them.
- *Up-to-date*: As security ages, security policies also age. A security policy may become outdated because there are new threats to address. The security policy must evolve correspondingly. A security policy may be outdated because its context has changed, e.g., the structure of the organization evolved. Adapting to the new context is important. For instance, there are two typical related issues. The titles and division names change, and the policy refers to older denominations. People change roles or move. The names of the responsible people and the contacts cited in the security policy must reflect the actual persons in charge. Users cannot take an outdated security policy seriously. Security policies should be reviewed once per year. The modifications should not be too frequent as the users may not keep up with them.
- *Supported by the management*: A security policy has no value if the management team does not support its application. The top managers must endorse it and accept the associated constraints. The top management should never approve risky exceptions to the security policies that would undermine the acceptance of security policies among employees. The middle-level and low-level management must enforce their applications. It is important that the management team drives by example when applying security policies.

Security policies have to be followed under all circumstances. They are efficient only under strict compliance. If a security policy is violated, it is often a good illustration of Chap. 6. This violation may become the weakest point. When writing a security policy, the editor assumes that users will comply with them. Hence, it is crucial that they be obeyed. This requires three conditions to be true:

- Users must be aware of security policies. Without knowledge of the security policies, it is not reasonable to expect employees to apply these policies. The security policies must be easily accessible by every employee. Furthermore, they must be understood by the potential readers.
- Security policies must not be obsolete. They must reflect the internals of the organization accurately and follow all evolutions of its organigram. The identities of the roles defined by the policies or procedures must be up-to-date. The status of departing and transferring employees must be modified if they are listed in these documents. A document that is not up-to-date has several issues:
  - Usually in the event of a security incident, people are under stress. If the reference documents are inaccurate, and the reader has to search for the missing information, this will only increase the pressure and the risk of errors. Furthermore, it may induce the reader to respond incorrectly.
  - If a document is not accurate, its credibility is endangered. Diminished credibility is a potential excuse for not complying with policy.
- Security policies must be applicable; if they are unrealistic or are not aligned with the business practices, then once more they offer a potential excuse for not complying with them.

10.3.4 Communicate Risks

Sometimes, people may prefer to hide known risks rather than disclose them. Many reasons may underlie such a decision. They may be afraid to frighten the management team or the users. For instance, salespeople may be afraid of losing a deal with customers who may believe the product is not secure. Other people may believe that disclosing a risk would give an edge to the attackers. It may also be primarily due to pusillanimity. In most cases, these reasons are not valid.

Rule 10.3: Be Transparent About Risks

Communicating risks is a good security practice. ISO 27005 insists on the need to communicate the risks. This is paramount, at least for the most important risks and the ones most likely.

Users should be aware of the significant risks. Identifying and understanding the risks helps users to accept constraints that security will necessarily impose. Without knowing why these restrictions are imposed upon them, users will resent them and may try to bypass them. If they understand the risks, they may be more willing to adopt security policies, security tools, and associated constraints as these constraints will be based on rational decisions rather than on arbitrary decisions. When knowing a risk, users are more likely to detect an ongoing corresponding attack. People are more likely to identify a known pattern than an unknown pattern. If they believe that strange behavior relates to a known threat, they are more liable to report it to the incident responding team (provided that they know whom to report the issue).

The management team needs to be aware of risks in advance. Without having a complete picture of the risk landscape, the management team cannot make rational decisions on the security strategy. It must know the major risks to forecast the necessary investment in new tools, to upgrade already deployed tools, and to size the security teams correctly. Risk management and analysis will allow prioritizing and estimating these investments depending on the company-defined security stance. When aware of a risk, the management team may react faster and more rationally in the case of its realization. This is essential in a situation of crisis. No necessary phase of education would slow down the response.

Customers need to be aware of the risks associated with the system they expect to purchase, or that they already have deployed. Any system has limitations and known weaknesses. It is better to communicate them to the customer than to hide them. First, it is a smart strategy that builds a trust relationship between designers and clients. It is better that the client knows beforehand of a known risk rather than learning it from a source other than the supplier. It is even worse if the customer discovers after a successful attack that its supplier was aware of the risk. Such a situation may destroy the customer's trust in the vendor. If the customer knows the risk, he can adapt his policy, the use of the product, and the environment to minimize this risk. The customer does not have the deep knowledge of the tool that the supplier has. Thus, he may not be able to identify the risk by himself. The client relies on the supplier to guide him.

The communication of the risk should be adapted to the targeted audience to be efficient. It must use the vocabulary and the context of the audience. Of course, the communicated risk must be relevant to the audience. The level of technical detail to convey must be adapted to the audience. Usually, the management team does not need as many technical details as the IT security team of a customer. In all cases, the consequences of the risk must be communicated together with the risk.

10.3.5 Think Out of the Box

When defending against an ongoing attack, it is usually better to follow defined procedures. Improvisation may quickly become a disaster. Procedure guides the defender step by step. It has several advantages.

- A safe procedure helps the defender even in domains he may not fully master. Experts incorporate their know-how in procedures. For example, a procedure may define all the steps needed to collect forensic information that may be used later in the case of litigation.

- When applying a procedure, the defender cannot miss a step ensuring that the recovery or the mitigation will be optimal. For instance, he cannot bypass the eventual collection of forensic data.
- Procedure reduces the stress inherent to a crisis phase. It diminishes the apprehension of navigating in unknown territories when a serious or severe attack occurs. Reducing stress limits the risk of errors.

On the other side, attackers have to think out of the box. If an attacker uses the same techniques and procedures relentlessly, she becomes predictable. A defender can build efficient measures against anticipated attacks. Thus, the attacker must vary her schemes.

To create new attacks, the attacker may think out of the box. She should look for conditions that the designer did not expect (Rule 1.1: Always Expect the Attacker to Push the Limits). Similarly, she should use the attacked system in ways that the designer did not envisage. Creativity is one of the major qualities of a successful attacker.

Therefore, defenders should always use the services of white-hat hackers. Designers must have an attacking team testing their designs. The most efficient testing procedures use white-box testing, i.e., the attacker has the same information as the designer. Defenders must regularly request pen testing to validate the robustness of the defense. These tests must not uniquely be based on the use of automated tools and tests. Automated penetration tools reproduce only known attack vectors. They are mandatory but not sufficient. Pen testing must also use human-driven attacks. It will mimick the operating mode of determined attackers. By using white-hat hackers, defenders benefit from the creativity of attackers.

10.4 Summary

Law 10: Security Is Not a Product, Security Is a Process
Security is not uniquely about technologies. Security is mostly about humans and procedures they have to apply. Without these elements, any system will be unsafe.

Rule 10.1: Tools Are Not Autonomous Security is not uniquely performed by a set of tools. Tools are good auxiliaries that may compile large amounts of data, analyze them, and detect patterns and anomalies. Nevertheless, only human experts can make the final decisions when the attack is complex or unforeseen.

Rule 10.2: Write Your Policies and Enforce Them Policies are the cornerstone of any effective security. They are useful only when applied and enforced. Upper management must support their enforcement and cost.

Rule 10.3: Be Transparent About Risks Risks must not be hidden from the management team, the customers, or the end users. Communicating risks may help to guarantee sufficient budget and the active collaboration of users for the defense.

Conclusions

In our increasingly connected world, security has become a mandatory feature. The threats are continuously growing. Hacking moved from a fun hobby activity of amateurs to a professional activity with strong financial incentives. A significant element of the security toolbox is awareness and knowledge. The ten laws and their associated rules are good starting points for educating, training and analyzing. These laws may be useful indicators of the soundness of a system or a service provider. They can be used as heuristics within a larger framework of risk management. They may be used as discussion topics to create a common ground and to start building mutual trust with new partners. They may be used for educational purposes when introducing security.

These rules may seem common sense. Security involves in most cases obeying common sense rules. Unfortunately, people have not imported common sense-based practices when migrating from the physical world to the digital world and embracing dematerialization. The only area of security that does not obey common sense is the interaction with users (for instance, through social engineering). Humans are predictably irrational [122].

The initial cautionary note of the introduction has to be repeated at the end of this book. Security is a complex problem. It does not fit well with a Manichean vision. Security is about equilibrium. Security is relative, never absolute. Thus, the rules must never be taken literally and considered as absolute rules. The reader will have to use his judgment when analyzing something through the prism of these ten laws.

I urge the reader to never forget that security is an extremely complex task. Security systems and the development of secure solutions should be handled exclusively by well-trained security experts. As illustrated by the inserts "The Devil Is in the Detail," it is far too easy to mess up the overall security with a small mistake or a minor detail in an implementation. Security is hard, and we cannot rely on amateurs.

I will maintain some resources about these ten laws on the website https://eric-diehl.com/ten-laws/, such as a poster with the ten laws, and new examples. The reader may contact me at book@eric-diehl.com.

E. Diehl, *Ten Laws for Security*, DOI 10.1007/978-3-319-42641-9

Appendix A
A Brief Introduction to Cryptography

A.1 Symmetric Cryptography

Symmetric cryptography is the oldest form of cryptosystem (Fig. A.1). Alice and Bob share the same key K. She applies to message m an algorithm E, called encryption, using the shared key K. The result ciphertext is $c = E_{\{K\}}(m)$.

To retrieve the clear text, Bob applies to message c an algorithm D, called decryption, using the shared key K_i such that $D_{\{K\}}(c) = D_{\{K\}}\left(E_{\{K\}}(m)\right) = m$.

The most deployed symmetric cryptosystem is AES. The key length of 128 bits is considered secure.

A.2 Asymmetric Cryptography

In 1976, two researchers, Whitfield Diffie and Martin Hellman, in a famous seminal paper [204] built the foundations of a new type of cryptosystem: asymmetric cryptography (Fig. A.2). Here, Bob owns a pair of keys: the public key K_{pub} and the private key K_{pri}. Both keys are linked with a mathematical relationship. To securely

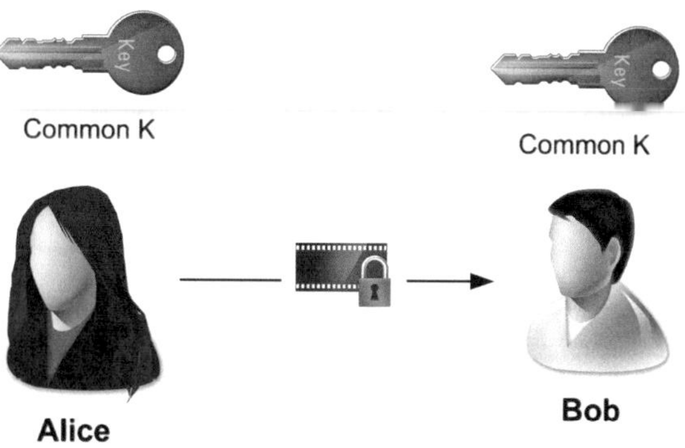

Fig. A.1 Symmetric cryptography

E. Diehl, *Ten Laws for Security*, DOI 10.1007/978-3-319-42641-9

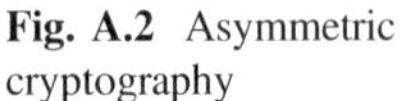

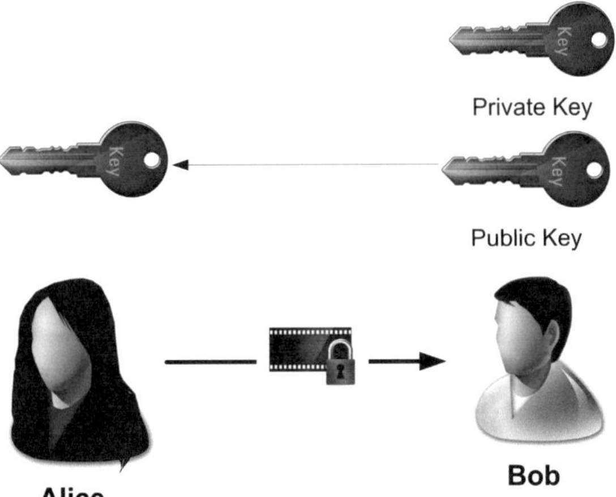

Fig. A.2 Asymmetric cryptography

send a message m to Bob, Alice obtains a copy of Bob's public key K_{pub}. She applies to message m an algorithm E, called encryption, using Bob's public key K_{pub}. The resulting ciphertext is $c = E_{\{K_{\text{pub}}\}}(m)$.

To retrieve the clear text, Bob applies to message c an algorithm D, called decryption, using his private key K_{pri} such that $D_{\{K_{\text{pri}}\}}(c) = D_{\{K_{\text{pri}}\}}\left(E_{\{K_{\text{pub}}\}}(m)\right) = m$.

The encryption and decryption algorithms are designed such that, at least in theory, without the knowledge of the private key K_{pri}, it would be infeasible for Eve to find m from c. More precisely, Eve does not find out any information about m except its length (except by breaking the underlying hard-to-solve mathematical problem). Therefore, it is paramount that Bob keeps his private key secret. However, Bob, as its name indicates, can safely distribute his public key to everybody, even to the malicious Eve.

Asymmetric cryptography can provide both integrity and authentication. Bob wants to send a message m to Alice. He does not look for confidentiality, but he wants Alice to be sure that the message has originated from him (authentication) and that Charlie did not tamper with the message (integrity). For that purpose, Bob signs the message m. Bob applies to m a mathematical function called a signature using his private key K_{pri}. The result, σ, called a signature, is $\sigma = S_{\{K_{\text{pri}}\}}(m)$. He sends m and σ together to Alice. Alice verifies the signature using Bob's public key K_{pub}. The signature σ is valid if $V_{\{K_{\text{pub}}\}}(\sigma) = m$, where V is an algorithm called Verification. In other words, the signature inverts the respective roles of the public and private keys. The main difference with MAC is that the signing entity and the verifying entity use different keys.

In fact, most implemented signature schemes use an additional cryptographic hash function (Sect. A.3). Using the cryptographic hash function H, we obtain the new signature scheme $\sigma = S_{\{K_{\text{pri}}\}}(H(m))$ and the signature σ is verified if

$V_{\{K_{pub}\}}(\sigma) = H(m)$. Signing the hash of the message rather than the full message has several advantages such as a fixed length of signature, regardless of the size of signed data, and efficiency. Calculating the hash of a large dataset is faster than signing the same large dataset with asymmetric cryptography.

The robustness of asymmetric cryptography relies on the difficulty of solving some difficult mathematical problems. Typical problems are the factorization of composite numbers made of two large prime numbers, or the difficulty of calculating discrete logarithms in finite fields or on elliptic curves.

Currently, the most widely used asymmetric cryptosystem is RSA. The name stems from the initials of its three inventors: Ron Rivest, Adi Shamir, and Leonard Adleman. RSA can be used for both encryption and signature. It was invented in 1977 [399] and is still considered secure. RSA Ltd. patented this algorithm. Since 2000, the algorithm is in the public domain; hence, it is not anymore protected by patents. It is currently estimated that RSA requires a key length of at least 2048 bits to be safe [52]. The security of a 2048-bit key of RSA is considered equivalent to the security of an 110-bit key for symmetric cryptography. In recent years, a new kind of asymmetric cryptosystem, Elliptic Curve Cryptosystems, has become mainstream. Here the size of the keys is smaller, and the footprint of the implementation is smaller than RSA; 256-bit ECC keys are considered secure today.

A.3 Hash Functions

A cryptographic hash function, also sometimes called a one-way hash function, has the following characteristics:

- The hash of any arbitrary size of data will always have a fixed size.
- Changing any bit of the hashed data generates a very different hash result, e.g., it may change every bit of the hash with a probability of 0.5.
- It is extremely difficult to find a message m such that its hash value is equal to an arbitrary value.
- It is extremely difficult to find two messages m_1 and m_2 such that their respective hash values are equal. This property is called collision resistance.

Currently, the most widely used hash function is SHA-1. However, in 2005, studies demonstrated that the robustness of SHA-1 was severely weakening [158, 400]. In October 2012, NIST selected KECCAK as the new SHA-3 [401] (Sect. 3.2.1). MD5 is another widely deployed hash function. MD5 hashes often serve as unique identifiers for torrents in P2P networks or file management. However, it was shown in 2004 that MD5 is not collision resistant.

Appendix B
Other Ten (or More) Laws of Security

B.1 Microsoft

Microsoft has published its ten laws of immutable security [219].

1. If a bad guy can persuade you to run his program on your computer, it's not your computer anymore.
2. If a bad guy can alter the operating system on your computer, it's not your computer anymore.
3. If a bad guy has unrestricted physical access to your computer, it's not your computer anymore.
4. If you allow a bad guy to upload programs to your website, it's not your website anymore.
5. Weak passwords trump strong security.
6. A computer is only as secure as the administrator is trustworthy.
7. Encrypted data is only as secure as the decryption key.
8. An out-of-date virus scanner is only marginally better than no virus scanner at all.
9. Absolute anonymity is not practical, in real life or on the Web.
10. Technology is not a panacea.

B.2 Building Secure Software

In 2001, John Viega and Gary McGraw published a book describing how to build secure software [402]. They listed ten guidelines.

1. Secure the weakest link
2. Practice defense in depth
3. Fail securely
4. Follow the principle of least privilege
5. Compartmentalize
6. Keep it simple

E. Diehl, *Ten Laws for Security*, DOI 10.1007/978-3-319-42641-9

7. Promote privacy
8. Remember that hiding secrets is hard
9. Be reluctant to trust
10. Use your community resources

Many of their guidelines overlap with the laws and rules of this book.

B.3 What Hackers Don't Want You to Know

In 2000, Jeff Crume published a book entitled *Inside Internet Security: What Hackers Don't Want You to Know* [403]. The narrative approach was interesting because it highlighted supposedly dirty little secrets. The supposed secrets are good rules or warnings. Sadly, fifteen years later, they are still valid.

1. Firewalls are just the beginning
2. Not all the bad guys are 'out there'
3. Humans are the weakest link
4. Passwords aren't secure
5. They can see you but you can't see them
6. Downlevel software is vulnerable
7. Defaults are dangerous
8. It takes a thief to catch a thief
9. Attacks are getting easier
10. Virus protection is inadequate
11. Active content is more active than you think
12. Yesterday's strong crypto is today's weak crypto
13. The back door is open
14. There's no such thing as a harmless attack
15. Information is your best defense
16. The future of hacking is bright

References

1. ISO/IEC 27002:2005 - Information technology – Security techniques – Code of practice for information security management (2005) (http://www.iso.org/iso/home/store/catalogue_ics/catalogue_detail_ics.htm?csnumber=50297)
2. Diehl, E.: Ten laws of security (http://eric-diehl.com/ten-laws/)
3. Diehl, E.: Content Security, Presented at the RESCOM 2006, Porquerolles, France (2006) (http://ericdiehl.x10.mx/wp-content/uploads/2012/05/ME060612-RESCOM06.ppt)
4. Diehl, E.: Securing Digital Video. Springer (2012)
5. Adams, D.: Mostly Harmless. Del Rey (2000)
6. Homer: The Iliad. Create Space Independent Publishing Platform (2010)
7. Hemanth, J.: DoSing Pebble SmartWatch and Thus Deleting All Data Remotely (2014)
8. Al-Kadit, I.A.: Origins of Cryptology: The Arab Contributions. Cryptologia. 16, 97–126 (1992)
9. Kahn, D.: The Code-Breakers. Macmillan (1976)
10. Hagelin (http://www.cryptomuseum.com/crypto/hagelin/)
11. Army Security Agency: Notes on German high level cryptography and cryptanalysis (1946) (http://www.nsa.gov/public_info/_files/european_axis_sigint/volume_2_notes_on_german.pdf)
12. Doom9.net - The Definitive DVD Backup Resource (http://www.doom9.org/)
13. Bilge, L., Dumitros, T.: Before we knew it, Presented at the 19th ACM Conference on Computer and Communications Security, Raleigh, NC, USA (2012)
14. Lanier, J.: You Are Not a Gadget: A Manifesto. Penguin UK (2010)
15. Greenberg, A.: Shopping for Zero-Days: A Price List For Hackers' Secret Software Exploits - Forbes (2012) (http://www.forbes.com/sites/andygreenberg/2012/03/23/shopping-for-zero-days-an-price-list-for-hackers-secret-software-exploits/)
16. Leyden, J.: Adobe Reader 0-day exploit surfaces on underground bazaars (2012) (http://www.theregister.co.uk/2012/11/08/adobe_reader_zero_day/)
17. Fisher, D.: ReVuln Emerges as New Player in Vulnerability Sales Market (2012) (https://threatpost.com/revuln-emerges-new-player-vulnerability-sales-market-101212/77112/)
18. Greenberg, A.: Meet the Hackers Who Sell Spies the Tools to Crack Your PC (And Get Paid Six-Figure Fees) (2012) (http://www.forbes.com/sites/andygreenberg/2012/03/21/meet-the-hackers-who-sell-spies-the-tools-to-crack-your-pc-and-get-paid-six-figure-fees/)
19. Vupen Contracts with NSA (https://www.muckrock.com/foi/united-states-of-america-10/vupen-contracts-with-nsa-6593/#787525-responsive-documents
20. McDougall, P.: Crowdsourcing War on Cybercrime (http://www.cruxialcio.com/crowd-sourcing-war-cybercrime-2162)
21. Google Vulnerability Reward Program (VRP) Rules (https://www.google.com/about/company/rewardprogram.html)
22. Mitigation Bypass and BlueHat Defense Guidelines (technet.microsoft.com/en-us/security/dn425049.aspx)

E. Diehl, *Ten Laws for Security*, DOI 10.1007/978-3-319-42641-9

23. Leyden, J.: Facebook coughs up $33.5k… its BIGGEST bug bounty EVER (2014) (http://www.theregister.co.uk/2014/01/24/facebook_bug_bounty_payout/)
24. Greene, C.: Bug Bounty Highlights and Updates (2014) (https://www.facebook.com/notes/protect-the-graph/bug-bounty-highlights-and-updates/1440732202833593)
25. Vulnerability Research Grant Rules (2015) (https://www.google.com/about/appsecurity/research-grants/)
26. Zetter, K.: United Airlines pays man a million miles for reporting bug (2015) (http://www.wired.com/2015/07/united-airlines-pays-man-million-miles-reporting-bug/)
27. Coordinated Vulnerability Disclosure (http://www.microsoft.com/security/msrc/report/disclosure.aspx#)
28. Disclosure Policy (http://www.zerodayinitiative.com/advisories/disclosure_policy/)
29. Freyssinet, E.: Threats spreading silently despite Java updates… (2013) (http://digitalcrime.wordpress.com/2013/04/28/threats-spreading-silently-despite-java-updates/)
30. Evans, C., Hintz, D.: Disclosure timeline for vulnerabilities under active attack (2013) (http://googleonlinesecurity.blogspot.ch/2013/05/disclosure-timeline-for-vulnerabilities.html)
31. Evans, C., Hawkes, B.: Feedback and data-driven updates to Google's disclosure policy (2015) (http://googleonlinesecurity.blogspot.com/2015/02/feedback-and-data-driven-updates-to.html)
32. Schneier, B.: The Vulnerabilities Market and the Future of Security, Forbes (2012) (http://www.forbes.com/sites/bruceschneier/2012/05/30/the-vulnerabilities-market-and-the-future-of-security/)
33. Renard, M.: Practical iOS Applications Hacking. In: Scribd. pp. 15–26 (2012) (https://www.scribd.com/doc/164094321/Practical-iOS-Applications-Hacking-WP)
34. Gligli, Tiros, Razkar, tuxuser: The Xbox reset glitch hack (2011) (https://raw.github.com/gligli/tools/master/reset_glitch_hack/reset_glitch_hack.txt)
35. Bar-El, H., Choukri, H., Naccache, D., Tunstall, M., Whelan, C., Rehovot, I.: The Sorcerer's Apprentice guide to fault attacks. Proceedings of the IEEE. 94, 370–382 (2006)
36. Kocher, P.C.: Timing Attacks on Implementations of Diffie-Hellman, RSA, DSS, and Other Systems. Lecture Notes in Computer Science, 1109, 104–113 (1996)
37. HackMii — Notes from inside your Wii (http://hackmii.com/)
38. Chaos Computer Club (http://www.ccc.de/en/)
39. Fildes, J.: iPhone hacker publishes secret Sony PlayStation 3 key (2011) (http://www.bbc.co.uk/news/technology-12116051)
40. Lawson, N.: DSA requirements for random k value (2010) (http://rdist.root.org/2010/11/19/dsa-requirements-for-random-k-value/)
41. The Three Musketeers: #5102182 (http://pastie.org/5102182)
42. Digital Millennium Copyright Act (1998) (http://www.copyright.gov/legislation/dmca.pdf)
43. US Copyright Office: Rulemaking on Exemptions from Prohibition on Circumvention of Technological Measures That Control Access to Copyrighted Works (http://www.copyright.gov/1201/)
44. Statement of the Librarian of Congress Relating to Section 1201 Rulemaking (http://www.copyright.gov/1201/2010/Librarian-of-Congress-1201-Statement.html)
45. JailbreakMe 3.0 (http://www.jailbreakme.com/#)
46. Pangu Jailbreak (http://en.pangu.io/)
47. Welcome to Cydia (http://cydia.saurik.com/)
48. Geohot: Towelroot v3 (https://towelroot.com)
49. Anderson, R.: Security Engineering: A Guide to Building Dependable Distributed Systems. Wiley (2008)
50. Auffret, P.: WPS, the new WEP? Technicolor Security Newsletter. 6 (2012) (http://ericdiehl.x10.mx/wp-content/uploads/2012/05/Security-Newsletter-21.pdf)

51. Lell, J.: CVE-2012-4366: Insecure default WPA2 passphrase in multiple Belkin wireless routers (http://www.jakoblell.com/blog/2012/11/19/cve-2012-4366-insecure-default-wpa2-passphrase-in-multiple-belkin-wireless-routers/)
52. Lenstra, A.K., Verheul, E.R.: Selecting Cryptographic Key Sizes. Journal of Cryptology. 14, 255–293 (2001)
53. Francillon, A., Danev, B., Capkun, S.: Relay Attacks on Passive Keyless Entry and Start Systems in Modern Cars (2010) (https://eprint.iacr.org/2010/332.pdf)
54. Corral, A., Mac, R.: More Criminals Using High-Tech Trick to Break Into Cars (2015) (http://www.nbclosangeles.com/investigations/LAPD-Warning-More-Criminals-Using-Hi-Tech-Trick-to-Break-Into-Cars-309644611.html)
55. Munilla, J., Peinado, A.: Distance bounding protocols for RFID enhanced by using void-challenges and analysis in noisy channels. Wireless Communications and Mobile Computing. 8, 1227–1232 (2008)
56. Boureanu, I., Vaudenay, S.: Challenges in Distance Boundings. IEEE Security and Privacy. 13 (2015)
57. Noga, M.C.: GetCodec Multimedia Trojan Analysis (2008) (www.hispasec.com/laboratorio/GetCodecAnalysis.pdf)
58. Yampolskiy, A.: Exploiting Media For Fun and Profit, Presented at the APPSEC DC 2010, Washington, USA (2010) (https://vimeo.com/20436133)
59. Update for Windows Media Player URL script command behavior (https://support.microsoft.com/en-us/kb/828026)
60. Hudson, T.: Thunderstrike: EFI bootkits for Apple MacBooks, Presented at the 31st Chaos Communication Congress (31C3), Hamburg, Germany (2014) (https://trmm.net/Thunderstrike_31c3)
61. Dalihun, D.: Malicious Code Execution in PCI Expansion ROM (http://resources.infosecinstitute.com/pci-expansion-rom/)
62. Xing, L., Pan, X., Wang, R., Yuan, K., Wang, X.: Upgrading Your Android, Elevating My Malware: Privilege Escalation Through Mobile OS Updating (2014) (http://www.informatics.indiana.edu/xw7/papers/privilegescalationthroughandroidupdating.pdf)
63. Mitnick, K., Simon, W.: Ghost in the Wires: My Adventures as the World's Most Wanted Hacker. Little, Brown and Company (2011)
64. Coviello, A.W.: Open letter to RSA customers (2011) (http://www.validian.com/pdfs/Open-Letter-to-RSA-Customers-Mar11.pdf)
65. Kalker, T., Samtani, R., Wang, X.: UltraViolet: Redefining the Movie Industry? IEEE MultiMedia. 19, 7 (2012)
66. Hypponen, M.: How We Found the File That Was Used to Hack RSA (2011) (http://www.f-secure.com/weblog/archives/00002226.html)
67. Rivner, U.: Anatomy of an Attack (2011) (http://blogs.rsa.com/rivner/anatomy-of-an-attack/)
68. Vulnerability Summary for CVE-2011-0609 (http://web.nvd.nist.gov/view/vuln/detail?vulnId=CVE 2011 0609)
69. Backdoor: W32/PoisonIvy (http://www.f-secure.com/v-descs/backdoor_w32_poisonivy.shtml)
70. Branco, R.: Into the Darkness: Dissecting Targeted Attacks (2011) (https://community.qualys.com/blogs/securitylabs/2011/11/30/dissecting-targeted-attacks)
71. Nevis Editor: Adobe Flash 0-day in the wild (2011) (http://nevis-blog.com/2011/03/adobe-flash-0-day-in-the-wild/)
72. Schwartz, M.: Lockheed Martin Suffers Massive Cyber Attack (2011) (http://www.informationweek.com/news/government/security/229700151)
73. When Advanced Persistent Threats Go Mainstrean. RSA (2011)
74. SinFP3 operating system fingerprinting and more (http://www.networecon.com/tools/sinfp/#.UUDnildQpEM)
75. Penetration Testing Software (http://www.metasploit.com/)

76. Wrightson, T.: Social Engineering – Scraping Data from Linkedin (2012) (http://twrightson.wordpress.com/2012/08/05/social-engineering-scraping-data-from-linkedin/)
77. Killing with a Borrowed Knife: Chaining Core Cloud Service Profile Infrastructure for Cyber Attacks (http://www.cybersquared.com/killing-with-a-borrowed-knife-chaining-core-cloud-service-profile-infrastructure-for-cyber-attacks/)
78. Symantec: Waterhole Attack (2012) (http://fr.slideshare.net/symantec/waterhole-attack)
79. McWhorter, D.: Mandiant Exposes APT1 – One of China's Cyber Espionage Units & Releases 3,000 Indicators (https://www.mandiant.com/blog/mandiant-exposes-apt1-chinas-cyber-espionage-units-releases-3000-indicators/)
80. Arkin, B.: Inappropriate Use of Adobe Code Signing Certificate (2012) (http://blogs.adobe.com/asset/2012/09/inappropriate-use-of-adobe-code-signing-certificate.html)
81. Tarzey, B., Fernandes, L.: The trouble heading for your business (2013) (http://www.quocirca.com/reports/797/the-trouble-heading-for-your-business)
82. Ten ways the IT department enables cybercrime (2010) (http://usa.kaspersky.com/resources/knowledge-center/10-ways-it-enables-cybercrime)
83. Platt, C.: Satellite Pirates (2004)
84. Lenoir, V.: EUROCRYPT, a successful conditional access system, In: 1991 IEEE International Conference on Consumer Electronics. pp. 206–207 (ieeexplore.ieee.org/iel1/30/2796/00085548.pdf)
85. Leduc, M.: Système de télévision à péage à controle d'accès pleinement détachable, un example d'implémentation: Videocrypt, In: Proceedings of the ACSA (1990)
86. McCormac, J.: European Scrambling System: Circuits, Tactics and Techniques: The Black Book. Baylin (1996)
87. Parker, D.: Cease and DeCSS: DVD's Encryption Code Cracked - Technology Information (1999) (http://connection.ebscohost.com/c/articles/2655184/cease-decss-dvds-encryption-code-cracked)
88. Kocher, P., Jaffe, J., Jun, B., Laren, C., Lawson, N.: Self-Protecting Digital Content: A Technical Report from the CRI Content Security Research Initiative. Whitepaper (2003)
89. X.509: Information technology - Open Systems Interconnection - The Directory: Public-key and attribute certificate frameworks (http://www.itu.int/rec/T-REC-X.509/en)
90. MDSEC: iOS passcode brute-forcing hardware (2015) (http://www.jwz.org/blog/2015/03/ios-passcode-brute-forcing-hardware/)
91. Ranum, M.J.: Thinking about firewalls, In: Proceedings of Second International Conference on Systems and Network Security and Management (SANS-II) (1993) (http://csrc.nist.gov/publications/secpubs/fwalls.pdf)
92. Khandelwal, S.: 100,000 refrigerators and other home appliances hacked to perform cyber attack (2014) (http://thehackernews.com/2014/01/100000-refrigerators-and-other-home.html)
93. Security PACE Book 2: Physical Security Concepts (http://www.simplexgrinnell.com/SiteCollectionDocuments/Training/PACEBook2.pdf)
94. The Critical Security Controls for Effective Cyber Defense Version 5.0. Council on Cyber Security (2014)
95. Cox, I., Miller, M., Bloom, J., Fridrich, J., Kalker, T.: Digital Watermarking and Steganography. Morgan Kaufmann (2007)
96. Lefebvre, F., Arnold, M.: Fingerprinting and filtering. Security newsletter (2006) (http://eric-diehl.com/newsletterEn.html)
97. Gazet, A.: Comparative analysis of various ransomware virii. J Comput Virol. 6, 77–90 (2010)
98. Thomson, I.: German ransomware threatens with sick kiddie smut (2013) (http://www.theregister.co.uk/2013/04/05/iwf_warning_smut_ransomware/)
99. O'Gorman, G., McDonald, G.: Ransomware: A Growing Menace (2012)
100. Pott, T.: Ransomware attack hits Synology's NAS boxen (2014) (http://www.theregister.co.uk/2014/08/05/synologys_synolocker_crisis_its_as_bad_as_you_think/)

101. RansomWeb: emerging website threat that may outshine DDoS, data theft and defacements? (2015) (https://www.htbridge.com/blog/ransomweb_emerging_website_threat.html)
102. Kassner, M.: The FBI locked your computer? Watch out for new spins on ransomware (2012) (http://www.techrepublic.com/blog/security/the-fbi-locked-your-computer-watch-out-for-new-spins-on-ransomware/8663)
103. Leyden, J.: Android ransomware demands 12x more cash, targets English-speakers (2014) (http://www.theregister.co.uk/2014/07/23/android_ransomware_simplocker_revamp/)
104. Ablon, L., Libicki, M.C., Golay, A.A.: Markets for Cybercrime Tools and Stolen Data (2014) (http://www.rand.org/pubs/research_reports/RR610.html)
105. Hernandez-Castro, J., Boiten, E., Barnoux: preliminary report: 2nd Kent Cyber Security survey (2014) (http://www.cybersec.kent.ac.uk/Survey2.pdf)
106. 2015 Trustwave Global Security Report. Trustwave (2015) (https://www2.trustwave.com/rs/815-RFM-693/images/2015_TrustwaveGlobalSecurityReport.pdf)
107. Burke, P., Craiger, P.: Assessing Trace Evidence Left by Secure Deletion Programs, In: Olivier, M.S., Shenoi, S. (eds.) Advances in Digital Forensics II. pp. 185–195. Springer (2006)
108. Kissel, R., Scholl, M., Skolochenko, S., Li, X.: Special for Publication 800-88: Guidelines for Media Sanitization (2012) (http://csrc.nist.gov/publications/drafts/800-88-rev1/sp800_88_r1_draft.pdf)
109. Wilhoit, K., Dawda, U.: Your Locker of Information for CryptoLocker Decryption (2014) (https://www.cinchit.com/your-locker-of-information-for-cryptolocker-decryption/)
110. Leyden, J.: Fiendish CryptoLocker ransomware: Whatever you do, don't PAY (2013) (http://www.theregister.co.uk/2013/10/18/cryptolocker_ransmware/)
111. McAllister, N.: Code Spaces goes titsup FOREVER after attacker NUKES its Amazon-hosted data (2014) (http://www.theregister.co.uk/2014/06/18/code_spaces_destroyed/)
112. Barcelo, M., Herzog, P.: The Open Source Security Testing Methodology Manual (2010)
113. Quisquater, J.-J., Quisquater, M., Quisquater, M., Quisquater, M., Guillou, L., Guillou, M. A., Guillou, G., Guillou, A., Guillou, G., Guillou, S.: How to Explain Zero-Knowledge Protocols to Your Children, In: Brassard, G. (ed.) Advances in Cryptology — CRYPTO' 89 Proceedings. pp. 628–631. Springer (1990)
114. Fiege, U., Fiat, A., Shamir, A.: Zero Knowledge Proofs of Identity, In: Proceedings of the Nineteenth Annual ACM Symposium on Theory of Computing. pp. 210–217. ACM (1987) (http://doi.acm.org/10.1145/28395.28419)
115. Anderson, R.H., Brackney, R.: Understanding the Insider Threat. RAND (2004) (http://www.rand.org/pubs/conf_proceedings/CF196.html)
116. Kadam, A.: Asset Classification and Control (http://www.networkmagazineindia.com/200212/security2.shtml)
117. Monnet, B., Véry, P.: Les nouveaux pirates de l'entreprise : Mafias et terrorisme. CNRS (2010)
118. Posthuma, R., Garcia, J.: Expatriate Risk Management: Kidnapping and Ransom. Center for Multicultural Management & Ethics (2011)
119. Leyden, J.: HBGary Chief Exec resigns over Anon hack (2011) (http://www.theregister.co.uk/2011/03/01/hbgary_ceo_resigns_over_anon_hack/)
120. Libicki, M.C., Ablon, L., Webb, T.: The Defender's Dilemma (2015) (http://www.rand.org/pubs/research_reports/RR1024.html)
121. Bell, D.E., LaPadula, L.J.: Secure Computer Systems: Mathematical Foundations (1973) (http://www.albany.edu/acc/courses/ia/classics/belllapadula1.pdf)
122. Ariely, D.: Predictably Irrational, Revised and Expanded Edition: The Hidden Forces That Shape Our Decisions. Harper Perennial (2010)
123. Tsu, S.: The Art of War. Dover Publications (2002)
124. Lasica, J.D.: Darknet: Hollywood's War Against The Digital Generation. Wiley (2005)

125. He, B., Patel, M., Zhang, Z., Chang, K.C.-C.: Accessing the Deep Web. Commun. ACM. 50, 94–101 (2007)
126. Abraham, D.G., Dolan, G.M., Double, G.P., Stevens, J.V.: Transaction security system. IBM Syst. J. 30, 206–229 (1991)
127. Anonymous Hackers (http://www.anonymoushackers.org/)
128. Lemos, R.: Dastardly Dozen: A Few APT Groups Carry Out Most Attacks (2011) (http://www.darkreading.com/vulnerabilities—threats/dastardly-dozen-a-few-apt-groups-carry-out-most-attacks/d/d-id/1136840)
129. Schneier, B.: Attack Trees. Dr. Dobb's Journal (1999)
130. Introduction to Return on Security Investment (2012) (https://www.enisa.europa.eu/activities/cert/other-work/introduction-to-return-on-security-investment)
131. Gordon, L.A., Loeb, M.P.: The economics of information security investment. ACM Trans. Inf. Syst. Secur. 5, 438–457 (2002)
132. Bistarelli, S., Fioravanti, F., Peretti, P.: Defense trees for economic evaluation of security investments, In: The First International Conference on Availability, Reliability and Security, 2006. ARES 2006 (2006)
133. VERIS (http://veriscommunity.net/index.html)
134. Hollnagel, P.E., Leveson, P.N., Woods, P.D.D.: Resilience Engineering: Concepts and Precepts. Ashgate Publishing (2012)
135. Stevens, M., Sotirov, A., Appelbaum, J., Lenstra, A., Molnar, D., Osvik, D.A., de Weger, B.: Short Chosen-Prefix Collisions for MD5 and the Creation of a Rogue CA Certificate, In: Halevi, S. (ed.) Advances in Cryptology - CRYPTO 2009. pp. 55–69. Springer (2009)
136. 24C3 Why silicon security is still that hard (2007) (http://www.youtube.com/watch?v=XtDTNnEvlf8)
137. The Open Kinect project – THE OK PRIZE (2010) (http://www.adafruit.com/blog/2010/11/04/the-open-kinect-project-the-ok-prize-get-1000-bounty-for-kinect-for-xbox-360-open-source-drivers/)
138. Terdiman, D.: Bounty offered for open-source Kinect driver (2010) (http://news.cnet.com/8301-13772_3-20021836-52.html#ixzz19zJmrX9F)
139. Martin, H.: git.marcansoft.com (http://git.marcansoft.com/?p=libfreenect.git)
140. AlexP: Windows Kinect Driver/SDK - Xbox NUI Audio, NUI Camera, NUI Motor and Accelerometer (2010) (http://nuigroup.com/forums/viewthread/11154/)
141. Thorsen, T.: Microsoft denies Kinect hack claims (http://www.gamespot.com/articles/microsoft-denies-kinect-hack-claims/1100-6283696/)
142. Carmody, T.: Hackers Take the Kinect to New Levels (2010) (http://www.technologyreview.com/news/421867/hackers-take-the-kinect-to-new-levels/)
143. Bradley, B.: What Is the True Cost of a Data Breach? It May Not Be That Easy (https://digitalguardian.com/blog/what-true-cost-data-breach-it-may-not-be-easy)
144. Rovi: RipGuard: Protecting DVD Content Owners from Consumer Piracy (http://www.rovicorp.com/products/content_producers/protect/ripguard.htm)
145. The Piracy Continuum (2012) (irdeto.com/documents/wp_piracy-continuum_en.pdf)
146. Chenoweth, N.: Murdoch's Pirates: Before the phone hacking, there was Rupert's pay-TV skullduggery. Allen & Unwin (2012)
147. Kerckhoffs, A.: La cryptographie militaire (1883)
148. GS2 Specs (http://www.gatekeepersystems.com/sup_cc_cc_gs2_specs.php)
149. Blender, N.: Reversing the Operation of CAPS Shopping Cart Wheel Locks (2000) (http://www.woodmann.com/fravia/nola_wheel.htm)
150. orthonormal_basis_of_evil: EMP shopping cart locker (http://www.instructables.com/id/EMP-shopping-cart-locker/)
151. Complaint for injunctive relief for misappropriation of trade secrets (1999) (http://cyber.law.harvard.edu/openlaw/DVD/filings/ca-complaint.html)
152. Schneier, B.: Memo to the Amateur Cipher Designer (1998) (http://www.schneier.com/crypto-gram-9810.html#cipherdesign)

153. Levy, S.: Crypto: How the Code Rebels Beat the Government Saving Privacy in the Digital Age. Penguin Books (2001)
154. Biham, E., Shamir, A.: Differential cryptanalysis of DES-like cryptosystems. J. Cryptology. 4, 3–72 (1991)
155. Coppersmith, D.: The Data Encryption Standard (DES) and its strength against attacks. IBM Journal of Research and Development. 38, 243–250 (1994)
156. Frequently Asked Questions (FAQ) About the Electronic Frontier Foundation's "DES Cracker" Machine (http://w2.eff.org/Privacy/Crypto/Crypto_misc/DESCracker/HTML/19980716_eff_des_faq.html)
157. Blaze, M., Diffie, W., Rivest, R.L., Schneier, B., Shimomura, T.: Minimal Key Lengths for Symmetric Ciphers to Provide Adequate Commercial Security. A Report by an Ad Hoc Group of Cryptographers and Computer Scientists (1996) (https://www.schneier.com/cryptography/paperfiles/paper-keylength.pdf)
158. Wang, X., Yin, Y.L., Yu, H.: Finding Collisions in the Full SHA-1 (2005) (http://citeseerx.ist.psu.edu/viewdoc/summary?doi=10.1.1.94.4261)
159. Glass, R.L.: Facts and Fallacies of Software Engineering. Addison-Wesley (2002)
160. Michele, B., Karpow, A.: Watch and be Watched: Compromising All Smart TV Generations, In: Proc. of 11th Consumer Communications and Networking Conference (CCNC). IEEE (2014)
161. Williams: Patch Bugzilla! Anyone can access your private bugs – including your security vulns (2015) (http://www.theregister.co.uk/2015/09/17/bugzilla_priv_esc/)
162. Dageron: AES encryption key extraction from RAGE games [reverse engineering, Xbox360] (2013) (http://dageron.com/?page_id=4723&lang=en)
163. Shamir, A., van Someren, N.: Playing 'Hide and Seek' with Stored Keys, In: Proceedings of Financial Cryptography (1999) (https://www.cs.jhu.edu/~astubble/600.412/s-c-papers/keys2.pdf)
164. IDA: Cross References/Xrefs (http://resources.infosecinstitute.com/ida-cross-references-xrefs/)
165. Chow, S., Eisen, P., Johnson, H., van Oorschot, P.C.: A White-Box DES Implementation for DRM Applications, In: Feigenbaum, J. (ed.) Digital Rights Management. pp. 1–15. Springer (2003)
166. Brecht, W.: White-box cryptography: hiding keys in software (2012) (http://whiteboxcrypto.com/files/2012_misc.pdf)
167. Clarke, R.: Trust in the Context of e-Business. Internet Law Bulletin. 4 (2002) (http://www.rogerclarke.com/EC/Trust.html)
168. Neme6: Reverse engineering du PSJailbreak (2010) (http://www.logic-sunrise.com/news-126726-reverse-engineering-du-psjailbreak-topic-technique.html)
169. defiler: Trojan Reversing part I (http://www.woodmann.com/fravia/defiler_TrojanRE.htm)
170. Guri, M., Monitz, M., Mirski, Y., Elovici, Y.: BitWhisper: Covert Signaling Channel between Air-Gapped Computers using Thermal Manipulations. arXiv (2015) (http://arxiv.org/abs/1503.07919)
171. Madhavapeddy, A., Sharp, R., Scott, D., Tse, A.: Audio networking: the forgotten wireless technology. Pervasive Computing, IEEE. 4, 55– 60 (2005)
172. Block, R.: W32.Wullik.B@mm worm burrows into shipping Zen Neeon (2005) (http://www.engadget.com/2005/08/29/w32-wullik-b-mm-worm-burrows-into-shipping-zen-neeon/)
173. Ricker, T.: McDonald's MP3 players ship with trojan horse (2006) (http://www.engadget.com/2006/10/16/mcdonalds-mp3-players-ship-with-trojan-horse/)
174. Our campaign prize of "MP3 player" with respect to virus infection (http://www.mcd-holdings.co.jp/news/2006/release-061013.html)
175. Small Number of Video iPods Shipped With Windows Virus (http://www.apple.com/support/windowsvirus/)

176. Hudson, T.: TomTom GO 910 = Virus Time! (http://gizmodo.com/232257/tomtom-go-910–virus-time)
177. Preston, T.: Virus Warning when connecting TomTom Go 910 (2006) (http://forum.avast.com/index.php?PHPSESSID=6flgg0itg7rd34c2kl2ibaq787&topic=25442.0;imode)
178. Patel, N.: Insignia photo frame virus much nastier than originally thought (2008) (http://www.engadget.com/2008/02/15/insignia-photo-frame-virus-much-nastier-than-originally-thought/)
179. HVACman: New computer virus from China (2008) (http://www.jeepforum.com/forum/f7/new-computer-virus-china-521660/)
180. Naraine, R.: Malware found in Lenovo software package (2008) (http://www.zdnet.com/blog/security/malware-found-in-lenovo-software-package/2203)
181. Kirk, J.: Pre-installed malware found on new Android phones (2014) (http://www.computerworld.com/s/article/9246764/Pre_installed_malware_found_on_new_Android_phones?pageNumber=1)
182. Henry, S.: Chip and pin scam "has netted millions from British shoppers" (2008) (http://www.telegraph.co.uk/news/uknews/law-and-order/3173346/Chip-and-pin-scam-has-netted-millions-from-British-shoppers.html)
183. Sawer, P.: Credit card scam: How it works (2008) (http://www.telegraph.co.uk/news/worldnews/asia/pakistan/3173161/Credit-card-scam-How-it-works.html)
184. Gorman, S.: Fraud Ring Funnels Data From Cards to Pakistan (2008) (http://online.wsj.com/article/SB122366999999723871.html)
185. mister.old.school: FBI Fears Chinese Hackers Have Back Door Into US Government & Military (2008) (http://www.abovetopsecret.com/forum/thread350381/pg1)
186. Rogers, M., Ruppersberger, D.: Investigative Report on the US National Security Issues Posed by Chinese Telecommunications Companies Huawei and ZTE (2012) (https://intelligence.house.gov/sites/intelligence.house.gov/files/documents/Huawei-ZTE%20Investigative%20Report%20%28FINAL%29.pdf)
187. Greenwald, G.: How the NSA tampers with US-made internet routers (2014) (http://www.theguardian.com/books/2014/may/12/glenn-greenwald-nsa-tampers-us-internet-routers-snowden)
188. Equation Group: Questions and Answers (2015) (https://securelist.com/files/2015/02/Equation_group_questions_and_answers.pdf)
189. Vulnerability Note VU#529496 (2015) (http://www.kb.cert.org/vuls/id/529496)
190. Cyber Supply Chain Risks, Strategies and Best Practices, In: Priorities for America's Preparedness: Best Practices from the Private Sector (2012)
191. Adee, S.: The Hunt for the Kill Switch. IEEE Spectrum. 45, 34–39 (2008)
192. Technion: HP D2D/StorOnce Storage unit backdoors (2013) (https://lolware.net/hpstorage.html)
193. HPSBST02896 rev. 2, HP StoreVirtual Storage, unauthorized remote access (2013) (http://h20564.www2.hpe.com/hpsc/doc/public/display?docId=emr_na-c03825537)
194. Krebs, B.: Security Firm Bit9 Hacked, Used to Spread Malware (2013) (http://krebsonsecurity.com/2013/02/security-firm-bit9-hacked-used-to-spread-malware/)
195. Morley, P.: Bit9 and Our Customers' Security (2013) (https://blog.bit9.com/2013/02/08/bit9-and-our-customers-security/)
196. Doherty, S., Gegeny, J., Baltazar, J., Spasojevic, B.: Hidden Lynx - Professional Hackers for Hire (2013) (http://www.symantec.com/content/en/us/enterprise/media/security_response/whitepapers/hidden_lynx.pdf)
197. Flanagan, K.: It's the Same Old Song: Antivirus Can't Stop Advanced Threats (2013) (https://blog.bit9.com/2013/02/08/its-the-same-old-song-antivirus-cant-stop-advanced-threats/)
198. Schneier, B.: NSA surveillance: A guide to staying secure (http://www.theguardian.com/world/2013/sep/05/nsa-how-to-remain-secure-surveillance)

199. Menn, J.: Exclusive: NSA infiltrated RSA security more deeply than thought - study (2014) (http://www.reuters.com/article/2014/03/31/us-usa-security-nsa-rsa-idUSBREA2U0TY20140331)
200. Fay, J.: So sad about the NSA web-spying bombshells - but think of the MONEY! (2013) (http://www.channelregister.co.uk/2013/10/02/nsa_scandal_business_opportunity/)
201. Paquette, E.: Cybersécurité: les ministres interdits de smartphones (2013) (http://lexpansion.lexpress.fr/high-tech/cybersecurite-les-ministres-interdits-de-smartphones_400697.html)
202. Sanders, J.: Japanese government warns Baidu IME is spying on users (2014) (http://www.techrepublic.com/blog/asian-technology/japanese-government-warns-baidu-ime-is-spying-on-users/)
203. Duo arrested for internet banking fraud (2013) (http://www.financialexpress.com/news/duo-arrested-for-internet-banking-fraud/1061205/1)
204. Diffie, W., Hellman, M.: New directions in cryptography. IEEE Transactions on Information Theory. 22, 644–654 (1976)
205. Borchers, D.: Loss of data has serious consequences for German electronic health card (2009) (http://www.h-online.com/security/news/item/Loss-of-data-has-serious-consequences-for-German-electronic-health-card-742441.html)
206. Microsoft Security Bulletin MS01-017: Erroneous VeriSign-Issued Digital Certificates Pose Spoofing Hazard (2001) (http://technet.microsoft.com/en-us/security/bulletin/ms01-017)
207. Linn, J.: Trust Models and Management in Public-Key Infrastructures (2000) (ftp://ftp.rsasecurity.com/pub/pdfs/PKIPaper.pdf)
208. Eckersley, P., Burns, J.: An observatory for the SSLiverse, DEFCON 18, Las Vegas, NV, USA (2010) (https://ngaytuyet.com/nph-vzh.s/en/20/https/www.eff.org/files/DefconSSLiverse.pdf)
209. ComodoHacker: Striking Back… (2011) (http://pastebin.com/1AxH30em)
210. Prins, J.: DigiNotar Certificate Authority breach "Operation Black Tulip," (2011) (http://www.rijksoverheid.nl/bestanden/documenten-en-publicaties/rapporten/2011/09/05/diginotar-public-report-version-1/rapport-fox-it-operation-black-tulip-v1-0.pdf)
211. VASCO Announces Bankruptcy Filing by DigiNotar B.V. (2011) (http://www.vasco.com/company/press_room/news_archive/2011/news_vasco_announces_bankruptcy_filing_by_diginotar_bv.aspx)
212. Schneier, B.: Forged Google Certificate (2011) (http://www.schneier.com/blog/archives/2011/09/forged_google_c.html)
213. Forristal, J.: Android Fake ID Vulnerability Lets Malware Impersonate Trusted Applications, Puts All Android Users Since January 2010 At Risk (2014) (https://bluebox.com/technical/android-fake-id-vulnerability/)
214. "Tor Stinks" (2012) (http://www.theguardian.com/world/interactive/2013/oct/04/tor-stinks-nsa-presentation-document)
215. Peeling back the layers of TOR with Guard-Egotistical Giraffe (2007) (https://www.eff.org/document/2013 10 04 guard egotistical giraffe)
216. Bonchi, F., Ferrari, E.: Privacy-Aware Knowledge Discovery. CRC Press (2010) (http://www.crcpress.com/product/isbn/9781439803653)
217. Clarke, R.: Privacy as a Strategic Factor in Social Media: An Analysis Based on the Concepts of Trust and Distrust (2012) (http://www.rogerclarke.com/DV/SMTD.html)
218. Mell, P., Grance, T.: The NIST Definition of Cloud Computing. NIST (2011) (http://csrc.nist.gov/publications/PubsSPs.html#800-145)
219. 10 Immutable Laws of Security (http://technet.microsoft.com/library/cc722487.aspx)
220. Chen, L., Franklin, J., Regenscheid, A.: Guidelines on Hardware-Rooted Security in Mobile Devices (Draft). NIST (2012) (http://csrc.nist.gov/publications/drafts/800-164/sp800_164_draft.pdf)
221. Trusted Platform Module Library Specification, Family "2.0", Level 00, Revision 01.16 (2014) (http://www.trustedcomputinggroup.org/resources/tpm_library_specification)

222. Anati, I., Gueron, S., Johnson, S., Scarlata, V.: Innovative Technology for CPU Based Attestation and Sealing (2013) (https://software.intel.com/en-us/articles/innovative-technology-for-cpu-based-attestation-and-sealing)
223. The Heartbleed Bug (heartbleed.com)
224. Willams, J.: DropSmack: How cloud synchronization services render you corporate firewall worthless, Black Hat Europe 2013, Amsterdam, The Netherlands (2013) (https://media.blackhat.com/eu-13/briefings/Williams/bh-eu-13-dropsmack-jwilliams-wp.pdf)
225. Vogel, D.: How to successfully implement the principle of least privilege (2013) (http://www.techrepublic.com/blog/security/how-to-successfully-implement-the-principle-of-least-privilege/9575)
226. Apple's SSL/TLS bug (22 Feb 2014) (2014) (https://www.imperialviolet.org/2014/02/22/applebug.html)
227. Haimes, Y.Y., Horowitz, B.M., Guo, Z., Andrijcic, E., Bogdanor, J.: Assessing Systemic Risk to Cloud Computing Technology as Complex Interconnected Systems of Systems. Systems Engineering. 18 (2014)
228. Collberg, C.S., Thomborson, C.: Watermarking, tamper-proofing, and obfuscation - tools for software protection. Transactions on Software Engineering. 28, 735–746 (2002)
229. Hudson, J.: Deciphering How Edward Snowden Breached the NSA (2013) (http://www.venafi.com/blog/post/deciphering-how-edward-snowden-breached-the-nsa/)
230. Byers, S., Cranor, L., Korman, D., McDaniel, P., Cronin, E.: Analysis of security vulnerabilities in the movie production and distribution process, In: Proceedings of the 3rd ACM Workshop on Digital Rights Management. pp. 1–12. ACM (2003) (http://lorrie.cranor.org/pubs/drm03-tr.pdf)
231. Insider Threat The CERT Division (http://www.cert.org/insider-threat/index.cfm)
232. The Insider Threat (http://www.fbi.gov/about-us/investigate/counterintelligence/the-insider-threat)
233. Edwards, J.: Tech Interns Confess To The Most Disastrous Mistakes They Ever Made (2013) (http://www.businessinsider.com/worst-mistakes-made-by-interns-at-tech-companies-2013-10)
234. Valeo: deux mois de prison ferme pour la stagiaire chinoise Li Li blanchie d'espionnage (2007) (http://www.rtl.be/info/monde/france/valeo-deux-mois-de-prison-ferme-pour-la-stagiaire-chinoise-li-li-blanchie-d-espionnage-29022.aspx)
235. Stempel, J.: Goldman says client data leaked, wants Google to delete email (2014) (http://www.reuters.com/article/2014/07/02/us-google-goldman-leak-idUSKBN0F729I20140702)
236. Andy: Leaked Doctor Who Episode Appears on The Pirate Bay (2014) (http://torrentfreak.com/leaked-dr-who-episode-appears-on-the-pirate-bay-140714/)
237. Oltsik, J.: 2013 Vormetric/ESG Insider Threats Survey (2013) (www.vormetric.com/sites/defaul/files/ap_Vormetric-Insider_Threat_ESG_Research_Brief.pdf)
238. An inside track on insider threats (2012) (https://www.imperva.com/lg/lgw.asp?pid=477)
239. Schneier, B.: Thwarting an Internal Hacker (2009) (http://online.wsj.com/article/SB123447990459779609.html)
240. To Increase Downloads, Instill Trust First (2012) (http://www.symantec.com/content/en/us/enterprise/white_papers/b-to_increase_downloads-instill_trust_first_WP.en-us.pdf)
241. Guignot, P.: Journal : Intrusion sur les serveurs Fedora/Red Hat (2008) (http://linuxfr.org/users/patrick_g/journaux/intrusion-sur-les-serveurs-fedorared-hat)
242. Forristal, J.: Android: One Root to Own Them All, Black Hat USA 2013, Las Vegas, NV, USA (2013)
243. Freeman (Saurik): Exploit (& Fix) Android "Master Key" (http://www.saurik.com/id/17)
244. DirecTV DSS Glossary of Terms (http://www.websitesrcg.com/dss/Glossary.htm)
245. Hunt, T.: Troy Hunt: Everything you need to know about the Shellshock Bash bug (2014) (http://www.troyhunt.com/2014/09/everything-you-need-to-know-about.html)
246. Lin, M., Bennett, J., Bianco, D.: Shellshock in the Wild (2014) (http://www.fireeye.com/blog/technical/2014/09/shellshock-in-the-wild.html)

247. Muncaster, P.: Shellshock Attackers Still Landing Punches on Unpatched Users (2015) (http://www.infosecurity-magazine.com/news/shellshock-attackers-landing/)
248. Mimoso, M.: Third-Party Software Library Risks To Be Scrutinized at Black Hat (2014) (http://threatpost.com/third-party-software-library-risks-to-be-scrutinized-at-black-hat/107319)
249. OWASP Top 10 2013 (https://www.owasp.org/index.php/Top_10_2013-Top_10)
250. Gonsalves, A.: Prices fall, services rise in malware-as-a-service market (2013) (http://www.csoonline.com/article/2133045/malware-cybercrime/prices-fall–services-rise-in-malware-as-a-service-market.html)
251. Durumeric, Z., Bailey, M., Halderman, J.A.: An Internet-wide view of Internet-wide scanning, In: USENIX Security Symposium (2014) (https://www.usenix.org/system/files/conference/usenixsecurity14/sec14-paper-durumeric.pdf)
252. N-Tron 702W Hard-Coded SSH and HTTPS Encryption Keys (2015) (https://ics-cert.us-cert.gov/advisories/ICSA-15-160-01)
253. Eric Diehl: Method and device for accessing content data (http://www.google.com/patents/EP2151999A1)
254. Hard-Coded Bluetooth PIN Vulnerability in LIXIL Satis Toilet (2013) (seclists.org/fulldisclosure/2013/Aug/18)
255. Pen Test Partners LLP: Infosecurity Europe 2015: Wifi Kettle SSID Hack Demo. (https://www.youtube.com/watch?v=GDy9Nvcw4O4)
256. Dhanjani, N.: Hacking Lightbulbs (2013)
257. Joux, A.: Multicollisions in Iterated Hash Functions. Application to Cascaded Constructions, In: Proc. Crypto 2004. pp. 306–316. Springer (2004)
258. Herodotus: The history of Herodotus - Volume 1
259. Boyette, C.: Sensitive documents found in Macy's Thanksgiving Day Parade confetti (2012) (http://www.cnn.com/2012/11/26/us/new-york-confidential-confetti/index.html)
260. Li, P., Fang, X., Pan, L., Piao, Y., Jiao, M.: Reconstruction of Shredded Paper Documents by Feature Matching. Mathematical Problems in Engineering. 2014 (2014) (http://www.hindawi.com/journals/mpe/2014/514748/abs/)
261. Unshredder - Document Reconstruction Software (http://www.unshredder.com/home/w1/i2/)
262. Retired JCG vessel "sold without data wipe" (2013) (http://the-japan-news.com/news/article/0000168249)
263. von Ahn, L. Blum, M., Hopper, N.J., Langford, J.: CAPTCHA: Using Hard AI Problems for Security, In: Biham, E. (ed.) Advances in Cryptology — EUROCRYPT 2003. pp. 294–311. Springer (2003)
264. Quantum Random Bit Generator Service: Sign up (http://random.irb.hr/signup.php)
265. Stiltwalker: Nucaptcha, Paypal, SecurImage, Slashdot, Davids Summer Communication (http://www.dc949.org/projects/stiltwalker/)
266. EC-Council takes the privacy and confidentiality of their customers very seriously (2014) (http://www.eccouncil.org/news/ec-council-takes-the-privacy-and-confidentiality-of-their-customers/)
267. Lydersen, L., Wiechers, C., Wittmann, C., Elser, D., Skaar, J., Makarov, V.: Hacking commercial quantum cryptography systems by tailored bright illumination. Nature Photonics. 4, 686–689 (2010)
268. Halderman, A., Schoen, S., Heninger, N., Clarkson, W., Paul, W., Calandrino, J., Feldman, A., Appelbaum, J., Felten, E.: Lest We Remember: Cold Boot Attacks on Encryption Keys (http://citp.princeton.edu/memory/)
269. Courtay, O., Karroumi, M.: AACS Under Fire. Security Newsletter. 2 (2007) (http://eric-diehl.com/newsletterEn.html)

270. Kim, Y., Daly, R., Kim, J., Fallin, C., Lee, J.H., Lee, D., Wilkerson, C., Lai, K., Mutlu, O.: Flipping bits in memory without accessing them: An experimental study of DRAM disturbance errors, Proceeding of the 41st annual international symposium on computer architecture. pp. 361–372. IEEE Press (2014) (http://users.ece.cmu.edu/~omutlu/pub/dram-row-hammer_isca14.pdf)
271. Evans, C.: Project Zero: Exploiting the DRAM rowhammer bug to gain kernel privileges (2015) (http://googleprojectzero.blogspot.com/2015/03/exploiting-dram-rowhammer-bug-to-gain.html)
272. Genkin, D., Shamir, A., Tromer, E.: RSA Key Extraction via Low-Bandwidth Acoustic Cryptanalysis (2013) (http://eprint.iacr.org/2013/857)
273. Aviv, A.J., Gibson, K., Mossop, E., Blaze, M., Smith, J.M.: Smudge Attacks on Smartphone Touch Screens, In: Proceedings of the 4th USENIX Conference on Offensive Technologies. pp. 1–7. USENIX Association (2010) (http://dl.acm.org/citation.cfm?id=1925004.1925009)
274. Moller, B.: This POODLE bites: exploiting the SSL 3.0 fallback (2014) (http://googleonlinesecurity.blogspot.com/2014/10/this-poodle-bites-exploiting-ssl-30.html)
275. FREAK: Factoring RSA Export Keys (https://www.smacktls.com/#freak)
276. Tracking the FREAK attack (https://freakattack.com/)
277. Chirgwin, R.: "Logjam" crypto bug could be how the NSA cracked VPNs (2015) (http://www.theregister.co.uk/2015/05/20/logjam_johns_hopkins_cryptoboffin_ids_next_branded_bug/)
278. Common Criteria: An introduction (http://www.niap-ccevs.org/Documents_and_Guidance/cc_docs/cc_introduction-v2.pdf)
279. Common Criteria Evaluation and Validation Scheme Validation Report: Microsoft Windows 2003 Server and XP Workstation. NIST
280. Arora, A., Telang, R., Xu, H.: Optimal Policy for Software Vulnerability Disclosure. Management Science. 54, 642–656 (2008)
281. Chirgwin, R.: KILL FLASH WITH FIRE until a patch comes: Hacking Team exploit is in the wild (http://www.theregister.co.uk/2015/07/08/hacking_teamderived_0day_is_now_in_the_wild/)
282. Security Updates available for Adobe Reader and Acrobat - APSB14-19 (2014)
283. Ion, L., Reeder, R., Consolvo, S.: "…No one Can Hack My Mind": Comparing Expert and Non-Expert Security Practices, Presented at the Symposium on Usable Privacy and Security (SOUPS2 2015), Ottawa, Canada (2015) (https://www.usenix.org/conference/soups2015/proceedings/presentation/ion)
284. Nicastro, F.M.: Security Patch Management. CRC Press (2011)
285. Souppaya, M., Scarfone, K.: Guide to Enterprise Patch Management Technologies. NIST (2013) (http://nvlpubs.nist.gov/nistpubs/SpecialPublications/NIST.SP.800-40r3.pdf)
286. Pauna, A., Moulinos, K.: Window of exposure … a real problem for SCADA systems? ENISA (2013)
287. Schneier, B.: The Internet of Things Is Wildly Insecure — And Often Unpatchable. WIRED (2014) (http://www.wired.com/2014/01/theres-no-good-way-to-patch-the-internet-of-things-and-thats-a-huge-problem/)
288. Thomas, D., Beresdorf, A., Rice, A.: Security Metrics for the Android Ecosystem, In: Proceedings of the 5th Annual ACM CCS Workshop on Security and Privacy in Smartphones and Mobile Devices. pp. 87–98. ACM (2015) (https://www.cl.cam.ac.uk/~drt24/papers/spsm-scoring.pdf)
289. Fleming, S.: Auto Safety: NHTSA Has Options to Improve the Safety Defect Recall Process. United States Government Accountability Office (2011) (http://www.gao.gov/assets/320/319698.pdf)
290. Cisco Security Advisory: Multiple Vulnerabilities in Cisco Unified Communications Domain Manager (http://tools.cisco.com/security/center/content/CiscoSecurityAdvisory/cisco-sa-20140702-cucdm)

291. FIPS 140-2 Security Requirements for Cryptographic Modules (2001) (http://csrc.nist.gov/publications/fips/fips140-2/fips1402.pdf)
292. Nie, Y.-Q., Huang, L., Liu, Y., Payne, F., Zhang, J., Pan, J.-W.: 68 Gbps quantum random number generation by measuring laser phase fluctuations. Review of Scientific Instruments. 86, 063105 (2015)
293. Gutmann, P.: Secure Deletion of Data from Magnetic and Solid-State Memory, In: Proceedings of the 6th USENIX Security Symposium. pp. 77–89 (1996)
294. McGraw, G., Felten, E.W.: Securing Java: getting down to business with mobile code. Wiley (1999)
295. Cluley, G.: Man who tricked women into taking hacked webcams into shower is jailed (2012) (http://nakedsecurity.sophos.com/2012/07/25/jail-hacked-webcams-shower/)
296. Everstine, B.: Carlisle: Air Force intel uses ISIS "moron's" social media posts to target airstrikes (2015) (http://www.airforcetimes.com/story/military/tech/2015/06/04/air-force-isis-social-media-target/28473723/)
297. Statistics highlight a growing number of data breaches and fines levied across multiple business sectors, with the highest percentage increases in Healthcare, Local Government, Education, Financial Services, Insurance and Telecoms (http://www.egress.com/ico-foi-data-breach/)
298. Grzonkowski, S.: Password recovery scam tricks users into handing over email account access (2015) (http://www.symantec.com/connect/blogs/password-recovery-scam-tricks-users-handing-over-email-account-access)
299. Cubrilovic, N.: Yahoo Axis Chrome Extension Leaks Private Certificate File (2008) (http://www.nikcub.com/posts/yahoo-axis-chrome-extension-leaks-private-certificate-file)
300. Davis, M.: Belkin WeMo Home Automation Vulnerabilities (2014)
301. Cowley, S.: How a lying "social engineer" hacked Walmart (2012) (http://money.cnn.com/2012/08/07/technology/walmart-hack-defcon/index.htm)
302. Los, R., Shackleford, D., Sullivan, B.: The Notorious Nine: Cloud Computing Top Threats in 2013 (2013) (https://downloads.cloudsecurityalliance.org/initiatives/top_threats/The_Notorious_Nine_Cloud_Computing_Top_Threats_in_2013.pdf)
303. Wenzel, S.: The real nightmare of today's CIO isn't BYOD, it is BYOC (2013) (http://caribtek.com/blog/2013/11/byoc-bring-your-own-cloud/)
304. New Webroot Survey Reveals Poor Password Practices That May Put Consumers' Identities at Risk (http://www.webroot.com/us/en/company/press-room/releases/protect-your-computer-from-hackers)
305. Das, A., Bonneau, J., Caesar, M., Borisov, N., Wang, X.: The Tangled Web of Password Reuse, Proceedings of NDSS (2014) (http://blogprod.dev.alligatorsneeze.com/sites/default/files/06_1_1.pdf)
306. Where to download a list of email accounts hacked from Adobe? (http://security.stackexchange.com/questions/45611/where-to-download-a-list-of-email-accounts-hacked-from-adobe)
307. franx47: Download Wordlist Password Collections (2013) (http://franx47.wordpress.com/2013/03/31/download-wordlist-password-collections/)
308. Oechslin, P.: Making a Faster Cryptanalytic Time-Memory Trade-off, In: Boneh, D. (ed.) Advances in Cryptology - CRYPTO 2003. pp. 617–630. Springer (2003)
309. Lost codes spark Haneda scramble (2014) (http://www.japantimes.co.jp/news/2014/04/22/national/lost-codes-spark-airport-scramble-eve-obama-trip/)
310. Drozhzhin, A.: Tell me who you are and I will tell you your lock screen pattern (https://blog.kaspersky.co.in/tell-me-who-you-are-and-i-will-tell-you-your-lock-screen-pattern/)
311. Herley, C.: Why do Nigerian Scammers Say They are from Nigeria?, Presented at the Workshop on the Economics of Information Security (WEIS 2012), Berlin, Germany (2012) (http://research.microsoft.com/apps/pubs/?id=167719)

312. Satnam, N.: Hacking Facebook: Scammers Trick Users to Gain Likes and Followers (2014) (http://www.symantec.com/connect/blogs/hacking-facebook-scammers-trick-users-gain-likes-and-followers)
313. Khandelwal, S.: Facebook Self-XSS Scam Fools Users into Hacking Themselves (2014) (http://thehackernews.com/2014/07/facebook-self-xss-scam-fools-users-into_28.html)
314. Cheng, N.: Hacker targets info on MH370 probe (http://www.thestar.com.my/News/Nation/2014/08/20/Hacker-targets-info-on-MH370-probe-Computers-of-officials-infected-with-malware/)
315. Moran, N., Lanstein, A.: Spear Phishing the News Cycle: APT Actors Leverage Interest in the Disappearance of Malaysian Flight MH 370 (2014) (http://www.fireeye.com/blog/technical/malware-research/2014/03/spear-phishing-the-news-cycle-apt-actors-leverage-interest-in-the-disappearance-of-malaysian-flight-mh-370.html)
316. Rika Joi, G.: Malaysia Airlines Flight 370 News Used To Spread Online Threats (2014) (http://blog.trendmicro.com/trendlabs-security-intelligence/malaysia-airlines-flight-370-news-used-to-spread-online-threats/)
317. Mitnick, K.D., Simon, W.L.: The Art of Deception: Controlling the Human Element of Security. Wiley (2003)
318. Milgram, S.: Obedience to Authority: An Experimental View. Harper Perennial Modern Classics (2009)
319. Wolf, P.: De l'authentification biométrique. Sécurité Informatique (2003) (http://www.sg.cnrs.fr/FSD/securite-systemes/revues-pdf/num46.pdf)
320. Objectif Sécurité - Ophcrack (https://www.objectif-securite.ch/ophcrack.php)
321. Silveira, V.: An Update on LinkedIn Member Passwords Compromised (2012) (http://blog.linkedin.com/2012/06/06/linkedin-member-passwords-compromised/)
322. Karnan, M., Akila, M., Krishnaraj, N.: Biometric personal authentication using keystroke dynamics: A review. Applied Soft Computing. 11, 1565–1573 (2011)
323. Stolerman, A., Fridman, A., Greenstadt, R., Brennan, P., Juola, P.: Active Linguistic Authentication Revisited: Real-Time Stylometric Evaluation towards Multi-Modal Decision Fusion, In: IFIP WG (2014) (http://www.stolerman.net/papers/active_auth_ifip-wg11.9-2014.pdf)
324. Derawi, M.O., Nickel, C., Bours, P., Busch, C.: Unobtrusive User-Authentication on Mobile Phones Using Biometric Gait Recognition, In: 2010 Sixth International Conference on Intelligent Information Hiding and Multimedia Signal Processing (IIH-MSP). pp. 306–311 (2010)
325. Ferro, M., Pioggia, G., Tognetti, A., Carbonaro, N., De Rossi, D.: A Sensing Seat for Human Authentication. IEEE Transactions on Information Forensics and Security. 4, 451–459 (2009)
326. Delac, K., Grgic, M.: A survey of biometric recognition methods, In: Electronics in Marine, 2004. Proceedings Elmar 2004. 46th International Symposium. pp. 184–193 (2004)
327. Zhang, Y., Cheng, Z., Xue, H., Wei, T.: Fingerprints On Mobile Devices: Abusing and Leaking, Black Hat 2015, Las Vegas, NV, USA (2015) (https://www.blackhat.com/docs/us-15/materials/us-15-Zhang-Fingerprints-On-Mobile-Devices-Abusing-And-Leaking-wp.pdf)
328. iPhone 5s: About Touch ID security (http://support.apple.com/kb/ht5949)
329. Chaos Computer Club breaks Apple Touch ID (http://www.ccc.de/en/updates/2013/ccc-breaks-apple-touchid)
330. Star, B.: hacking iPhone 5S Touch ID (2013) (https://www.youtube.com/watch?v=HM8b8d8kSNQ)
331. Reddy, P.V., Kumar, A., Rahman, S., Mundra, T.: A New Method for Fingerprint Antispoofing using Pulse Oxiometry, In: First IEEE International Conference on Biometrics: Theory, Applications, and Systems, 2007. BTAS 2007. pp. 1–6 (2007)
332. Samsung Galaxy S5 Finger Scanner also susceptible to ordinary spoofs (2014) (http://www.youtube.com/watch?v=sfhLZZWBn5Q&feature=youtube_gdata_player)

333. Statement by OPM Press Secretary Sam Schumach on Background Investigations Incident (2015) (http://www.opm.gov/news/releases/2015/09/cyber-statement-923/)
334. Akhawe, D., Felt, A.P.: Alice in Warningland: A Large-Scale Field Study of Browser Security Warning Effectiveness (2013) (http://research.google.com/pubs/archive/41323.pdf)
335. Felt, A.P., Ha, E., Egelman, S., Haney, A., Chin, E., Wagner, D.: Android Permissions: User Attention, Comprehension, and Behavior, In: Proceedings of the Eighth Symposium on Usable Privacy and Security. pp. 3:1–3:14. ACM (2012) (http://doi.acm.org/10.1145/2335356.2335360)
336. Felt, A.P., Egelman, S., Finifter, M., Akhawe, D., Wagner, D.: How to Ask for Permission. HotSec (2012) (https://www.usenix.org/system/files/conference/hotsec12/hotsec12-final19.pdf)
337. Felt, A.P., Ainslie, A., Reeder, R.W., Consolvo, S., Thyagaraja, S., Bettes, A., Harris, H., Grimes, J.: Improving SSL Warnings: Comprehension and Adherence, Proceedings of 33rd Annual ACM Conference on Human Factors in Computing Systems. pp. 2893–2902. ACM (2015) (http://doi.acm.org/10.1145/2702123.2702442)
338. Reeder, R., Kowalczyk, E., Shostack, A.: Helping Engineers Design NEAT Security Warnings (2011)
339. Shneiderman, B., Plaisant, C., Cohen, M., Jacobs, S.: Designing the User Interface: Strategies for Effective Human-Computer Interaction. Prentice Hall (2009)
340. Kark, K.: Articulating The Business Value Of Information Security (2009) (http://www.forrester.com/Research/Document/Excerpt/0,7211,54908,00.html)
341. Wash, R.: Folk Models of Home Computer Security, In: Proceedings of the Sixth Symposium on Usable Privacy and Security. pp. 11:1–11:16. ACM (2010) (http://doi.acm.org/10.1145/1837110.1837125)
342. Pauli, D.: Kids hack Canadian ATM during LUNCH HOUR (2014) (http://www.theregister.co.uk/2014/06/12/kids_hack_canuck_bank_atm_during_lunch_break/)
343. Gayer, O., Atias, R., Zeifman, I.: Lax Security Opens the Door for Mass-Scale Abuse of SOHO Routers (2015) (https://www.incapsula.com/blog/ddos-botnet-soho-router.html)
344. Wikholm, Z.: CARISIRT: Yet Another BMC Vulnerability (And some added extras) (http://blog.cari.net/carisirt-yet-another-bmc-vulnerability-and-some-added-extras/)
345. Wireless Key Calculator (http://www.gredil.net/tech-stuff/12-other/50-wpa-calc)
346. Varian, H.: System Reliability and Free Riding, In: Camp, L.J. and Lewis, S. (eds.) Economics of Information Security. pp. 1–15. Springer (2004)
347. Moriarty, T.: Crime, Commitment and the Responsive Bystander (1972) (http://eric.ed.gov/?id=ED076923)
348. Guéguen, N., Dupré, M., Georget, P., Sénémeaud, C.: Commitment, crime, and the responsive bystander: effect of the commitment form and conformism. Psychology, Crime & Law. 21, 1–8 (2015)
349. Kuhn, J.: The Dyre Wolf Campaign Stealing Millions and Hungry for More (2015) (http://securityintelligence.com/dyre-wolf/)
350. Wilson, M., de Zafra, D.E., Pitcher, S.I., Tressler, J.D., Ippolito, J.B.: SP 800 16. Information Technology Security Training Requirements: A Role- and Performance-Based Model. National Institute of Standards and Technology (1998)
351. Wilson, M., Hash, J.: SP 800-50. Building an Information Technology Security Awareness and Training Program. National Institute of Standards and Technology (2003)
352. US Mobile Device Security Survey Report (2015) (http://www.absolute.com/en/resources/research/mobile-device-security-survey-report-us)
353. Ernesto: Busted: BitTorrent Pirates at Sony, Universal and Fox (2011) (https://torrentfreak.com/busted-bittorrent-pirates-at-sony-universal-and-fox-111213/)
354. Abrams, L.: Your browser has been locked, Ransomware Removal Guide (2013) (http://www.bleepingcomputer.com/virus-removal/remove-your-browser-has-been-locked-ransomware)

355. maurizio: Ransomwares. gendarmerie constaté sur ubuntu (2013) (http://forum.ubuntu-fr.org/viewtopic.php?id=1413081)
356. Kelion, L.: Russian site lists breached webcams (2014) (http://www.bbc.com/news/technology-30121159)
357. Shekyan, S., Harutyunyan, A.: Turning your surveillance camera against you. Hack In The Box 2013, Amsterdam (2013) (http://www.slideshare.net/SergeyShekyan/d2-t1-sergey-shekyan-and-artem-harutyunyan-turning-your-surveillance-camera-against-you)
358. Ackerman, S., Ball, J.: Optic Nerve: millions of Yahoo webcam images intercepted by GCHQ (2014) (http://www.theguardian.com/world/2014/feb/27/gchq-nsa-webcam-images-internet-yahoo)
359. Greenberg, A.: Hackers Remotely Kill a Jeep on the Highway—With Me in It (2015) (http://www.wired.com/2015/07/hackers-remotely-kill-jeep-highway/)
360. Greenberg, A.: After Jeep Hack, Chrysler Recalls 1.4M Vehicles for Bug Fix (2015) (http://www.wired.com/2015/07/jeep-hack-chrysler-recalls-1-4m-vehicles-bug-fix/)
361. Bhat, R.: How to bypass Zeus Trojan's self-protection mechanism (2014) (http://int0xcc.svbtle.com/how-to-bypass-zeus-trojans-self-protection-mechanism)
362. Kujawa, A.: You Dirty RAT! Part 2 – BlackShades NET (2012) (http://blog.malwarebytes.org/intelligence/2012/06/you-dirty-rat-part-2-blackshades-net/)
363. I Know Where Your Cat Lives (http://iknowwhereyourcatlives.com/about/)
364. Mowery, K., Shacham, H.: Pixel Perfect: Fingerprinting Canvas in HTML5, W2SP Web 2.0 Security and Privacy (2012)
365. Acar, G., Eubank, C., Englehardt, S., Juarez, M., Narayanan, A., Diaz, C.: The Web never forgets: Persistent tracking mechanisms in the wild (2014) (https://securehomes.esat.kuleuven.be/~gacar/persistent/the_web_never_forgets.pdf)
366. Young, S.: Designing a DMZ (2001)
367. Bauer, M.: Paranoid Penguin: Designing and Using DMZ Networks to Protect Internet Servers. Linux J. 2001 (2001) (http://dl.acm.org/citation.cfm?id=364764.364780)
368. Carr, N.: The Big Switch: Rewiring the World, From Edison to Google. W.W. Norton (2008)
369. Cook, J.: Hackers Access At Least 100,000 Snapchat Photos And Prepare To Leak Them, Including Underage Nude Pictures (2014) (http://www.businessinsider.com/snapchat-hacked-the-snappening-2014-10?op=1)
370. Allen, J.: Pennsylvania teen killed classmate, took "selfie" with body: police (2015) (http://www.reuters.com/article/2015/02/09/us-usa-crime-selfie-idUSKBN0LD2C320150209)
371. Agreement Containing Consent Order Snapchat (2014)
372. Singel, R.: You Deleted Your Cookies? Think Again (2009) (http://www.wired.com/epicenter/2009/08/you-deleted-your-cookies-think-again/)
373. Schneier, B.: A Taxonomy of Social Networking Data (2010)
374. Bartlett, J.: The Dark Net: Inside the Digital Underworld. Melville House (2015)
375. Anderson, C.: Free: The Future of a Radical Price. Hyperion Books (2009)
376. Barbaro, M., Jr., T.Z.: A Face Is Exposed for AOL Searcher No. 4417749 (2006) (http://www.nytimes.com/2006/08/09/technology/09aol.html)
377. Sweeney, L.: K-anonymity: A Model for Protecting Privacy. International Journal on Uncertainty Fuzziness and Knowledge-based Systems. 10, 557–570 (2002)
378. Machanavajjhala, A., Kifer, D., Gehrke, J., Venkitasubramaniam, M.: L-diversity: Privacy Beyond K-anonymity. ACM Trans. Knowl. Discov. Data. 1 (2007) (http://doi.acm.org/10.1145/1217299.1217302)
379. Ernesto: Which VPN Services Take Your Anonymity Seriously? 2015 Edition (2015) (https://torrentfreak.com/anonymous-vpn-service-provider-review-2015-150228/3/)
380. Syverson, P.: A Peel of Onion, Proc. 27th Annual Computer Security Applications Conference. pp. 123–137. ACM (2011) (http://doi.acm.org/10.1145/2076732.2076750)
381. Kubrick, S.: Dr. Strangelove or How I Learned to Stop Worrying and Love the Bomb (1964)

382. Blair, B.: Keeping Presidents in the Nuclear Dark (Episode #1: The Case of the Missing "Permissive Action Links") - CDI
383. Brodkin, J.: iOS apps hijack Twitter accounts, post false "confessions" of piracy (2012) (http://arstechnica.com/tech-policy/2012/11/ios-apps-hijack-twitter-accounts-post-false-confessions-of-piracy/)
384. Andrew, V.: Case Study: Pro-active Log Review Might Be A Good Idea (2013) (https://securityblog.verizonenterprise.com/?p=1626#more-1626)
385. McAllister, N.: NSA: NOBODY could stop Snowden – he was A SYSADMIN (2013)
386. Allen, J.: NSA to cut system administrators by 90 percent to limit data access (2013) (http://www.reuters.com/article/2013/08/09/us-usa-security-nsa-leaks-idUSBRE97801020130809)
387. Legere, J.: T-Mobile CEO on Experian's Data Breach (2015) (http://www.t-mobile.com/landing/experian-data-breach.html)
388. COBIT 4.1 (2007) (http://www.isaca.org/Knowledge-Center/cobit/Documents/CobiT_4.1.pdf)
389. Epstein, J.: Security lessons learned from Société Générale. IEEE Security & Privacy. 80–82 (2008)
390. Peikari, C., Chuvakin, A.: Security Warrior. O'Reilly Media (2004)
391. Marty, R.: Applied Security Visualization. Addison-Wesley (2008)
392. Riley, M., Elgin, B., Lawrence, D., Matlack, C.: Missed Alarms and 40 Million Stolen Credit Card Numbers: How Target Blew It (2014) (http://www.businessweek.com/articles/2014-03-13/target-missed-alarms-in-epic-hack-of-credit-card-data)
393. Target Reports Fourth Quarter and Full-Year 2014 Earnings (http://www.businesswire.com/news/home/20150225005513/en/Target-Reports-Fourth-Quarter-Full-Year-2014-Earnings#.VXCp0c9VhBc)
394. Poulsen, K.: Hacker Disables More Than 100 Cars Remotely (2010) (http://www.wired.com/2010/03/hacker-bricks-cars/)
395. McCumber, J.: Information systems security: A comprehensive model, In: Proc. 14th National Computer Security Conference (1991) (http://trygstad.rice.iit.edu:8000/Government%20Documents/NSTISS/NSTISSI4011Annex.rtf)
396. Kaplan, S., Garrick, J.: On the quantitative definition of risk. Risk Analysis. 1, 11–22 (1985)
397. Manuele, F.A.: Acceptable Risk. Professional safety (2010) (http://www.asse.org/professionalsafety/docs/F1Manuel_0510.pdf)
398. Ionita, D.: Current established risk assessment methodologies and tools (2013) (http://eprints.eemcs.utwente.nl/23767/01/D_Ionita_-_Current_Established_Risk_Assessment_Methodologies_and_Tools.pdf)
399. Rivest, R.L., Shamir, A., Adleman, L.: A method for obtaining digital signatures and public-key cryptosystems. Communications of the ACM. 21, 120–126 (1978)
400. Biham, E., Chen, R., Joux, A., Carribault, P., Lemuet, C., Jalby, W.: Collisions of SHA-0 and Reduced SHA-1, In: Advances in Cryptology – EUROCRYPT 2005. Springer (2005)
401. Boutin, C.: NIST Selects Winner of Secure Hash Algorithm (SHA-3) Competition (2012) (http://www.nist.gov/itl/csd/sha-100212.cfm)
402. Viega, J., McGraw, G.: Building Secure Software: How to Avoid Security Problems the Right Way. Addison-Wesley (2001)
403. Crume, J.: Inside Internet Security: What Hackers Don't Want You To Know. Addison-Wesley Professional (2000)

Printed in the United States
By Bookmasters